design essentials in earthquake resistant buildings

Edited by

Architectural Institute of Japan

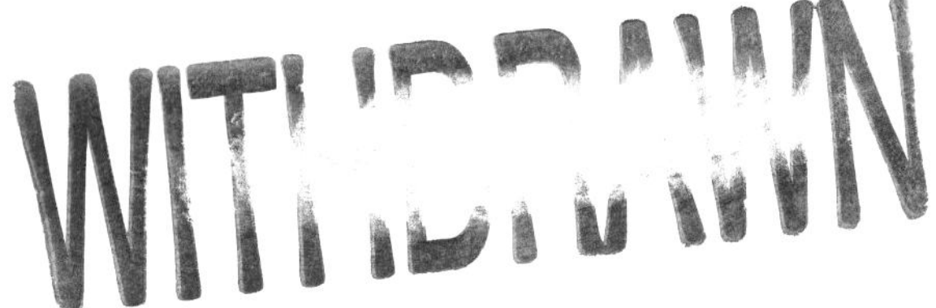

Published jointly by

Architectural Institute of Japan, TOKYO

ELSEVIER PUBLISHING COMPANY
Amsterdam, London, New York
1970

Sole Distributor for JAPAN:
MARUZEN COMPANY, LTD.
P.O. Box 5050, Tokyo International, 100-31 Japan

Sole Distributor for the USA and Canada
AMERICAN ELSEVIER PUBLISHING COMPANY, INC.
52, Vanderbilt Avenue
New York, New York 10017

Sole Distributor for the British Commonwealth excluding Canada
ELSEVIER PUBLISHING CO., LTD.
Barking, Essex, England

For all remaining areas:
ELSEVIER PUBLISHING COMPANY
335 Jan van Galenstraat
P.O. Box 211, Amsterdam, The Netherlands

STANDARD BOOK NUMBER: 444-99954-X

Printed in Japan

PREFACE

Since Japan is seismically active, all buildings must be earthquake resistant so that this consideration occupies an important position in structural design. The problem of earthquake resistant construction saw considerable development due to the efforts of pioneers in this field and after the Kanto earthquake of 1923 and experiences gained from subsequent earthquakes, the methods of earthquake resistant calculation and/or structural requirements were included in the AIJ structural standards for different types of construction that were published from time to time. However, problems other than those of earthquake resistance were included for different types of construction so that the back-bone of anti-seismic construction was in danger of being lost sight of.

In 1941, the Japan Council for the Promotion of Science published a book entitled "The Principles of Earthquake Resistant Construction of Buildings" with Dr. Riki Sano as chairman. It was a very valuable document at that time and contained many fine papers of unchanging value but it is now out of print and difficult to obtain. Additional experiences have been accumulated from destructive earthquakes since 1941 and due to social developments and technical advancements, new types of structures are being constructed at present.

The Architectural Institute of Japan, on the occasion of the eightieth anniversary of its establishment, decided to amplify the above mentioned book and to compile a new comprehensive book pertaining to the present status of earthquake resistant construction methods and appointed a special subcommittee within the Structural Standards Committee to undertake this project in April 1963 and it commenced work immediately. The experiences gained from the unusual Niigata earthquake of 1964 are also included in this work. In compiling this book, the policy of adhering to the Building Standard Laws, effective in 1966, and related building regulations has been adopted and portions relating to earthquake resistance in the above and AIJ Structural Standards have been excerpted and explanatory comments added to them. Researches under process at present will be published in the future and revisions made in the present work as new standards are adopted.

Each chapter was written by designated members of the sub-committee and this committee as a whole endeavored to obtain consistency throughout without infringing on the opinions of the authors which should be respected. Inspite of some inconsistencies and imperfections in the present work, if it can contribute to the understanding and development of earthquake resistant construction methods, the Committee would be most fortunate. Finally, deep appreciation is expressed to each member of the Committee.

July 1966

Kiyoo Matsushita, Chairman
Special Sub-Committee on Compilation of
Earthquake Resistant Construction Principles,
Structural Standards Committee of AIJ

Committee Members Participating in Compilation of This Book, as of April 1966

Structural Standards Committee

Chairman	Takeo Naka
Secretaries	Toshihiko Hisada
	Kiyoo Matsushita
(since April 1966)	
Chairmam	Yoshikatsu Tsuboi
Secretaries	Yorihiko Ohsaki
	Morihisa Fujimoto

Special Sub-Committee on Compilation of Design Essentials in Earthquake Resistant Buildings

Chairman	Kiyoo Matsushita
Secretaries	Hajime Umemura
	Toshihiko Hisada
Members	Shinichi Aikawa
	Masanori Izumi
	Makoto Ueno
	Kazuo Uchiyama
	Akira Enami
	Koichiro Ogura
	Yorihiko Ohsaki
	Yutaka Ohsawa
	Tsutomu Kato
	Kiyoshi Kaneta
	Hiroyoshi Kobayashi
	Seiji Kokusho
	Yoshio Sakai
	Toshio Shiga
	Hideo Sugiyama
	Shuzo Takada
	Masahide Tomii
	Kyoji Nakagawa
	Kiyoshi Nakano
	Yoshio Baba
	Minoru Makino
	Akira Matsumura

Contents

1. GENERAL CONSIDERATIONS

1.1 Earthquake Motion

1.1.1 Earthquake and Earthquake Motion

Earthquake is a phenomenon produced by a sudden change inside the earth. The origin of an earthquake is called the hypocenter, and the point vertically above it on the earth's surface is called the epicenter. The wave motions radiated from the epicenter and the ground motion of the earth's surface produced by these waves are called seismic waves and the earthquake ground motion, respectively.

The hypocenter is not limited to a point but sometimes it has considerable length or volume. The depth of the hypocenter could be several hundred kms, but in severe earthquakes which cause damage to buildings, it can be less than 50 km.

The wave motions transmitted from the hypocenter through the interior of the earth are classified as the P-wave (longitudinal wave) and the S-wave (transverse wave), and those transmitted along the earth's surface are called the surface waves. As the earth's material is not homogeneous and the various layers having different characteristics are near the earth's surface, various kinds of waves are produced by reflections and refractions at the boundary of the layers even if the wave motion near the hypocenter is very simple, and in addition, various kinds of surface waves are also produced after those waves reach the earth's surface. Thus, the earthquake motion of a point at a certain distance from the hypocenter becomes very complicated as the result of superposition of various wave motions.

Such complicated earthquake motions can be divided into vertical and horizontal components. The horizontal motion is usually much larger than the vertical except at places near the hypocenter.

According to the seismograms recorded with seismographs, the earthquake motion starts with P-wave having small amplitudes and relatively short periods and then suddenly change to S-wave having large amplitudes and longer periods. Sometimes the large, long-period surface wave follows, but this is not clearly observed near the hypocenter. The tremor between the beginning of the P-wave and the S-wave is called the preliminary tremor, and the large amplitude portion after the beginning of the S-wave is called the principal motion. It is considered that buildings suffer damage during the principal motion which consists of the S-waves and the surface waves. In some cases the principal motion ceases in a very short time like a shock wave.

The velocities of P and S-wave have been measured in various soil layers, and the results are listed in Table 1.1.

Table 1.1 Velocity of seismic waves

	P-wave (km/sec)	S-wave (km/sec)
Tertiary	1.9~3.2	1.3~2.5
Diluvium	1.4~2.6	0.8~1.5
Alluvium	0.3~1.8	0.1~0.2
Loam	0.4~1.0	0.1~0.3

1.1.2 Earthquake Magnitude

The earthquake magnitude scale has been used to express the magnitude of an earthquake. The fundamental definition, which is accepted throughout the world, is given by C.F. Richter as follows: "Value to be calculated taking the common logarithmus of the recorded maximum amplitude in terms of micron taken at a distance of 100 km from the epicenter with the torsion seismometer developed by Anderson and Wood having a static magnification of 2,800 with a pendulum period of 0.8 sec and the damping constant of 0.8". This magnitude scale is a quantitative expression of the energy which is radiated from the hypocenter, and it is estimated that every upward step of one magnitude unit corresponds to an increase in the energy by about 30 times.

In general, the damage to structures occurs from an earthquake with a magnitude of more or less than 6. The damaged area will be considerable if M=7 to 8. The largest scale ever experienced is around M=8.5, and the recent examples are the 1960 Chilean earthquake (M=8.5) and the 1964 Alaskan earthquake (M=8.5). The magnitude scale of recent earthquakes in Japan is shown in Table 1.2 together with the outline of damage.

Table 1.2 Recent Large Earthquakes in Japan

Name of earthquake	Date	Location of epicenter	Outline of earthquake damage	M
Nohbi Earthq.	Oct. 28, 1891	35.6°N, 136.6°E	Felt by persons all over Japan except northern part from Sendai. Persons killed 7,273; dwelling houses with total damage 80,000 and 142,177 including other kind of buildings; remarkable fault. After shocks continued for ten years or more at Gifu.	8.4
Kanto Earthq.	Sep. 1, 1923	35.2°N, 139.3°E	Houses with total damage 128,266; half damage 126,233; houses destroyed by fire 447,128; houses flooded 868. Persons killed 99,331; persons wounded 103,733; persons missing 43,476.	7.9
Tango Earthq.	Mar. 7, 1927	35.6°N, 135.0°E	Houses with total damage 4,974; houses destroyed by fire 2,651. Persons killed 3,017.	7.4

Kitaizu Earthq.	Nov.26, 1930	35.0°N, 138.9°E	Houses with total damage 2,141. Persons killed 259.	7.0
Sanriku Earthq.	Mar. 3, 1933	39.15°N, 144.4°E	Ayari bay was struck by big Tsunami, whose maximum wave height was 24 m. Houses flooded 4,086. Persons killed 2,986.	8.5
Shizuoka Earthq.	Jul. 11, 1935	35.0°N, 138.4°E	Houses with total damage 363. Persons killed 9.	6.6
Tottori Earthq.	Apr.10, 1943	35.5°N, 134.2°E	Remarkable natural calamity, such as fault, crack or fissure in the ground and land slides were evident. Persons killed 1,083; persons heavily wounded 6,153. Houses with total damage 7,485; houses with half damage 5,158; houses destroyed by fire 254.	7.3
Tohnankai Earthq.	Jul. 7, 1944	33.7°N, 136.2°E	The wave height of Tsunami was 6 m at Owase. Persons killed 998; persons heavily wounded 2,135. Houses with total damage 26,130; houses with half damage 46,950; houses flooded 3,059; houses destroyed by fire 11.	8.3
Mikawa Earthq.	Jan. 13, 1945	34.7°N, 137.2°E	Fault (uplift 2 m). Persons killed 1,961; persons heavily wounded 896. Houses with total damage 5,536; houses with half damage 11,706; unoccupied houses with total damage 6,603; houses with half damage 9,976.	6.9
Nankaidoh Earthq.	Dec.21, 1946	33.0°N, 135.7°E	Earthquake damage was extended over a part of Shikoku, Kyushu, Chugoku and Chubu districts. Persons killed 1,330. Houses with total damage 9,060; houses with half damage 19,204; unoccupied houses with total damage 2,521; unoccupied houses with half damage 4,283; houses flooded 1,451; houses destroyed by fire 2,598. The southern extremity of Kii peninsula was struck by big Tsunami. (wave height is 6.6 m). Crustel deformation was found at various districts. At Tosa, ground about 15 km^2 sunk below the sea surface.	8.1
Fukui Earthq.	Jun. 28, 1948	36.1°N, 136.3°E	Persons killed 3, 895; persons wounded 16,375. Houses with total damage 35,420; houses with half damage 11,449; houses destroyed by fire 3,960. A fault with 50~60 cm vertical difference was found by surveying.	7.2

Off Tokachi Earthq.	Mar. 4, 1952	41.8°N, 144.1°E	Very disastrous at Tokachi, Hidaka and Kushiro with tsunami. Houses with total damage 815; houses with half damage 1,324; houses flooded 91; houses destroyed by fire 14. Persons killed 28; persons wounded 287; persons missing 5.	8.2
Niigata Earthq.	Jun. 16, 1964	38.2°N, 139.1°E	Niigata, Yamagata damaged. Houses with total damage 1,960; houses with half damage 6,640. The heighest tidal wave was 6 m. Ground movement accompanied by lequifaction of soil caused remarkable damage.	7.3

1.1.3 Intensity of the Earthquake Motion (Seismic Intensity)

a) Seismic Intensity

The seismic intensity is a quantitative expression of the intensity of the earthquake motion at a certain place, while the earthquake magnitude expresses the whole energy radiated from the epicenter. The seismic intensities take different values at the places having different soil layers and different epicentral distances for same earthquake or the earthquakes having same magnitudes. The following description will be given to the seismic intensity of ground motion. There are other kinds of seismic intensities such as "design seismic coefficient" which appears in the Building Standard Law, and the "seismic coefficient for buildings".

The seismic intensity of ground motion is defined as the maximum acceleration expressed in terms of the acceleration of gravity. It is used, for instance, that "it is estimated that during the A-earthquake the seismic intensity of the B-city was 0.2 in the down town and 0.1 in the up town.

Consider a rigid body which moves together with the ground, as shown in Fig. 1.1 and denote the weight W, the mass M, and the maximum acceleration of the ground motion α. In this case the rigid body is subjected to the seismic force F in the opposite direction to the ground movement.

$$F=\alpha M=\frac{\alpha}{g}W=kW$$

Where $\quad k=\dfrac{\alpha}{g}$

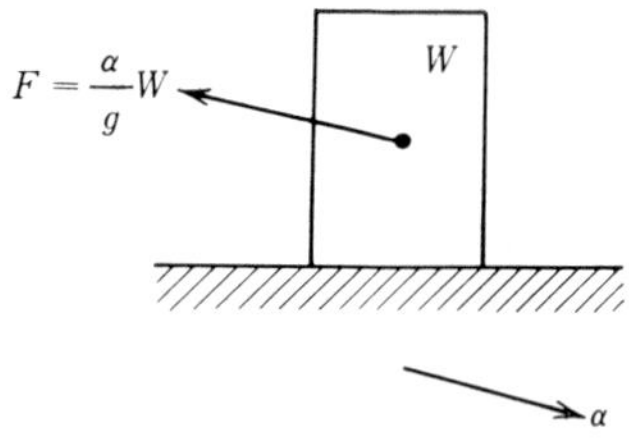

$F=\frac{\alpha}{g}W$, W, α

Fig. 1.1

It is clear from the above that the largest force subjected to the rigid body is obtained by multiplying its weight by the seismic coefficient.

The expression of the seismic coefficient, therefore, is effectively used to directly related to the earthquake intensity and the earthquake force, and consequently, to have better understanding of the destruction force of the earthquake. Nevertheless, it should be noted that for the investigation of earthquake damage to buildings the seismic coefficient is not sufficient to express the nature of the earthquake effect, being preferable to know the variation of period and amplitude of the ground motion with the time.

As described in section 1.1.6, the special type seismographs, which are capable of recording the acceleration wave forms even in case of destructive earthquake, have been installed in many places and the observation with them have been continued. It is expected that the quantitative relation between the intensity of the earthquake ground motion and the resulting damage to buildings becomes clear.

b) Seismic Intensity Scale

The seismic intensity can be classified by the human reactions or the degree of damage to buildings without the direct record of seismographs.

Japan Meteorological Agency has been using the seismic intensity scale which is devided into eight classes. The scale is described in Table 1.3.

Seismic intensity scales of this kind are widely used in foreign countries. The United States and many other countries are using the Modified-Mercalli Intensity Scale which is described in Table 1.3. It is similar to the JMA Intensity Scale except that the former is subdivided in the high intensity range. Recently, the MSK intensity of 12 classes was proposed to establish the commonly used scale throughout the world.

As mentioned above the intensity scale does not express the recorded acceleration, and therefore, it is somewhat vague. However it has been used considerably in the research of the nature of the earthquake and for the earthquake resistant construction as it is possible to understand the intensity of the earthquake motion by use of the intensity scale.

1.1.4 Predominant Period of the Ground

It has been recognized for many years that the ground has a proper period in the earthquake motion. In the 1930's it was found that the specific vibration period was predominant at the specific place as the results of the observation of the earthquake motion by the acceleration seismographs, and this period is called the predominant period.

Table 1.3 Seismic Intensity Scale

Japan Meteorological Agency Intensity Scale of 1949	
0	Not Felt : Shocks too weak to be felt by human beings, registered only by a seismograph. Acceleration < 0.8 gal (cm/sec^2)
1	Slight : Extremely feeble shocks only felt by persons at rest or by those who are very close to the epicenter. 0.8~2.5 gal
2	Weak : Shocks felt by most persons, slight shaking of doors and Japanese latticed sliding doors (Shoji). 2.5~8.0 gal
3	Rather Strong : Slight shaking of houses and buildings, rattlings of doors and Japanese latticed sliding doors (Shoji), swinging of hanging objects like electric lamps, moving of liquids in vessels. 8.0~25 gal
4	Strong : Strong shaking of houses and buildings, overturning of unstable objects, spilling of liquids out of vessels. 25~80 gal

5	Very Strong : Cracks in the walls, overturning of gravestones, stone lanterns, etc., damage to chimneys and mud-and-plaster warehouses. 80～250 gal
6	Disastrous : Demolition of houses less than 30%, landslips, fissures on the roads, in the ground, etc. 250～400 gal
7	Very Disastrous : Demolition of houses more than 30%, intense landslips, large fissures in the ground. $>$400 gal

	Modified Mercalli Intensity Scale of 1931 (Abridged and rewritten by C.F. Richter)
I	Not felt. Marginal and long-period effects of large earthquakes (for details see text).
II	Felt by persons at rest, on upper floors, or favorably placed.
III	Felt indoors. Hanging objects swing. Vibration like passing of light trucks. Duration estimated. May not be recognized as an earthquake.
IV	Hanging objects swing. Vibration like passing of heavy trucks; sensation of a jolt like a heavy ball striking the walls. Standing motor cars rock. Windows, dishes, doors rattle. Glasses clink. Crockery clashes. In the upper range of IV wooden walls and frame creak.
V	Felt outdoors; direction estimated. Sleepers wakened. Liquids disturbed, some spilled. Small unstable objects displaced or upset. Doors swing, close, open. Shutters, pictures move. Pendulum clocks stop, start, change rate.
VI	Felt by all. Many frightened and run outdoors. Persons walk unsteadily. Windows, dishes, glassware broken. Knickknacks, books, etc., off shelves. Pictures off walls. Furniture moved or overturned. Weak plaster and masonry D cracked. Small bells ring (church, school). Trees, bushes shaken (visibly, or heard to rustle CFR).
VII	Difficult to stand. Noticed by drivers of motor cars. Hanging objects quiver. Furniture broken. Damage to masonry D, including cracks. Weak chimneys broken at roof line. Fall of plaster, loose bricks, stones, tiles, cornices (also unbraced parapets and architectural ornaments—CFR). Some cracks in masonry C. Waves on ponds; water turbid with mud. Small slides and caving in along sand or gravel banks. Large bells ring. Concrete irrigation ditches damaged.
VIII	Steering of motor cars affected. Damage to masonry C; partial collapse. Some damage to masonry B; none to masonry A. Fall of stucco and some masonry walls. Twisting, fall of chimneys, factory stacks, monuments, towers, elevated tanks. Frame houses moved on foundations if not bolted down; loose panel walls thrown out. Decayed piling broken off. Branches broken from trees. Changes in flow or temperature of springs and wells. Cracks in wet ground and on steep slopes.
IX	General panic. Masonry D destroyed; masonry C heavily damaged, sometimes with complete collapse; masonry B seriously damaged. (General damage to foundations—CFR). Frame structures, if not bolted, shifted off foundations. Frames racked. Serious damage to reservoirs. Underground pipes broken. Conspicuous cracks in ground. In alluviated areas sand and mud ejected, earthquake fountains, sand craters.

X	Most masonry and frame structures destroyed with their foundations. Some well-built wooden structures and bridges destroyed. Serious damage to dams, dikes, embankments. Large landslides. Water thrown on banks of canals, rivers, lakes, etc. Sand and mud shifted horizontally on beaches and flat land. Rails bent slightly.
XI	Rails bent greatly. Underground pipelines completely out of service.
XII	Damage nearly total. Large rock masses displaced. Lines of sight and level distorted. Objects thrown into the air.

Some of the predominant periods are illustrated as follows.

Place	Predominant period	
Hongo, Tokyo (Campus of the University of Tokyo)	about	0.3 sec
Marunouchi, Tokyo	〃	0.6 sec
Yamashita Park, Yokohama	〃	0.5 sec and 1.2 sec
Mt. Tsukuba	〃	0.05 sec

The predominant period of the ground has also been found by the observed results of the microtremor which is always produced by traffic and other vibrational sources. The existence of the predominant period is interpreted that the vibration having the proper period of the surface layer of the ground is induced by earthquakes or by traffic.

The predominant period of the ground is considered to be of great importance in the earthquake resistant construction because the resonance phenomenon may occur in the building the natural period of which is equal to the predominant period of the ground on which the building stands. It should be noted that the predominant periods indicated above are observed during relatively small earthquakes and it is not certain that these periods are the same during destructive earthquakes. Furthermore, the natural period of buildings will change with the development of damage.

1.1.5 Expectancy of the Seismic Intensity

It is important in the anti-seismic design of buildings to know the expectancy of the seismic intensity of the ground on which the building is to be built, although it is very difficult at the present time to predict accurately the intensity of the future earthquakes.

H. Kawasumi calculated the expectancy of the seismic intensity of the diluvial layer with average hardness by means of the probability theory using the data of earthquake activity in Japan over 1,000 years, and the results have been plotted on maps. In Fig. 1.2 the maximum accelerations of the earthquake ground motion expected for 75 years (A) and 100 years (B) are shown by the isolines. The total number of earthquakes used in this analysis is 343, which occurred before 1949, and the periods considered are 1,350 years in the west of Kanto district, 1,120 years in Tohoku district and 160 years in Hokkaido.

1.1.6 Observation of Strong Earthquakes

Since the ordinary seismograph is not capable of recording strong earthquake motions which produce certain damage to buildings, a special type of seismograph for recording strong earthquakes was designed and manufactured. This is called the SMAC (strong motion accelerograph

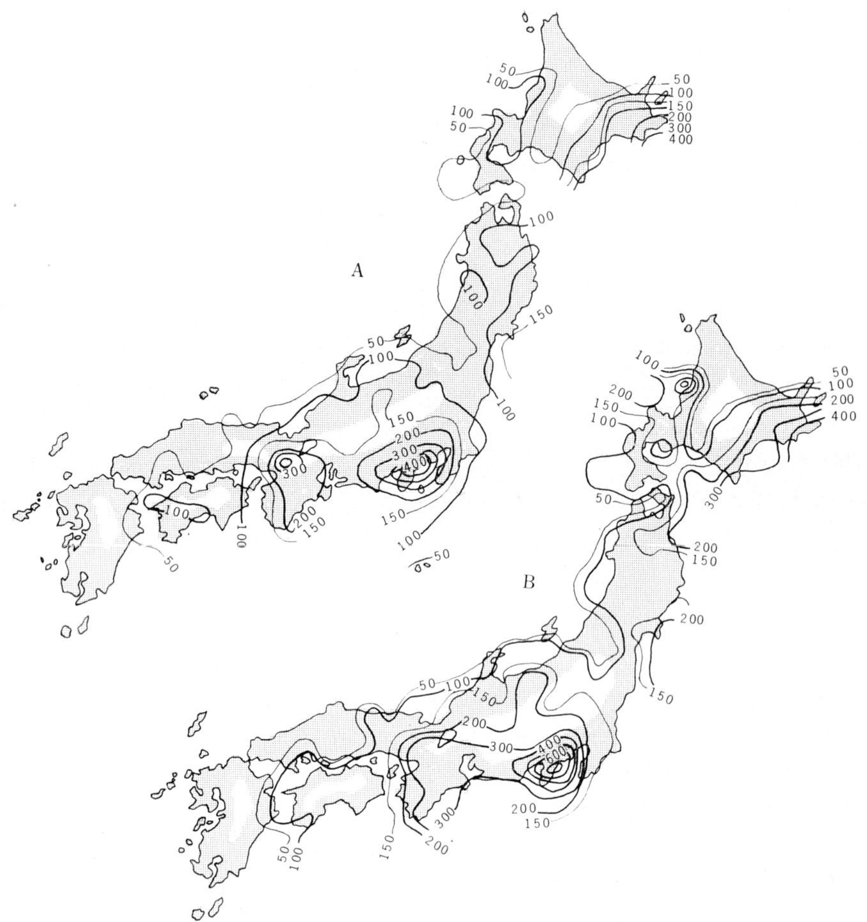

Fig. 1.2 Expectancy of Maximum Seismic Intensity for (A) 75 years (B) 100 years (unit in gal)

committee) accelerograph and the first SMAC was made in 1952.

The SMAC starts antomatically to record earthquake motions greater than 10 gals (cm/sec^2) and is capable of recording up to 1,000 gals. It can be kept in good condition through inspection by a technician once a year and by a keeper of the buildings and others in which the SMAC is installed, once or twice a month. There is another type of accelerograph for the same purpose which is called DC.

The number of SMAC and DC type strong motion accelerographs which were installed before 1965 in the buildings, civil engineering structures, ground, etc. are about 250 and 15, respectively. They are distributed all over the country but most of them are located in large cities, mainly Tokyo and Osaka. In case of buildings the accelerographs are installed not only on the basement floor but also on the upper floors so that the acceleration subjected to the buildings during a big earthquake can be recorded directly. As a successful example, during the Niigata Earthquake of June 1964, the acceleration records were obtained by the strong motion accelerographs installed at the apartment house in Kawagishi-cho, Niigata city, where the damage was most severe among the affected area.

In the United States the strong motion acceleration measurements were started earlier in

1930's, and the accelerographs of destructive earthquakes since the El Centro Earthquake of 1940 have been obtained successfully.

1.2 Outline of Earthquake Damage

1.2.1 Outline of Earthquake Damage to Buildings

The various types of earthquake damage to buildings have been observed depending on the nature of the earthquake, the composition of the subsoil, and the scale and the type of construction of the building. In this section, the outline of the damage is simply described on the basis of past experiences, being classified and compared with the type of construction of the building.

The damage to steel buildings of light weight such as factories with steel plate or slate was relatively few in case that much attention was paid to the foundation construction, and even in case of poor workmanship the damage was usually not severe. In buildings having steel frames and brick walls, the cracks and the falling of the brick walls, and the distortion of the steel frames were observed, but there were no buildings which collapsed. The damage to steel buildings having reinforced concrete slabs was in general, very slight.

The reinforced concrete buildings rarely collapsed but in some cases they failed when the construction of the superstructure or foundation was inadequate. In the past severe earthquakes some of the buildings that were insufficient in the framing system and that had small amount of walls, suffered fatal damage. In case of the Niigata Earthquake many reinforced concrete buildings either inclined or subsided due to liquefaction of the ground without any damage to the superstructure, and in some buildings the cracks in concrete were induced not by vibration but by unequal settlement of the footings caused by liquefaction.

The steel reinforced concrete buildings have experienced but a few severe earthquakes, and it can be said that the damage has been very slight.

The wooden buildings often suffer minor damage such as sliding off of roof tiles, cracks in the walls, disconnection of small pieces of timber, etc. even in case of not so severe earthquakes. In case of severe earthquakes the wooden buildings are easily inclined or shifted, and sometimes are fell down with the failure of main frames. In the heavily shaken area, the number of collapsed buildings is usually considerable. In general the lateral resisting capacity of traditional wooden buildings is comparatively small. The buildings having heavy roof or large rooms were severely damaged because the heavy roof brings large destructive force and the large room diminishes the resisting capacity.

The "dozo" construction made of heavy plaster walls often suffer damage such as cracks in the walls but such buildings rarely collapsed. In buildings having wood frames and brick or stone walls a piece of brick or stone is easy to fall down if they are in the heavily shaken area, and many such buildings have fallen down during past severe earthquakes.

Brick and stone buildings, in general, are extremely weak against earthquakes, especially those with wooden roof trusses and wooden slabs are easy to suffer crushing failure of the wall. Generally speaking, the earthquake damage to buildings of any type of construction is closely related to the foundation and ground conditions. This problem is discussed in Chapter 3.

1.2.2 Ground Movements Accompanied by Earthquakes

a) Fault

A fault is the discontinuity of the ground in both horizontal and vertical directions. The biggest one extends over 100 km long. The buildings on the ground with thin surface layer or directly on the exposed in the mountain area near the fault may suffer serious damage. And the buildings on the relatively thick alluvial or diluvial layer in the plain area sometimes suffer the additional damage due to ground fissure and boiling sand if the fault occurs under these layers.

b) Upheaval and Subsidence

The ground is lifted up or settles down during earthquakes, and sometimes this happens in the large area in case of severe earthquakes. In Japan more than several decades cm of upheaval and subsidence was ever observed. It was also observed in a certain area that the ground had gradually been lifted up for about 100 years after the earthquake and then suddenly settled down when another earthquake occurred, and upheaval and subsidence have been alternately repeated. The damage to buildings directly due to upheaval and subsidence was rarely observed but the secondary damage of submarged houses in the settling area due to the collapse of the bank was experienced.

c) Ground Fissure, Depression and Landslide

At a steep slope, a cliff, a bank and a soft ground like reclaimed land the ground fissure and/or the depression of the ground are easy to take place due to the earthquake ground motion. In the mountain area it frequently happens that the earthquake motion causes the landslide and the mass of soil without cohesion flows down on the slope. Due to the movement of the ground mentioned above lots of houses have been damaged.

d) Liquefaction of Sandy Ground

Some sandy ground is subjected to liquefaction by the earthquake and it is accompanied by water and sand eruption. One of the most typical examples is that in and around the Niigata city during the Niigata Earthquake where the 10 m thick surface layer of liquefaction and as the results the heavy structures like reinforced concrete buildings were severely damaged due to settling and tilting.

1.2.3 Other Damage Caused by Earthquakes

a) Earthquake Fire

The earthquake damage increases with the fire following earthquakes in addition to the damage to buildings directly due to earthquakes. Especially, the cities are in danger of wide-spread fire. And they are menaced by the increasing danger in explosion and burning of fuels, explosion of gas, etc. When the fire starts, it is difficult to extinguish even at ordinary times. In case of big earthquakes much less activity in extinguishing the fire is expected because the big earthquake accompanies the lack of water supply, the traffic jam in streets and the simultaneous occurence of fires in the collapsed houses and by the special sources for fires mentioned above. Therefore, in order to minimize earthquake disaster it is necessary to pay particular attention to the fire.

The wooden houses crowded in the city are most dangerous to the fire. In case of the

Kanto Great Earthquake the firing occurred at more than 100 places in old Tokyo city and 50% of its area and 70% of the total number of houses were burnt down and more than 100,000 people were killed by the fire. The measure against such a disastrous earthquake fire must be taken not only by making individual building earthquake and fire resistant but also from the standpoint of city planning.

b) Tsunamis

The tsunami is produced by a sudden movement of submarine blocks of the earth's crust caused by earthquakes. It is propagated through the sea as a long wave, and when it reaches at the coast the sea water floods on the land and sweeps the materials on it. There are many instances that the buildings were either collapsed or swept away by such a flood water. As Japan is surrounded by the sea, she has been suffered in the past by many tsunamis such as Sanriku, Nankaido, Chilean Earthquake, and will be attacked by that in the future. Therefore, the counter measure against tsunami is also very important. The damage due to tsunami is closely related with the topography of the coast and in general the tsunami surges to extreme hights at the inner of V shape bay. From the history of tsunami disaster it can be said that the region that have been attacked by tsunami in the past has large possibility to be attacked again in the future.

1.3 General Principles for Earthquake Resistant Design of Buildings

1.3.1 General

a) Philosophy of Earthquake Resistant Design

The purpose of earthquake resistant design of buildings is to protect human lives from the earthquake disaster and to prevent the damage to buildings during the earthquake. However, it is almost impossible from the economical viewpoint to construct so rigid and strong buildings that they will never be damaged during any destructive earthquake. In general, it is considered that the damage to buildings to a certain extent is inevitable during the biggest earthquake expected at the site, although the buildings should never collapse. There is, of couse, an exceptional case in which the perfect earthquake resistant design is required to have no damage during the biggest earthquake expected in the future considering the danger caused by slight damage like in case of the reactor buildings.

The most important thing in the earthquake resistant design is to protect human lives, and this is the minimum requirement. For earthquakes which are not so big and encountered rather frequently, the structural elements of buildings should not be damaged.

b) Earthquake Resistivity Characterized by the Earthquake Damage

From the experiences in the past severe earthquakes it can be said that the buildings having sufficient strength, ductility and rigidity have shown excellent earthquake resistivity, and those having uniform and simple structure have also well resisted earthquakes.

Both reinforced concrete and steel reinforced concrete constructions have enough strength, ductility and rigidity to prevent earthquake damage, although they are of heavy weight. The steel construction has also good quality as the earthquake resistant construction. The wooden construction having light weight roof has high safety for earthquakes. The masonry construction made of brick or stone has heavy weight and poor ductility, and much difficulty has been found to make it earthquake resistant.

The reinforced concrete block construction and precast reinforced concrete construction are similar to ordinary reinforced concrete construction but it is necessary to give sufficient ductility and deformability. There are many instances that the damage was caused by poor construction work.

It should be noted that the remarkable damage to such construction as concrete block fenses has been reported in recent earthquakes resulted from the careless design and construction, and many persons have been killed or injured even in moderate earthquakes.

c) Inspection of Building

The buildings that suffered slight damage should be carefully inspected after severe earthquakes. The cracks in the structures are especially important as they indicate that the structure was stressed considerably during the earthquake, and if they are extended to the most part of the structure, the repair of the structure may be considered. The buildings, which inclined or unequally settled during the earthquake, should be investigated even when they suffered only slight damage. When the remarkable difference is found between the vibration characteristics before and after the earthquake, the suitable reinforcement must be made. Sometimes the cracks are covered by painting merely to make a fine appearance, but this is most dangerous, and therefore, should be avoided.

There are several methods to investigate the soundness of the building such as vibration test, observation of the microtremor, observation of the earthquake motion, etc. It is desirable to make vibration test when the construction work of the building is completed, because the vibration characteristics after the earthquake must be compared with those before the earthquake. The continuous observation of the earthquake motion is most helpful to get better understanding of the vibration characteristics of the building.

It is important to always inspect every part of the building at an ordinary time, because the condition of the building should be kept to be same as that just after the completion of the construction work.

1.3.2 Consideration for the Site Selection of Buildings

It is recommended, in general, to select the hard soil layer as the building site, and this is particularly important when considering the earthquake resistant design. The bed rock, the hard clay layer, the clay layer with gravel and the gravel layer are classified as the hardest layers, and the sand or clay layer with low water content ranks next. The sand or silt layer with high water content and the reclaimed soil layer are the most undesirable. Recent development in the national land utilization is so rapid and the city area is expanding so fast that it is not easy in some cases to get satisfactory ground conditions mentioned above. As it is observed that the buildings on the unsatisfactory ground conditions suffered serious damage due to recent earthquakes, the site selection of buildings should be carefully made considering its importance in the economical effect in case of the design of the new industrial city, expansion of the existing city, construction of buildings, etc.

Recent examples of damage to buildings on the soft ground are the Tohnankai, the Nankaido, the Fukui, the Off Tokachi and the Niigata Earthquakes. It is no exaggeration to say that in these earthquakes damaged buildings were concentrated only in the soft ground area. The damage of this kind is frequently observed in other countries, too.

The problems related to the soft ground may be pointed out as follows from the antiseismic point of view.

i) The earthquake motion at the surface of soft soil layer is much larger than that of hard soil layer. Because of the existence of soft soil layer the ground motion is amplified several times between the bed rock and the surface of soft layer. The effect of this amplification is serious to wooden houses because their collapse is closely related to the amplitude of ground motion. The bed rock means the hard soil layer such as gravel or soft rock underlain by alluvial and others on which the natural dynamic properties depend.

ii) In the soft soil layer the unusual ground movement due to earthquake easily takes place. The fissures in the ground, landslides, liquefaction of sand, sand eruption, depression of the sand, etc. are considered to occur only in loose or soft soils.

Especially, attention should be paid to the places that are often subjected to ground fissure or the landslides. Though it is rather difficult to distinguish those places to avoid such secondary damage when considering the site selection of buildings, it should be kept in mind that those places where the damage due to unusual ground movement occured in the past, have suffered similar type of damage repeatedly.

iii) Recently, the ground subsidence has occurred in some coastal cities in Japan, and sometimes this has made the damage due to tsunami or flood tide more serious. It should be kept in mind in the site selection that the ground subsidence is especially serious in soft soil layers.

It is also important to pay attention to the fact that in the place where the land slide is easy to take place many houses have been damaged due to the flow of soils or ground movement under their footings during past earthquakes. The degree of damage due to tsunami depends not only on the height of the land but also on the shape of bay and sea-bottom. As its behavior is similar to that in case of flood tide, in case of city or town planning as well as site selection of buildings it is necessary to make plan carefully to prevent the tsunami disaster under sufficient data.

1.3.3 Notes on Structural Planning

a) Superstructure

As the vibration of buildings during an earthquake is very complicated, it is very difficult to analyze its whole aspect mathematically by means of present technique of dynamic analysis. Especially, it is almost impossible to know in advance the vibration of specific building on the specific ground during the earthquake which will occur in the future. However, recent remarkable development in the dynamic analysis enables us to put some notes on the structural planning of superstructure against earthquakes referring to the experience of building damage in the past earthquakes.

i) Plan

It is desirable that any structure has a simple layout in plan. The building having a wing or wings in plan such as L-, T-, U-, H-, Y-shape will vibrate in complicated way mainly due to the dominant torsional vibrations. In such a case it should be noted that the abnormal stresses are easy to occur at the joint of wing and main part of the building. It is preferable to have uniform lateral rigidity in every part of the building in the same story.

ii) Elevation

It is desirable that each floor has the same shape in plan. Where a sudden change in the size of plan of the building exists such as in a set-back building and in a building with appendage on the roof, the extraordinarily large deflection may occur in that part during an earthquake. Uniformity in the distribution of masses, rigidities and strengths is also desirable in the vertical direction of the building. As for the columns, beams, and shear walls, they should be arranged in a simple configuration so as to avoid disturbing the uniformity as a whole, and it is desirable that their locations do not differ between floors.

Summarizing the above, the building with simple, uniform and compact configuration will be safe against earthquake motions. Judging from the fact that many buildings were damaged by the past severe earthquakes because of insufficient attention in the structural design, it must be effective for reducing unexpected damage to make the building frames more analyzable as mentioned above.

iii) Shear Wall

The shear wall is effective to give sufficient rigidity and strength to the structure, and it is required to be uniformly arranged in both directions. Between two shear walls it is necessary to arrange a floor slab or a horizontal frame with sufficient rigidity in the horizontal plane to transmit the earthquake force through it.

In case of wooden construction, damage to intermediate frame between shear walls was often observed during the past earthquakes because of the lack of rigidity in the floor slab or horizontal frame. As the shear walls arranged in a box shape and connected tightly to each other are expected to take most of the earthquake force, they should be designed to have sufficient strength and ductility.

iv) Attached Structure, Lean-to, Connecting Corridor, etc.

Attention should be called to these appendage, because their dynamic characteristics are different from those of main structures. The connections of appendant and main structures are subjected to large stresses during an earthquake. There are many instances where the appendages having light weight compared with main structures such as fire walls, chimneys and connecting corridors in the wooden structures were either thrown off or crushed during the past earthquakes. Attached structures at the roof are subjected to large earthquake force, especially, in case that the natural period of these structures is close to that of main structure a special caution is needed for large vibration due to quasi-resonance phenomenon.

v) Free Joint

Free joint may have to be provided in some cases such as a building of large scale, two buildings connected to each other, a building with complicated shape in plan and a building subjected to thermal stresses. The determination of the distance at a free joint should be carefully made to avoid the collision of two buildings or portions during an earthquake due to the difference in their dynamic characteristics. It is desirable to make dynamic analysis considering that they may vibrate in a different phase.

b) Foundations

It is required to design structures to diminish the settlment as a whole, and especially, to avoid unequal settlements, because there are many uncertainties in the dynamic behavior of the soil in addition to uncertainties in the vibration characteristics of the superstructure. For this reason the contact pressure or pile stress for permanent loading should be uniform in every part, and those for the loading during an earthquake should be kept to low values to possess sufficient safety factor. Special caution is needed for soft ground because large amount of earthquake acceleration and displacement or unusually remarkable ground movement may be anticipated.

To construct a building on two or more different kinds of ground should be avoided, but if it is unavoidable, special considerations in design and construction are needed paying attention to unequal settlement due to permanent loading and to unusual deformation and vibration during a severe earthquake. It is also desirable to avoid using two or more different types of foundation construction in one building.

c) Footing Slab

Footing slab, which connects the foundation to superstructures, is one of the most important elements in the antiseismic construction. The sliding of foundation sill in wooden construction, the settlement and movement of independent footing in steel buildings and the unequal settlement of the foundation in reinforced concrete buildings caused serious damage to superstructure in past severe earthquakes. It was revealed by careful inspection that the cause of damage in some cases, which had been believed to be due to vibration, was in fact due to sliding of footing slab or unequal settlement of foundation. The raft foundation is most advantageous against earthquakes, especially, in case of having basement floor because the footing slab and basement walls form a rigid box. The continuous footing and the independent footing with tie-beams rank next, and the independent footing without tie-beams easily suffer earthquake damage. It should be noted that the earthquake resistivity will be insufficient if considerations are not given to the design and construction of footing slabs even though the superstructure and foundation are made earthquake resistant.

1.3.4 Notes on Construction Details and Finish Materials

a) Considerations for Acceleration

In some cases the upper part of buildings is subjected to an acceleration several times larger than that of the ground during earthquakes. In the Niigata Earthquake of 1964, the strong motion accelerograph installed on the second floor in penthous of the Akita prefectural buildings recorded about 300 gals (more than 300 gals is estimated at the roof), and this is about three times the acceleration recorded in the basement. Similar value of the ratio between top and bottom accelerations of the building is found in some steel reinforced concrete buildings in Tokyo for rather small earthquakes, and also in the results of response analysis for 15 to 20 story buildings represented as a multi-mass vibratory system. Therefore, the top of the building will be subjected to more than gravity acceleration, if the ground accelerations of 300 gals or more is expected. For this reason it is necessary to consider the large amount of horizontal acceleration for the design of appendages at the roof, attached chimneys, parapets, water tanks, curtain walls, furnitures, etc. for preservation of public peace.

b) Considerations for Deformation

The relative displacement of a story during an earthquake is expected to be in the range of 1/200~1/150 of story height for steel buildings and 1/500~1/300 for reinforced concrete buildings. These displacements may cause the deformation of non-structural elements of the building such as covering for fire protection, partition walls, external walls, piping units and elevator shafts. The earthquake damage to ordinary buildings shows that in many cases the finishings, partition walls, doors, windows and other non-structural elements suffer damage prior to structural elements.

From the economical point of view the damage to non-structural elements may be allowed to some extent during severe earthquakes. However, the falling of exterior finishings and window glasses often kill or injure the people who are just outside the buildings, especially, pieces of stone or tile plastered on the exterior walls and masonry curtain walls are most frequent to fall down, and therefore, it is important to prevent their falling down. The window glass of large size is also dangerous having the large possibility of falling down as a result of cracking, while those of small size are considered to be almost harmless.

There is an instance in a foreign country that the partition elements are either self-standing on each floor or hung from the above floor slab so that they are able to withstand large story displacements.

On the effect of earthquake motion on the damage to non-structural elements of the building special attention should be paid to both displacements and accelerations in the process of structural design as explained in present and previous sections.

1.3.5 Specially Designed Structures

The ideas of earthquake isolation structures or those of decreasing earthquake forces acting on structures have been presented by many engineers, although most of them are limited to experimental stage except few cases which have been put into practice. It may be possible in the future to develop such ideas, for instance those to have an energy absorbing story in the building in order to isolate the earthquake ground motion, particularly for an important part of building or the machineries which are especially weak against vibrations.

The following are the ideas which have been presented to date.

a) On the Idea for Wooden Structure in which the Sill Slides Freely on the Foundation Having no Connection between Them

This kind of structure has been actually built without the intention of earthquake isolation, and it gave satisfactory results in past severe earthquakes. However, its displacement is very large, sometimes amounting to one meter and it is not easy to prevent slipping off. Furthermore, it is difficult to arrange wiring, piping, bath room and toilet properly, and there may be some trouble with wind forces.

b) Isolation by Means of Cylinders or Spheres Placed between Foundation and Superstructure

In the idea a) the earthquake force up to frictional force between foundation and superstructure can be transmitted but the maximum earthquake force to be transmitted is reduced by using cylinders or spheres which have less friction than in idea a). The weak points of this idea is similar to those in a).

c) Isolation by Elastic Elements such as Springs

This idea is to reduce the earthquake force which acts on structures by using flexible members or frames. This has the advantage of preventing the occurance of permanent deformation.

d) Control of Earthquake Force by Mechanical Equipment

Recently, the development in the method of automatic control makes it possible to manufacture the oil pressure system of large capacity controlled by electric signal through servo-valves, and this is applied to the method to isolate from earthquake forces in which the structure is subjected to oil pressure controlled by the amount of measured vibration in order to diminish the displacement of structures during an earthquake.

The ideas mentioned above have some questionable points such as allowing local failure or partial large displacement which may cause some trouble with the other functions of buildings, and this should be kept in mind. Furthermore, the practical study for large scale structures like actual buildings is needed for further development of these ideas.

e) Dynamic Control System for Vibration

To hang a body having far different natural period from that of the structure similar to dynamic control system which is popular in mechanical engineering. This idea was applied to the structure of a steam power station in which the boiler was hung to give the effect of vibration control to the structure.

2. EARTHQUAKE RESISTANT CALCULATIONS

2.1 Basic Principles in Earthquake Resistant Design of Buildings

In general, earthquake motion has both horizontal and vertical components but in the present antiseismic design method in Japan, only the horizontal movement is considered. However, the effect of the vertical movement is considerable for large span structures, so it cannot be neglected in all cases.

With the recent progress in dynamic analysis techniques, the effect of the horizontal earthquake motion on structures has been taken into consideration in the actual design even in the case when statical method is employed. The present design codes used in many countries of the world may be classified into the following three types:

i) The seismic forces which act on the building shall be determined as the equivalent static loads using the statically evaluated seismic coefficients or the shear force coefficients. The section of members shall be determined from the stresses and strains caused by these equivalent loads.

ii) The seismic forces which act on buildings shall be determined from the response spectrum, considering various factors such as the ground characteristics, natural periods and damping coefficients of buildings. Then the deflection and story shear force can be calculated and the stresses and sections of each member determined[1),2)].

iii) Considering subsoil conditions, earthquake motions shall be selected from past strong earthquake records and by means of structural response analysis for them, the sections shall be revised after considering the amount of deflection and the stresses[3)].

Among these three methods, the intensity of the ground motion has to be given in the form of maximum acceleration in methods ii) and iii). However, some problems such as the energy dissipation between the ground and foundation, amplification factor of the ground, and the relationship between the previously recorded ground motions and statistically expected ones are still unsolved. Therefore, it is rather difficult to determine the earthquake motion which acts at the bottom of buildings.

At present, the method being used in Japan is as follows:

i) Considering the seismicity, the whole country is divided into three zones and the maximum reduction of design seismic force in the least dangerous area is 20%[4)].

ii) According to the soil conditions, the soil classification is defined. The seismic coefficient shall be increased or decreased according to the type of construction.

The above ideas are based on the seismic risk decided from past statistical distribution of earthquake damage, recognizing the possible difference in the maximum ground acceleration of possible future earthquakes. The latter modifies the seismic coefficient method considering the general seismic hazard as the combination of amplification factor of the ground and the type of construction. At present, the stresses in the structure are calculated for one horizontal component of the earthquake motion, while the effect of the other component is neglected. However, the actual earthquake ground motion is a complicated movement having vertical and two horizontal components. Considering the wave velocity, the movement of each part in a plane is not always equal and will possibly produce torsional vibration of the building. In many foreign codes, calculation for torsion is required. It should be noted that beams, slabs, parapets of cantilever type are often in danger unless their vertical movement is considered as they are statically determinate members.

In the Building Standard Law, the structures for which antiseismic calculation is required are as follows: (Building Standard Law, Chapter 6)

i) Buildings of more than 100 m^2 in floor area used for schools, hospitals, clinics, theaters, cinema houses, markets, public buildings, meeting halls, department stores, dormitories, and garages.

ii) Wooden structures of more than 3 stories or 500 m^2 in floor area.

iii) Any structure other than wooden structures which has more than two stories or 200 m^2 in floor area.

iv) Except those described in i), ii), and iii), the structures existing in the city-planned area or in the area designated by the governors of prefectures.

The buildings described above are generally required to follow the antiseismic calculations described in Chapter 2.2. Chimneys, advertising towers, elevated water tanks, retaining walls, etc. of large scale (Building Standard Law, Construction Art. 38) are also required to follow the antiseismic calculations, as described in Chapter 2.2. High-rise buildings of over 45 m high are not required to follow the antiseismic calculation method described in Chapter 2.2 but their structural strength must be checked to be more than the equivalent. For high-rise buildings, the Architectural Institute of Japan recommends the antiseismic calculation indicated in "The Guide for Design of High-Rise Buildings". This calculation is based on the dynamic response analysis method.

2.2 Current Earthquake Resistant Calculation Methods

2.2.1 General

Article 20 of the Building Standard Laws require that structures must be safely designed against dead load, live load, snow load, wind pressure, soil pressure, water pressure and seismic load. In Japan, the structural design and calculations mean exclusively antiseismic calculations, in the process of which all the other forces as described above should be considered.

It seems quite reasonable that buildings should be safely designed against seismic forces. However, the meaning of "safely designed" is not perfectly clear. Does it mean that the buildings must bear seismic forces without any damage, or with small damage to the buildings but without any loss of human lives? This different understanding depends not only on the difficulty in accurate estimation of the earthquake intensity but also on the fact that "safely designed" gives different definitions according to the type of the structure and importance of the building under consideration. Generally speaking, some effective means are needed to make the damage slight and deflection produced by seismic forces small. These considerations, of course, are related to the economy of the building.

According to past earthquake damage investigations, wooden buildings, steel structures and reinforced concrete structures, although damaged by past strong earthquakes, are still regarded as safe structures. On the other hand, as the basic concept of earthquake resistant design, it is considered that the masonry structures for which the deflection limit is very small cannot be allowed to suffer even slight damage in the event of earthquakes. In the earthquake resistant calculation of ordinary buildings, only the horizontal component of ground motion is taken into consideration as the external load while the vertical component is neglected. This is based on the concept that the safety factor for the permanent vertical loads may cover the safety of the building for the vertical dynamic loads. However, in the case of important structures such as atomic power generators, the vertical motion of the earthquake should be considered and the vertical seismic coefficient of about 1/2 of the horizontal seismic coefficient should be taken into account in the design.

In structural calculations, the loads and external forces that act on the structure are the following.

i) Dead load
ii) Live load
iii) Snow load
iv) Wind pressure
v) Seismic force
vi) Soil pressure, water pressure or impact force which shall be applied to each building according to its actual condition of use.

The design stresses shall be determined from the temporary stress and permanent stress for the main structural parts of the building for the above mentioned loads and external forces, according to Table 2.1, and the combination of the forces that makes the structural member most disadvantageous shall be used in the design. At the same time, the deflection of the

Table 2.1 Combination of Loads

Kind of stress	Conditions estimable regarding loads and external forces	General case	Case of snow districts
Permanent stress	Normal time	$G+P$	$G+P+S_2$
Temporary stress	Snow season	$G+P+S_1$	$G+P+S_1$
	Storm	$G+P+W_1$	$G+P+W_1$ $G+P+S_3+W_2$
	Earthquake	$G+P+K$	$G+P+S_3+K$

Note: For checking the overturning of a building or pulling out of a column, live load shall be reduced in accordance with actual condition of the building.

structural members shall, in necessary cases, be checked to make sure not to affect the convenient usage of the building.

In Table 2.1, symbols G, P, S_1, S_2, S_3, W_1, W_2, K represent the following stresses (axial stress, bending stress, shearing stress)

G : stresses due to dead load

P : stresses due to live load

S_1 : stresses due to snow load

S_2 : in heavy snow district (70% of the maximum vertical depth of the snow fall in that district)

S_3 : in heavy snow district (35% of the maximum vertical depth of the snow fall in that district)

W_1 : stresses due to typhoon

W_2 : stresses due to cyclone

K : stresses due to seismic loads which are determined by consideration of the soil condition and the type of construction. The quantative values of these loads are indicated in 2.2.3.

There are some exceptions in the Building Standard Laws that legally defines the structural calculation of certain buildings in Japan. For instance, "Guide for the Design of High-Rise Buildings" (March 1964) shall be (tentatively) refered to for the structural calculation of high-rise buildings exceeding the height limit specified in Article 57 of the Building Standard Law.

2.2.2 Design Seismic Coefficient K, and the Basic Value K_0

In the antiseismic calculations for buildings, the fundamental problem is to determine the design seismic loads. In the present antiseismic calculations, the design seismic force is obtained as the sum of the dead and live loads for the design seismic force (also the snow load in the case of heavy snow districts) multiplied by the horizontal seismic coefficient.

The Building Standard Laws require that the value of the design horizontal seismic coefficient shall be determined by considering the strength and safety factor of the allowable unit stresses for the temporary loads. However, as the destructive effect of earthquakes on the buildings depends on the ground condition, the type of construction and also the seismicity which is known from experience in some districts, the design horizontal seismic coefficient K for the superstructure must be more than the value obtained from the following formula.

$$K = K_0 \times \alpha \times \beta \quad \cdots\cdots (2.1)$$

provided $\alpha \times \beta \geqq 0.5$

where K : the standard value of the design horizontal seismic coefficient

α : the modification factor of the seismic coefficient owing to the ground condition and the type of construction

β : reduction factor of the seismic coefficient for the district

The so-called "Standard Seismic Coefficient K" is equal to 0.2 for the superstructure up to 16 m in height, and for the portion over 16 m it shall be increased by 0.01 for each 4 m in height.

The values of α and β are shown in Tables 2.2 and 2.3.

It is considered that the seismicity differs even in a small country like Japan. The original plan for defining the coefficient β was made by zoning the country into 3 areas——where an

Table 2.2 Modification Factor (α) of the Seismic Coefficient According to the Ground Condition and the Type of Construction

Kind of ground \ Type of construction	Wooden construction	Steel framed construction	Reinforced concrete construction, steel reinforced concrete construction, or steel-concrete construction	Masonry construction
I. Ground consisting of rock, hard sandy gravel, etc. classified as tertiary or older strata over a considerable area around the structure.	0.6	0.6	0.8	1.0
II. Ground consisting of sandy gravel, sandy hard clay, loam, etc. classified as diluvial, or gravelly alluvium, about 5 meters or more in thickness, over a considerable area around the structure.	0.8	0.8	0.9	1.0
III. Standard ground except Kind II, consisting of sand, sandy clay, clay, classified as alluvium.	1.0	1.0	1.0	1.0
IV. Bad and soft ground which falls in one of the following categories : 1) Alluvium consisting of soft delta deposits, topsoil, mud or the like (including heaping up, if any), whose depth is about 30 meters or more. 2) Filled ground which falls in one of the following items. A. Land obtained by reclamation of a marsh, muddy sea bottom etc. B. Earth filling is bad and soft such as topsoil and mud. C. The depth of earth filling is about 3 meters or more. D. Thirty years have not yet elapsed since the time of reclamation.	1.5	1.0	1.0	1.0

earthquake of more than 200 gal is expected to occur in a century, where more than 100 gal earthquake is expected to occur in a century, and where no earthquake of more than 100 gal is expected in the same period. The data were provided by the Earthquake Research Institute of the University of Tokyo (see Fig. 1.2).

This chart was drawn to indicate the expected scale and frequency of future earthquakes in different areas, statistically arranging the past earthquake records. Many opinions were proposed on making the zoning for the areas where the seismicity and the design seismic coefficient differs, though this method of zoning should be permitted under the present circumstances where earthquakes can be hardly predicted. The main opinions for adopting the original zoning plan were as follows :

(1) Future earthquakes cannot be predicted from the past data alone. For instance, Fukui

Earthquake was not expected from the statistical data of the past. An earthquake which occurs in the unexpected area often brings fatal damage to the district. Therefore, the seismic coefficient of the area along the "Inner Earthquake Zone" which runs along the coast of Sea of Japan should not be made less than that of Tokaido area although the statistical expectation of a big earthquake is small.

(2) As an objection to (1), it has been argued that, if the seismic coefficient is made large, the increased section of the structural members would make the construction uneconomical, and that in the area along the Japan Sea heavy snow load must be taken into account in the calculation of the seismic force so that proper size of sections is retained. Therefore, the seismic coefficient on the coast of Sea of Japan should not be the same as that of Tokaido area.

Table 2.3 Reduction Factor (β) of Seismic Coefficient

	Districts	Numerical value
(1)	In Hokkaido : Asahikawa-shi, Kitami-shi, Abashiri-shi, Rumoi-shi, Wakkanai-shi, Uryū-gun, Nakagawa-gun (Teshio-koku), Kamikawa-gun (Teshio-koku and Ishikari-koku), Monbetsu-gun, Tokoro-gun, Abashiri-gun, Shari-gun, Soya-gun, Esashi-gun, Rishiri-gun, Rebun-gun, Teshio-gun, Tomamae-gun, Rumoi-gun and Mashike-gun ; Yamaguchi Pref., Fukuoka Pref., Saga Pref., Nagasaki Pref., Kumamoto pref., Oita Pref., Miyazaki Pref., Kagoshima Pref.	0.8
(2)	In Hokkaido ; districts other than those mentioned in (1) ; Aomori Pref., Iwate Pref., Akita Pref., Miyagi Pref., Yamagata Pref., Fukushima Pref., Ibaragi Pref., Tochigi Pref., Gumma Pref., Niigata Pref., Toyama Pref., Ishikawa Pref., Fukui Pref., Tottori Pref., Shimane Pref., Okayama Pref., Hiroshima Pref., Tokushima Pref., Kagawa Pref., Ehime Pref., Kochi Pref.	0.9

(3) In the area where a large earthquake has occured, expectation of another severe earthquake may be great, and an earthquake of very large intensity might be dormant. For this reason, the seismic coefficient of the area from Oita district to Unzen in Nagasaki district where "Setouchi" earthquake zone exists should be somewhat greater than other areas in Kyushu.

(4) For a structure such as a reinforced concrete building that has a long life expectancy and is supposed to have many chances to encounter large earthquakes, it is not right to deal with it in the same way as those of wooden construction which have a shorter period of life. That is, for the wooden structures the danger zone should be limited as much as possible, while for reinforced concrete structures the danger zone must be enlarged. In this manner, the danger zones should be varied according to the type of construction.

(5) In the same manner, the seismic coefficient for office buildings should differ from that of theaters or cinemas where a large number of people gather together, even though the type of construction is the same. The seismic coefficient of cities should also vary according to their importance. For instance, expectation of large earthquakes is small in northern Kyushu area but the seismic coefficient should be increased, because of the reason that heavy industry is developing there and the terrible consequences of damage from a severe earthquake must be considered.

(6) An objection to (5) is that the principles of the Building Standard Laws require the minimum safety of all structures and different safety factors for various areas should not be allowed.

(7) The argument is advanced that due to lack of decisive data for seismic zoning, it should be reasonable to make Hokkaido and Kyushu zoning into one group and the other areas into another group.

Considering the original plan and the opinions listed above and for administrative convenience, the plan to divide the whole country into three areas, using county boundaries in Hokkaido and prefectural boundaries in the other areas, was finally adopted and is as shown in Table 2.3 (see Fig. 2.1).

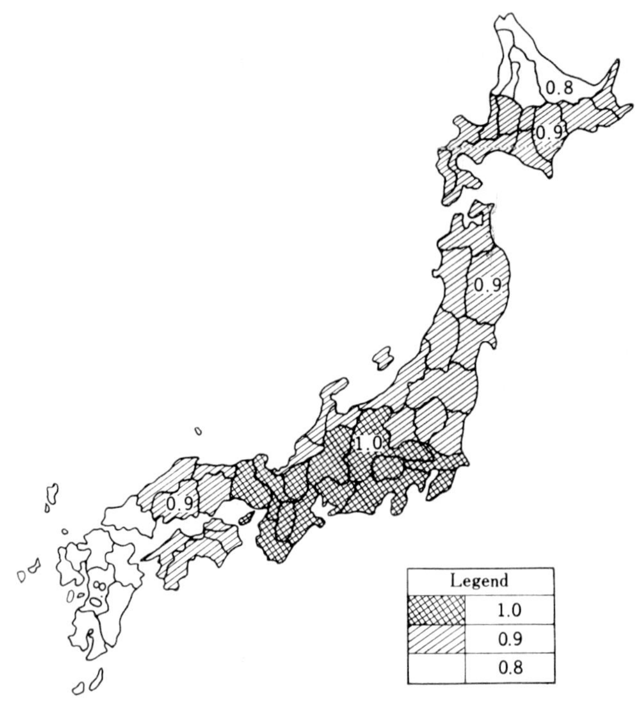

Fig. 2.1 Diagram Showing the Rate of Reduction of Lateral Seismic Coefficient

The value of K determined by the formula (2.1) is the minimal value of the lateral seismic coefficient that should be applied to the superstructure of ordinary buildings. For the substructure, values of the seismic coefficient are not specified. Under present circumstances, the accelerations applying to the underground structure and its behavior are not well known. However, for practical design purposes, the reduction of the underground seismic coefficient to 0.1 or linear interpolation between zero at the bottom of the foundation and 0.2 at the ground surface are often assumed.

For parapets of buildings, interior and exterior ornaments, water tanks on the roof level, chimneys and the like, the standard design seismic coefficient is required by the Building Standards Law to be more than 0.3.

The "Guide for the Design of High-Rise Buildings" published by the Architectural

Institute of Japan has taken great account of things attached to the buildings, and proposed to take especialy large seismic forces for them (For instance, a proposal is that the seismic coefficient of the parapets and ornaments be 0.5 while that of water tanks, chimneys etc. be more than 1.5 times the dynamic shear force coefficient of the top floor). For the lower buildings as well as the high-rise buildings, it is desirable to assume adequate lateral forces in antiseismic calculations.

2.2.3 Loads

a) Dead Load

Table 2.4

Part of Building	Classification			Weight (kg/m²)		Remarks
Roof	Tile roof	Without soil		For face of roof	65	Including roofing and rafters but not including purlins
		With soil			100	
	Asbestos slate roof	Placed directly over purlins			25	Not including purlins
		Other cases			35	Including roofing and rafters but not including purlins
	Corrugated iron sheet roof directly on purlins				5	Not including purlins
	Iron sheet roof				20	Including roofing and rafters but not including purlins
	Glass roof				30	Including steel frame but not including purlins
	Thick type slate roof				45	Including roofing and rafters but not including purlins
Wooden purlins	In case of span length of purlins 2 m or less			For face of roof	5	
	In case of span length of purlins 4 m or less				10	
Ceiling	Common moulding and strip			For face of ceiling	10	Including wood hangers, strips and other grounds.
	Fiber sheet, wood board, ply wood or metal sheet				15	
	Wood fiber-cement sheet				20	
	Cross framed moulding				30	
	Plaster finish				40	
	Mortar finish				60	
Floor	Floor for wooden building	Wood floor		For floor face	15	Including joists
		Tatami (Japanese mat)			35	Including floor boards and joists
		Floor beam	In case of span 4 m or less		10	
			In case of span 6 m or less		17	
			In case of span 8 m or less		25	

Part		Item		Value	Remarks
Floor	Finish on concrete floor	Wood boards	For floor face	20	Including joists and beams
		Flooring blocks		15	For every 1 cm of thickness it shall be multiplied by the number equal to that value in cm
		Mortar finish, artificial stone finish and tile finish		20	
		Asphalt waterproof layer		15	For every 1 cm of thickness it shall be multiplied by the number equal to that value in cm
Wall	Wall frames for wooden buildings		On wall face	15	Including column, strut and bracing
	Finish of walls in wooden building	Clapboards, shingle or fiber board		10	Including the base material but not including frame works
		Plaster on wood lath		35	
		Mortar on metal lath		65	
	Soil plastering wall (Komaikabe) in wooden building			85	Including frame-work
	Finish of concrete wall	Plaster finish		17	For every 1 cm of thickness of the finish, it shall be multiplied by the number equal to that value in cm
		Mortar and artificial stone finish		20	
		Tile finish		20	

(1) The dead load of parts of a building is specified in Article 84 of the Building Standard Law Enforcement Order. For the dead load of each part of the building, the values in the following table must be taken or it must be calculated by considering the actual conditions of the building.

(2) The weight of reinforced concrete depends on actual conditions in each case. When no special investigation is made, the values given in Table 2.5 may be used.

Table 2.5 Dead Weight of Reinforced Concrete Construction

Kind of concrete		Dead weight of reinforced concrete construction (t/m^3)
Ordinary concrete		2.4
Light weight aggregate concrete	LC 150	2.0
	LC 120	1.8

(3) In addition to the loads mentioned above, special loads, if any, shall be considered.

b) Live Load (the Article 85 of the Building Standard Law Enforcement Order)

(1) The live loads of each part of the building, to be used for structural calculations may be determined from Table 2.6, except when special calculations have been made.

(2) In the calculation of the axial forces in the section of foundation and columns, the live loads may be reduced to the values given in Table 2.7, according to the number of stories that this foundation and columns support.

(3) Live loads for warehouses shall be more than 400 kg/m^2 and shall be calculated in each case.

Table 2.6 Live Loads

Classification of rooms	For structural calculation in different cases : kg per sq.m.		
	(a) On calculating floor strength	(b) On calculating strength of girders, columns or footings	(c) On calculating for seismic forces
1) Ordinary room of residential building, sleeping room or patient's room of other than residential buildings	180	130	60
2) Office room	300	180	80
3) Class room	230	210	110
4) Sales room in department store or shop	300	240	130
5) Seatings or assembly room of theatre, cinema, entertainment hall, grand-stand, public hall, assembly hall or building available for other similar use	300	270	160 fixed seats
Do, other than fixed seats	360	330	210
6) Garage or passageway for automobiles	550	400	200
7) Corridor, vestibule or stairways	For room connected to those item (3) to item (5) inclusive, the value of other than fixed seats of item (5) is to be taken		
8) Open space on roof or balcony	Value of item (1) to be used. However, for buildings used as school or department store, take the value in item (4)		

Table 2.7 Live Loads for Calculation of Axial Stresses in Column and Foundation

Number of floors supported	1	2	3	More than 4
Valuc to bc multiplicd for reducing live load	100%	95%	90%	5% reduction shall be added for each floor, unless the value becomes less than 60%

c) Snow Load

(1) Snow load is specified in Article 86 of the Building Standard Law Enforcement Order as follows:

Snow load shall be calculated from the unit snow load multiplied by the maximum depth of snow fall in the district.

(2) Heavy show district is defined, by the Administrative Construction Order Paragraph 1074, to be "the area where the vertical maximum snow depth exceeds 1 m".

(3) As a reference to know the actual amount of snow in the provinces, see the "Order and Its Explanation on Load and External Force" (published by AIJ, Nov., 1952) page 13, Fig. 1

which can be regarded as the standard for refining the heavy snow district.

(4) Snow load on the roofs having the slope over 30 deg. but less than 60 deg. shall be figured, excepting there are strips to prevent snow from sliding down from the roof, by multiplying the snow load prescribed in paragraph 1 with the values shown in the following table according to its slope (When the administrative agency concerned determines a different value from those in the table in consideration of the roof materials, property of snow and so forth by its regulation, that determines value). However, if the slope is steeper than 60 deg., the snow load can be neglected.

Slope	(In degrees) Case for 30∼40	Case for 40∼50	Case for 50∼60
Value to be multiplied for the snow load	0.75	0.50	0.25

(5) Snow load in the heavy snow district prescribed in paragraph (2) treated as a semi-permanent load can be reduced to 70% of the value obtained by calculating according to the provisions of the preceding four paragraphs and snow loads counted to act together with wind pressure of seismic force can be reduced to 35% thereof.

(6) If there is possibility of snow piling on the roof in an unbalanced way, its influence shall be considered in the structural calculation.

(7) In the locality where the snow is removed from roofs from time to time, even in case the snow depth in that locality exceeds 1 m., snow depth can be reduced, according to the actual condition of removing snow, to depth of 1 m. for the calculation of snow load.

d) Wind Pressure, Soil Pressure, and Other Special Loads

Structures must be safely designed and constructed for earthquakes. Much consideration should also be paid to wind pressure and soil pressure which are applied to the building as horizontal forces.

It is apparent that there are many wooden or very light steel structures capable of resisting seismic forces that are damaged by wind.

Several cases have been reported that a serious problem for the building was induced from lack of enough consideration of soil pressure working on the underground structure. Even though remarkable researchs on soil dynamics during an earthquake are being conducted, importance of these loads as well as of the seismic load shall be only indicated herein. As for their quantative values, it is desirable to refer on the "AIJ Standard for Structural Calculation of Steel Structure and Commentary" and the "AIJ Standard for Structural Design of Building Foundation".

Excepting the loads prescribed in the preceding paragraphs, the special loads which shall be considered are as follows :

i) Load of travelling cranes or of other conveyance machines.

ii) In case of the structure which supports motors and power conveying machines, tensile forces of belts as well as the dead load and also vibration of the rotating shaft shall be considered.

iii) The load of the equipment attached to the building.

iv) Temperature stress, in case of the structure to which a large temperature stress works according to the type and shape of the structure.

v) Stress produced during operation and construction.

These loads and stresses have different importance according to the condition of the design and construction in each case. Especially for items i), ii), and iii), enough consideration shall be paid not only to the vertical and horizontal loads during the operation of the equipments and machines, but also to those in the event of an earthquake. Many cases have been reported that during an earthquake, the damage to building has become greater because of inadequate connection of each attached equipment and the structural frame or insufficient consideration for horizontal earthquake load. The loads of travelling cranes are found in the "AIJ Standard of Structural Calculation of Steel Structures and Commentary".

2.2.4 Elastic Moduli and Allowable Unit Stresses

a) Physical Constants of Materials

(1) Physical Constants of Steel

Young's modulus of structural steel is about 2.1×16^6 kg/cm^2. Young's modulus of P.C. steel lies between $1.9\times10^6\sim2.1\times10^6$, and the value of 2.0×10^6 kg/cm^2 can be used for the calculation on design and construction of P.C. steel structures when tensile tests are omitted.

Table 2.8 Constants for Structural Steel

Material	Young's modulus (t/cm^2)	Modulus of rigidity (t/cm^2)	Poisson's ratio	Coefficient of linear expansion
Steel, cast steel	2,100	810	0.3	0.000012/°C
cast iron	1,000	380	—	—

(2) Young's Modulus of Concrete

Since Young's modulus of the concrete used for reinforced concrete structure varies according to the magnitude of stresses, properties and mixing of cement and aggregate, in the "AIJ Standard for Structural Calculation of R.C. Structures and Commentary" the value $2.1\times10^6/n$ (kg/cm^2) for Young's modulus is determined, where n is determined from Table 2.9 according to the kind of concrete, stress condition, permanent or temporary, etc.

Table 2.9 Ratio of Modulus of Elasticity of Steel and Concrete

Kind of concrete		For stress computation	For design of sections	
			Permanent	Temporary
Normal weight concrete	C 225	10	21	14
	C 180	10	24	16
	C 135	10	30	20
Light weight concrete	LC 150	15	35	25
	LC 120	12	40	30

(3) Young's Modulus of the Concrete Used for Prestressed Concrete Structure

Young's modulus of the concrete used for the P.S. concrete structures shall be determined from the compression strength test defined in chapter 2 of the "AIJ Standard for Structural

Design and Construction of Prestressed Concrete Structure and Commentary".

The moduli for calculation without any test results can be taken from Table 2.10 for both tension and compression.

Table 2.10 Young's Moduli of Concrete

Compressive strength of concrete (kg/cm^2)	Young's modulus (kg/cm^2)
300	270,000
400	320,000
500	360,000

(4) Young's Modulus of Lumbers

In general, Young's modulus shall be determined in accordance with Table 2.11 by taking the following provisions into consideration.

i) In the calculation of deformation for principal members or for single members in which the deformation is considered to be most important, the values of E shall be 80% of that shown in Table 2.11.

ii) In the calculation of deformation for the permanent loading, the values in Table 2.11 shall

Table 2.11 Young's Moduli of Lumbers (10^3 kg/cm^2)

			Species	$E_{\parallel}$	$E_{\perp}$
Ordinary structural lumbers	Softwoods	Group I	Japanese red pine, Japanese black pine, Japanese larch, Japanese cedar, Japanese cypress, Southern Japanese hemlock, Douglas fir, Port Orford cedar	90	2.5
		Group II	Cryptomeria[1], Momi fir, Yezo spruce, White fir, Western red cedar, Western hemlock	70	
	Hardwoods	Group I	Live oak	100	4.0
		Group II	Japanese chestnut, Japanese white oak[2], Japanese beech, Zelkova[2], Apitong	80	
		Group III	Lauan	70	3.0
Higher grade structural lumbers	Softwoods	Group I	Japanese red pine, Japanese black pine, Japanese larch, Japanese cedar, Japanese cypress, Southern Japanese hemlock, Douglas fir, Port Orford cedar	100	3.0
		Gorup II	Cryptomeria[1], Momi fir, Yezo spruce, White fir, Western red cedar, Western hemlock	80	
	Hardwoods	Group I	Live oak	100	4.5
		Group II	Japanese chestnut, Japanese white oak[2], Japanese beech, Zelkova[2], Apitong	90	
		Group III	Lauan	80	4.0

Note :

(1) For crytomeria of inferior quality (specific gravity less than 0.3, or average width of annual ring more than 6 mm), the values shown in the table shall be reduced by 30 percent.

(2) For Japanese white oak and for zelkova, average width of annual ring shall be more than 1 mm.

be reduced to one-half.

iii) For the bearing capacity perpendicular to the grain, the values in Table 2.11 may be doubled according to the conditions. In the case of bearing on bastard grain, the values in Table 2.11 shall be reduced to one-third.

b) Allowable Unit Stress

(1) Steel

Allowable unit stresses of structural materials and allowable stresses of wire rope shall be the values given in Tables 2.12~2.16.

i) Materials

Table 2.12 Allowable Unit Stresses of Structural Materials (t/cm²)

Material	Type of Stress	Tension	Compression	Bending	Shear	Bearing	Contact
Structural Steel	Common Steel	1.4	1.4	1.4	0.8	2.6	4.0
	SS 41, SM 41	1.6	1.6	1.6	0.9	3.0	4.6
	SS 50	2.0	2.0	2.0	1.2	3.8	5.8
	SM 50	2.2	2.2	2.2	1.3	4.1	6.3
Rivet SV 34, SV 41		1.6	—	—	1.2	—	—
Bolt : SS 41	Unfinished Bolt	0.8	—	—	—	—	—
SS 41	Finished Bolt	1.0	—	—	1.2	—	—
Cast Iron		—	1.0	—	—	—	2.8

Note :

(1) Allowable unit stresses listed above are for the case of permanent loading. The values in the above table should be multiplied by 1.5, for a temporary loading.

(2) When large normal and shearing stresses act simultaneously, the combined effect of these two stresses shall be taken into consideration.

(3) For structural parts which are subjected to high temperatures for prolonged periods, the allowable stresses shall be reduced in consideration of temperature and creep.

ii) Connection

Table 2.13 Allowable Unit Stresses of Rivet and Bolt Connection (t/cm²)

Materials / Type of Stresses	For permanent stress: Tension	For permanent stress: Shearing	For temporary stress
Rivet SV 34, SV 41	1.6	1.2	1.5 times the values for permanent stress
Bolt SS 41 Black bolt	0.8	—	
SM 41 Medium bolt	1.0	1.2	

Note :

(1) Allowable stress of rivets shall be calculated for the diameter of rivet hole.

(2) Allowable stress of bolts shall be calculated for the sectional area of bolts.

iii) Thin Plate

Table 2.14 Allowable Unit Stress of Thin Plate (t/cm²)

	For permanent stresses				For temporary stresses
	Tension	Compression	Shear	Bearing	
SSC 41	1.4	1.4	0.8	2.4	1.5 times the values for permanent stress

iv) Tubular Steel

Table 2.15 Allowable Unit Stresses of Structural Tubular Steel (t/cm²)

Grade of Steel	Thickness	Permanent loading				Temporary loading
		Tension	Compression	Shear	Bearing	
STK 41	Under 4 mm	1.4	1.4	0.8	2.4	1.5 times the values for permanent load-ing
	4 mm or over	1.6	1.6	0.9	3.0	
STK 50	Under 4 mm	2.0	2.0	1.2	3.8	
	4 mm or over	2.2	2.2	1.3	4.1	

v) Allowable strength of steel wire and wire ropes, for the permanent and temporary loadings, shall be one-fourth of the values given in Table 2.16.

Table 2.16 Limitation of Strand and Wire Ropes

Materials	Standard Specifications
High tension rope	JIS G 3506, JIS G 3521 JIS G 3534, JIS G 3537 JIS G 3525
Wire rope	

vi) Allowable Tensile Unit Stresses of P.C. Steel Bars

Allowable tensile unit stresses of P.C. steel bars, σ_p, shall be the values in Table 2.17 ; before getting into final anchoring, it shall be within 1.05 times the values of the allowable unit stress.

Table 2.17 Allowable Unit Stresses of P.C. Bars

Kind of P.C. bar	Allowable unit stress (kg/cm²)
Kind 1.	4,500
Kind 2.	5,500
Kind 3.	6,500
Kind 4.	7,500

vii) Allowable Unit Stresses of Arc-Welding

Allowable unit stresses of arc-welding for throat depth shall be determined from Table 2.18.

Table 2.18 Allowable Unit Stress of Welded Joints (t/cm^2)

Steel	Operational* Conditions	Permanent loading: Butt joint: Tension	Butt joint: Compre-ssion	Butt joint: Bending	Butt joint: Shear	Permanent loading: Fillet joint	Temporary loading	Remarks
Common Steel**	(1) (2)	1.2	1.2	1.2	0.7	0.7	The figures will be 1.5 times those for compression, tension, bending and shear of permanent loading	**Limited to materials suitable for welding. (Include cast steel and forged steel)
SS 41	(1) (2)	1.4	1.4	1.4	0.8	0.8		
SM 41	(2)	1.4	1.4	1.4	0.8	0.8		
	(1)	1.6	1.6	1.6	0.9	0.9		
SM 50	(2)	2.0	2.0	2.0	1.2	1.2		
	(1)	2.2	2.2	2.2	1.3	1.3		

Note :

* (1) The case of work positioned for flat welding, by means of universal jig, positioner, etc. as provided for in Art. 92 of the Building Standard Law Enforcement Order.

(2) The other case than the above case (1).

Attention should be paid to the type of welded joints and the sequence of welding ; otherwise a large amount of shrinkage stress will develop due to the welding. If an undesirable type or process of welded joints has to be taken, the design unit stresses should be reduced.

Table 2.19 Stress Increase Coefficients

Type of welding	Number of repetition	$\frac{N_2}{N_1}$	γ: Common steel, SS 41, SM 41	γ: SM 50
Butt welding: In case where only one side is welded, coefficients for the fillet welding shall be used	$>2\times10^6$	1.0～0.5	1	1
		0.5～−1	$1.40-0.8\frac{N_2}{N_1}$	$1.67-1.34\frac{N_2}{N_1}$
	$>6\times10^5$	1.0～0.3	1	1
		0.3～−1	$1.21-0.7\frac{N_2}{N_1}$	$1.37-1.24\frac{N_2}{N_1}$
	$>1\times10^5$	1.0～0.0	1	1
		0.0～−1	$1.00-0.5\frac{N_2}{N_1}$	$1.00-1.06\frac{N_2}{N_1}$
Fillet welding : In case of the web-to-flange joint of plate girders by means of continuous fillet welding, coefficients for the butt welding shall be used.	$>2\times10^6$	1.0～0.5	1	1
		0.5～−1	$1.80-1.6\frac{N_2}{N_1}$	$2.30-2.60\frac{N_2}{N_1}$
	$>6\times10^5$	1.0～0.3	1	1
		0.3～−1	$1.30-\frac{N_2}{N_1}$	$1.54-1.80\frac{N_2}{N_1}$
	$>1\times10^5$	1.0～0.0	1	1
		0.0～−1	$1.00-\frac{N_2}{N_1}$	$1.00-1.90\frac{N_2}{N_1}$

Notes :

N_1 : absolute value of the larger stress

N_2 : absolute value of the smaller stress

1. When stresses N_1 and N_2 are tension and compression or vice versa, N_2/N_1 shall be taken as negative, and when they are of the same sign, N_2/N_1 shall be taken as positive.
2. Stress increase coefficients for plates without welded joints take same values as fatigue coefficients for riveted joints subjected to repeated stresses.
   In this case, the permanent allowable unit stress of steel members in Table 2.18 are to be used.

As to the welded structures which receive repeated stresses, the working stresses shall be multiplied by the factors listed in Table 2.19 and shall be checked by the allowable stresses for permanent loads in Table 2.18.

When the number of stress cycle is below 1×10^5, it is not required to consider the effect of fatigue.

(2) Concrete

i) Allowable unit stresses of concrete and reinforcing bars shall be in accordance with the values shown in Tables 2.20 and 2.21.

Table 2.20 Allowable Unit Stresses in Concrete and Reinforcing Bars (kg/cm²)

Material		For permanent stresses			For temporary stresses
		Tension	Compression	Shear	
Concrete		$\frac{1}{30}F_c$	$\frac{1}{3}F_c$	$\frac{1}{30}F_c$	2 times the values for permanent stresses
Reinforcing bar	Common steel	1,400	1,400	—	1.5 times the values for the permanent stresses
	SS 39, SSD 39 SRB 39, SRD 39	1,600	1,600		
	SS 49, SSD 49 SRB 49, SRD 49	2,000	2,000		

F_c : 4 week compressive strength of concrete for structural design. For the ordinary concrete with natural aggregates, $F_c=135\sim250$ kg/cm², and for the light weight aggregate concrete using light weight gravel and ordinary sand, $F_c=120\sim150$ kg/cm².

Table 2.21 Allowable Bond Stresses per Unit of Surface Area of Bars in Concrete (kg/cm²)

Material	For permanent stresses			For temporary stresses
	Top bar in *1) flexural members	General bar in flexural members	Anchorage and lap joint	
Plate, rolled steel	$0.03\,F_c$ *2)	$0.03\,F_c$	$0.02\,F_c$	2 times the values for the permanent stresses
Plain bar	$0.04\,F_c$	$0.06\,F_c$	$0.04\,F_c$	
Deformed bar	$0.07\,F_c$	$0.10\,F_c$	$0.07\,F_c$	

* Note :

1) It shall not be greater than 250 kg/cm², if $F_c=250$ kg/cm².

2) In case of the reinforced light weight aggregate concrete, LC 150 shall be used if $f_t,\ _rf_c=2{,}000$ kg/cm² is permitted for SRB 49, SD 30.

ii) Allowable Unit Stresses of Light Weight Aggregate Concrete

Table 2.22

Allowable unit stresses for permanent load (kg/cm²)			Allowable unit stresses for temporary load (kg/cm²)		
Compression	Tenssion, Shear	Bond	Compression	Tension, Shear	Bond
One third of 4 week comp. strength of concrete, as well as less than 40.	One tenth of allowable unit stress.	6	2 times the allowable unit stress (compression, tension, shear, bond) for the permanent stresses.		

Allowable unit stresses of light weight aggregate concrete with $F_c = 120 \sim 150$ kg/cm^2 are listed in Table 2.19 while those of light weight aggregate concrete with $F_c = 90 \sim 120$ kg/cm^2 shall be determined from Table 2.22.

iii) Allowable Unit Stress of Reinforced Concrete for Prefabrication

Allowable unit stresses of reinforced concrete for prefabrication shall be in accordance with the Building Standard Law Enforcement Order Chapter 90 and Chapter 91. If the forming, curing and other factory equipment are satisfactory, if the complete checking of the materials and products is conducted and if the uniformity of the products is obtained, the allowable unit stresses of concrete can be taken, only for the products of that factory, up to 100 kg/cm^2 for permanent compressive loading, one-tenth of it for shearing, double of it for temporary loading, considering the conditions. For the joint mortar to join the masonry unit or members, the allowable unit stress for permanent shear forces shall be not greater than 4 kg/cm^2 and 40 kg/cm^2 for permanent compression. For the temporary loading the values shall be doubled in each case. If the mortar is especially well made, these values of allowable unit stresses can be multiplied by 1.5.

The allowable unit stresses for the wall in masonry construction shall be the values indicated in Table 2.23.

Table 2.23

Allowable unit stress / Kind of masonry const.	Allowable unit stress for permanent load		Allowable unit stress for temporary load
	Compression	Tension and shear	Compression, tension and shear
Brick and concrete block without hollow	Less than 1/8 of compressive strength of concrete block or less than 15 kg/cm^2	Less than 1/80 of compressive strength of concrete block or less than 1.5 kg/cm^2	1.5 times the value of allowable unit stress for permanent load
Hollow brick and hollow concrete block	Concerning net section, less than 1/12 of compressive strength of concrete block or less than 10 kg/cm^2	Concerning net section, less than 1/80 of compressive strength of concrete block or less than 1.5 kg/cm^2	

(3) Concrete Used for Prestressed Concrete Structures

i) Allowable Unit Compressive Stresses

Allowable unit compressive stress of concrete due to any possible combination of load at each period shown in Table 2.24 shall not exceed the values shown in Table 2.24, and 180 kg/cm^2 for in-situ concrete or 210 kg/cm^2 for concrete cast in the factory.

Table 2.24 Allowable Unit Compressive Stresses of Concrete

At transfer	Dead load is not affecting	$0.45\,F_{28}$*
	Dead load is affecting	$0.40\,F_{28}$*
At Design load		$0.35\,F_{28}$*

Note :

* F_{28} : compressive strength of concrete at 28 days

ii) Allowable Unit Tensile Stresses of Concrete

Allowable unit tensile stresses of concrete due to any possible combination of loads at each period as shown in Table 2.25 shall not exceed the values listed in Table 2.25, and 18 kg/cm^2 for in-situ concrete or 21 kg/cm^2 for concrete cast in the factory.

Table 2.25 Allowable Unit Tensile Stresses of Concrete

Full prestressing	At transfer	Dead load is not affecting	$0.07 f_c$*
		Dead load is affecting	0
	At design load		0
Partial prestressing	At transfer		$0.15 f_c$*
	At design load		$0.10 f_c$*

Note :
* f_c : allowable unit compressive stress at each period described in Table 2.24.

(4) Allowable Unit Stresses of Lumbers

i) Quality of Structural Lumbers

Grades of Structural Lumbers

Structural lumbers shall be classified into two groups as follows :

"Ordinary structural lumbers"

"Higher grade structural lumbers" which have higher strength and rigidity and excellent quality.

Provisions for Grades

Quality of ordinary structural lumbers shall conform to the corresponding provisions specified in JASS 11.

Quality of higher grade structural lumbers shall conform to the follwing items :

Sawn squares, small squares and boards of softwoods shall be better in quality specified in the "Japanese Agriculture and Forestry Standard for Lumber".

Specific gravity, width of annual ring and defects shall be in accordance with the provisions specified in JASS 11, and shall satisfy the requirements shown in Table 2.26.

ii) Allowable Unit Stresses Parallel to Grain

Ordinary Structural Lumbers :

The allowable unit stresses parallel to grain for ordinary structural lumbers shall be in accordance with Table 2.27 (a). For the species not shown in Table 2.27(a), allowable unit stresses of the species having approximately the same specific gravity as the species shown in Table 2.27 (a) shall be applicable.

Higher Grade Structural Lumbers :

The allowable unit stresses parallel to grain for softwoods (needle leaf trees) shall be in accordance with Table 2.27 (b). For hardwoods (broad leaf trees) proved to be of good quality, the allowable unit stresses shown in Table 2.27 (a) may be increased by 30 per cent. The allowable unit stresses for hardwoods of superior quality to be used as dowels may be double the value in Table 2.27 (a).

Table 2.26 Specifications for the Specific Gravity, Width of Annual Ring and Defects for Higher Grade Lumber

		Species	Minimum value of specific gravity at air-dried state (moisture content 15%)	Maximum value of average width of annual ring	Maximum value of ratio of knot diameter	Maximum value of wane	Maximum value of inclination of grain	Maximum value of initial deflection
Softwoods	Group I	Japanese red pine, Japanese black pine, Japanese larch, Japanese cedar Japanese cypress, Southern Japanese hemlock, Douglas fir, Port Orford cedar	0.5	6 mm	0.2 for tension and flexural members 0.3 for compression members	10% for tension and flexural members 15% for compression members, 10% for 1/3 middle of the span for bending members	0.02 for tension and flexural members	0.002 for compression members
	Group II	Cryptomeria, Momi fir, Yezo spruce, White fir, Western red cedar, Western hemlock	0.43	6 mm			0.04 for compression members	

Table 2.27 (a) Allowable Unit Stresses Parallel to Grain for Ordinary Structural Lumbers (unit : kg/cm^2)

		Species	The value for permanent loading			The value for temporary loading
			${}_Lf_c$	${}_Lf_t$, ${}_Lf_b$	${}_Lf_s$	${}_sf$
Softwoods	Group I	Japanese red pine, Japanese black pine, Japanese larch, Japanese cedar, Japanese cypress, Southern Japanese hemlock, Douglas fir, Port Orford cedar	80	90	7	Double the value for permanent loading
	Group II	Cryptomeria[1], Momi fir, Yezo spruce, White fir, Western red cedar, Western hemlock	60	70	5	
Hardwoods	Group I	Live oak	90	130	14	
	Group II	Japanese chestnut, Japanese white oak[2], Japanese beech, Zelkova[2], Apitong	70	100	10	
	Group III	Lauan	70	90	6	

Note :

(1) For cryptomeria of inferior quality (specific gravity less than 0.3, or average width of annual ring more than 6 mm), the value shown in the table shall be reduced by 30 per cent.

(2) For Japanese white oak and for zelkova, average width of annual ring shall be more than 1 mm.

Table 2.27 (b) Allowable Unit Stresses Parallel to Grain for Higher Grade Structural Lumbers (unit : kg/cm²)

			The value for permanent loading			The value for temporary loading
			$_{L}f_{c}$	$_{L}f_{t}$, $_{L}f_{b}$	$_{L}f_{s}$	$_{s}f$
Softwoods	Group I	Japanese red pine, Japanese black pine, Japanese larch, Japanese cedar, Japanese cypress, Southern Japanese hemlock, Douglas fir, Port Orford cedar	100	120	9	Double the value for permanent loading
	Group II	Cryptomeria, Momi fir, Yezo spruce, White fir, Western red cedar, Western hemlock	80	90	7	

iii) Allowable Unit Stresses in Compression Perpendicular to Grain

Allowable unit stresses in bearing and compression perpendicular to grain shall be in accordance with Table 2.28.

Table 2.28 Allowable Unit Stresses in Bearing and Compression Perpendicular to Grain

Species	Allowable unit stress in bearing perpendicular to grain		Allowable unit stress in compression
Softwood	$0.20\,f_c$	$0.16\,f_c$	$0.125\,f_c$
Hardwood	$0.33\,f_c$	$0.25\,f_c$	$0.200\,f_c$
Conditions of loads	(A) Bearing at the middle part of member	(B) Bearing at the end of member $a<d$	(C) Full compression

iv) Allowable Unit Stresses in Compression at Angle to Grain

The allowable unit stresses in compression on the lumber surfaces inclined to the grain shall be in accordance with the values obtained by multiplying f_c in Table 2.27 by the coefficients given in Table 2.29.

Table 2.29 Coefficient for Allowable Unit Stresses in Compression at Angle to Grain

Conditions of force \ Angle θ		0~10°	20°	30°	40°	50°	60°	70°~90°
(A) in Table 2.28	Softwood	1.00	0.87	0.73	0.60	0.47	0.33	0.20
	Hardwood	1.00	0.89	0.78	0.67	0.56	0.44	0.33
(B) in Table 2.28	Softwood	1.00	0.86	0.72	0.58	0.45	0.31	0.16
	Hardwood	1.00	0.88	0.75	0.63	0.50	0.38	0.25
(C) in Table 2.28	Softwood	1.00	0.85	0.71	0.56	0.42	0.27	0.125
	Hardwood	1.00	0.87	0.73	0.60	0.47	0.33	0.20

v) Increment and Reduction of Allowable Unit Stresses

Use in Wet Conditions :

For lumbers used under continuously wet conditions, the values of allowable unit stresses given in Tables 2.27 and 2.28 shall be reduced by 30 percent.

Exposed Structures :

For exposed structures, the allowable unit stresses shall be reduced down to 80 percent of the values given in Tables 2.27 and 2.28, according to the conditions concerned.

Temporary Structures :

For the temporary structures in which the duration of loads is to be less than three months, allowable unit stresses for permanent loading in normal condition and for snow season may be increased, according to the periods of structures in service, as follows :

within one week	130% of the values in Table 2.27
one week to one month	125% of the values in Table 2.27
one month to 3 months	120% of the values in Table 2.27

where the maximum snow depth shall be determined for the snow load if the deflection calculation is omitted.

The preceding provision shall not apply to the allowable unit stresses for temporary loading and to the moduli of elasticity.

Shear :

The allowable unit stresses in shear not accompanied by cleavage may be increased up to 1.5 times the allowable unit stresses in shear given in Table 2.27.

Compression Perpendicular to Grain :

For the structures in which a little bearing deformation is not objectionable, the allowable unit stresses in bearing perpendicular to grain may be increased up to 1.5 times the values shown in Tables 2.28.

For the structures in which the deformation is a serious question, the allowable unit stresses in bearing perpendicular to grain shall be decreased, according to the conditions concerned. In the case of compressive stresses imposed on the bastard grain, the values in Table 2.28 shall be reduced to two-thirds.

(5) Allowable Strength of Pile Materials

i) The allowable unit compressive stress for permanent stresses of a wooden pile shall not exceed 50 kg/cm^2 irrespective to the kind of wooden materials, and the allowable bearing capacity of the piles shall be calculated with respect to the minimum cross-sectional area thereof.

ii) The allowable unit compressive stresses for permanent loads of a precast reinforced concrete pile shall not exceed 80 kg/cm^2 nor 1/4 of the compressive strength of concrete material at the age of 28 days. The allowable bearing capacity of the pile shall be calculated with respect to the minimum cross-sectional area thereof, and the compressive strength of the concrete at the age of 28 days shall not be less than 180 kg/cm^2.

iii) The allowable unit compressive stress for permanent stresses of a cast-in-place concrete pile shall comply with the appropriate requirements in the following, according to the structural type of the proposed pile. The allowable bearing capacity of the pile shall be calculated with respect to the minimum cross-sectional area thereof, and the compressive strength of the concrete at the age of 28 days shall not be less than 180 kg/cm^2.

The allowable unit compressive stresses for the permanent loading of a cast-in-place concrete pile without a shell shall not exceed 50 kg/cm^2 nor 1/5 of the compressive strength of the concrete at the age of 28 days.

The allowable unit compressive stresses for the permanent loading of a cast-in-place concrete pile with a steel shell shall not exceed 60 kg/cm^2 nor 1/4 of the compressive strength of the concrete at the age of 28 days.

iv) The allowable unit compressive stresses for the permanent loading of a steel pile shall comply with the appropriate provisions of the AIJ Standard for Structural Calculations of Steel Structures. For purposes of stress calculations, the effective thickness of load-bearing steel shall be taken as 2 mm less than actual thickness on all surfaced to the soil.

v) The allowable unit tensile stresses for the permanent loading of pile materials shall comply with the appropriate provisions of the AIJ Standard for Structual Calculations of Timber Structures, the AIJ Standard for Structural Calculation of Reinforced Concrete Structures or the AIJ Standard for Structural Calculation of Steel Structures, and the allowable pulling capacity of the pile shall be calculated with respect to the cross-sectional area prescribed in the appropriate items i) to iv) of this Article.

Table 2.30 Allowable Bearing Capacities for Permanent Loading

Soils		Permanent allowable bearing capacity (t/m^2)	Remarks	
			N-Value	q_u (kg/cm^2)
Rock		100	over 100	
Sand bedrock		50	over 50	
Hard bedrock		30	over 30	
Gravel	Dense	60		
	Loose	30		
Sandy soil	Dense	30	30~50	
	Medium	20	20~30	
		10	10~20	
	Loose	5	5~10	
	Very loose	0	below 5	
Clayey soil	Very stiff	20	15~30	over 2.50
	Stiff	10	8~15	1.00~2.50
	Medium	5	4~ 8	0.50~1.00
	Soft*	2	2~ 4	0.25~0.50
	Very soft*	0	0~ 2	below 0.25
Kanto-loam	Stiff	15	over 5	over 1.50
	Medium	10	3~ 5	1.00~1.50
	Soft	5	below 3	below 1.00

Note :
* Not suitable as bearing material

Allowable Bearing Capacity of Soils

Unless the allowable bearing capacity of soil is determined in accordance with the appropriate requirements of Chapter 9-4, it may be estimated on the basis of the plate loading test results or the standard penetration test results according to the soil conditions and the type and scale of the proposed building.

i) If a plate loading test is conducted and a load-settlement curve is obtained in accordance with the requirements of Art. 9. 4, the allowable bearing capacity for the permanent loads shall be 1/2 of the yield point, 1/3 of the ultimate bearing capacity or the intensity of load corresponding to the settlement of loading plate equal to 20 mm, whichever gives the smallest value.

ii) If the soil conditions are already known to a certain extent but a plate loading test is not carried out, the allowable bearing capacity for the permanent loads may be empirically estimated on the basis of well-established and successful local experience and practice according to Table 2.30. However, the allowable bearing capacity thus estimated shall be reduced, if necessary, according to the conditions of underlying strata.

iii) As the allowable bearing capacity for the temporary loading, twice the bearing capacity for the permanent loading shown in Table 2.30 may be employed.

2.3 Earthquake Resistant Calculations for High Rise Buildings

2.3.1 Dynamic Analysis of Structures

a) General

Recent progress in the technique of numerical analysis is great with the aid of remarkable development of electronic computors, which has solved various kinds of problems whose practical computation has been regarded impossible although the method of solving these problems had been known.

Earthquake response of structures has been computed, idealizing the structures into simple models. For structures having very complicated plans, since the earthquake has three dimensional movement, the precise analysis can not be expected with the present numerical technique. However, the analysis on this stage may greatly help us to understand the aspect of the vibration of structures during an earthquake.

The structures have been substituted by simple vibration models in the previous analysis. For instance, the multi-storied spring-mass system with linear or non-linear characteristics and in addition with viscous friction of the structure and for another instance, the system which represents the whole structure as a continuous shear beam. The foundation of the building is assumed to be supported on the solid or semi-solid ground.

There are many instances in which the responses of vibratory systems mentioned above have been computed for the acceleration of recorded earthquake motion at the bottom of the structure. Although it is possible to find out some qualitative characteristics of earthquake response of structures from the results of these calculations, many quantative problems of structural responses are still left unsolved. It should be kept in mind that the assumptions employed in the dynamic analysis are far simpler than those in the static analysis of structures.

The fact that the dynamic analysis is being used for the design of high-rise buildings will contribute toward the development of the method of dynamic analysis itself. However, in the process of actual design of buildings much caution should be paid for understanding and evaluating the results of dynamic analysis which are based on various assumptions.

b) Earthquake Ground Motion

Severe earthquakes which bring disastrous damage to the structures have been known in the history of the earthquake. But very few of them have been recorded by the seismic apparatus. Recently the strong motion accelerographs have been installed all over Japan but their distribution is not so satisfactorily complete from the earthquake engineering point of view. The ground motion of Niigata earthquake of 1964 was recorded only at Niigata, Akita and Shiogama, and no accelerograph was installed in the area between Niigata and Akita where the seismic motion was supposed to have been most severe, while considerable numbers of strong motion accelerographs are installed in Tokyo, Osaka, and Nagoya.

The maximum acceleration of the recorded earthquake in Niigata is 159 gals in the East-West direction and 155 gals in the North-South direction while the record obtained in Akita is 82 gals in EW and 91 gals in NS. Both records were obtained in the basement of these reinforced concrete buildings. With regard to the record in Niigata, the building itself sank considerably during the earthquake and the record shows some motions of very long-period due to sinking of the building. For this reason it is difficult to consider it as very accurate.

As for the seismic measurement in the United States, a strong earthquake record was obtained at El Centro, California in 1940 with the maximum acceleration of 319 gals and has been used in various studies, but its damage to structures was not so serious. As described before the scheme of earthquake measurement is being advanced and the recorded data are being accumlated, but these data are not yet enough for practical purposes.

The recent studies concerning the earthquake ground motion which is to be applied for actual design of buildings, will be introduced below.

(1) Relation between Strong Earthquake Records and Microtremors of the Ground by K. Kanai.

K. Kanai reported that the characteristics of microtremors of the ground are very similar to those of the strong earthquake record of that place, in the frequency domain.

Therefore, the strong earthquake records of other places, where the characteristics of microtremors are similar to those in the ground on which the building is to be built, can be adopted for design purposes.

(2) Average Response Spectrum by G. W. Housner

G. W. Housner proposed the average response spectrum as the result of rearrangement of numerous response spectra for recorded earthquake motions. It cannot be directly applied for calculating the earthquake response of a structure but in case of calculating the elastic vibration of multimass vibratory system this spectrum can be conveniently used for the method of modal analysis which gives approximately the maximum response values as a resultant of each modal response. However, Housner gathered the accelerograms recorded only in the west coast of the United States, and therefore, almost no record on the soft ground is included.

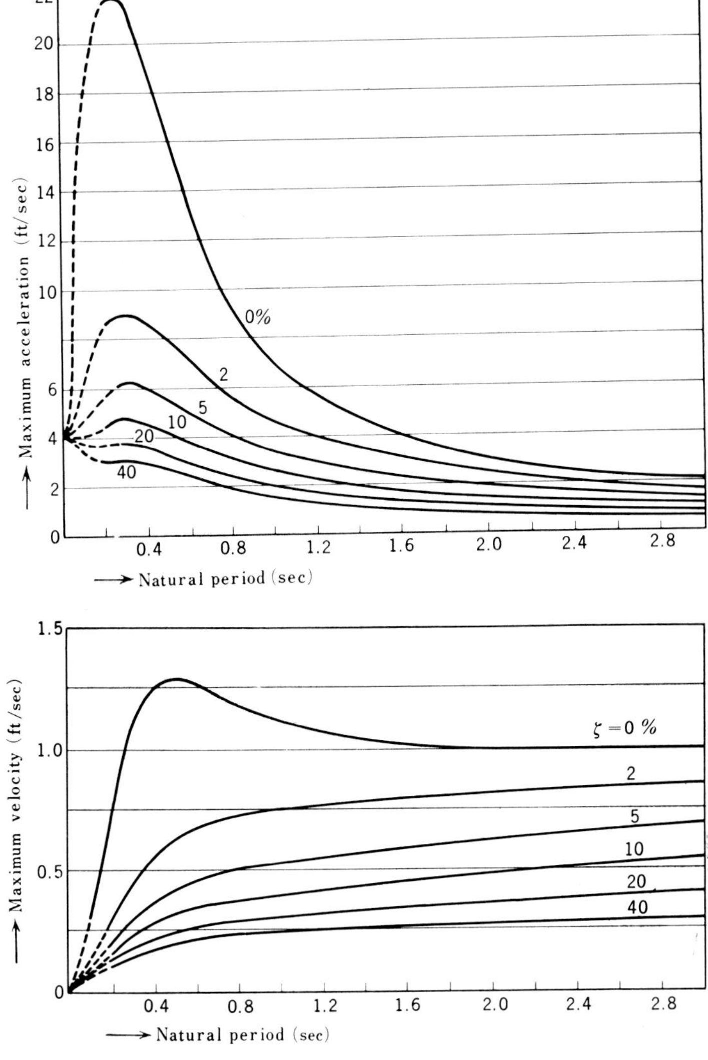

Fig. 2.2 Average Response Spectra (G.W. Housner)

(3) Response to Random Oscillations

The characteristics of response spectra to random oscillations are similar to those to earthquake ground motions. Several people have recently pointed out that by filtering those random oscillations based on the dynamic characteristics of the ground, a more similar movement to that of earthquake ground motion can be obtained, and it is argued that to use these filtered random oscillations is more general than to use a specific earthquake ground motion for design purposes.

(4) Response to Idealized Ground Motions.

R. Tanabashi and others proposed that a few number of rectangular waves which are obtained as a group of band limited noise, should be adopted as a design movement of earthquake ground motion, after they developed an idea of random oscillations.

The proposed methods of determining the design earthquake motion introduced above are mostly concerned with the characteristics in frequency domain but the question still remains in the intensity of ground motion for design purposes.

H. Kawasumi presented the maps of expectancy of earthquake intensity in Japan based on the statistics of past severe earthquakes. According to his paper, the maximum acceleration of more than 600 gals is expected in a certain area having standard soil conditions, and this value is much greater than that expected in Tokyo area based on the experience in the Kanto earthquake of 1923 or recorded with the instrument at El Centro in 1940. Moreover, considering the remaining buildings and simple bodies near the epicenter the maximum ground acceleration is estimated as 450～500 gals in Fukui and 300 gals in Tottori earthquake. Therefore, although there are some differences in the intensity depending on the district, rather big ground acceleration must be expected.

On the other hand, according to the comparative study on the observed earthquake motions on and under the ground, the amplitude of the earthquake motion at the place which is under the surface soil layer and on the so-called "bed rock", is considerably smaller than that on the surface of the ground. Judging from the facts mentioned above, it seems the maximum acceleration of 300 gals is too small as the expected earthquake intensity for buildings with shallow foundations. In case of buildings with deep foundations resting on the "bed rock", different consideration is needed. The problems on the earthquake intensity are not yet thoroughly solved, and they must be seriously considered as the major research objectives in the future, as well as those of frequency characteristics of earthquake ground motion.

2.3.2 Guide for Design of High-rise Buildings

In Japan two design procedures are being adopted according to the height of a building from the ground surface, the boundary of which is about 45 m. For buildings lower than 45 m, the structural design is made by the static method in accordance with the Building Standard Law which prescribes the seismic coefficient of 0.2 for the part of the building under 16 m high and of increasing value for the increase of the height, and the seismic forces to be obtained by multiplying the weight of that part by the corresponding seismic coefficients.

On the other hand, for high-rise buildings the amendment of the Building Standard Law, which mainly consists of considerations for the volume limitation, was published in July, 1963. Then, the "Guide for the Design and Construction of High-rise Buildings" was published by the Architectural Institute of Japan.

According to this guide the ordinary high-rise buildings over 45 m high can be designed as follows provided that appropriate caution is paid to the structural design.

i) Consider a design base shear coefficient C_B. The value of C_B will decrease in the limited band as described below, as the natural period of the building increases.

$$0.2 \geq C_B \geq 0.05$$

But in the case where the earthquake response is expected to be small from the consideration of the intensity of the ground motion and the natural period of the building, the lower boundary can be removed.

ii) Determine the distribution of the design shear force coefficient in each story.

iii) Design the member and joint to certify the stiffness, yield strength, yield deformation and the other characteristics after yielding. It is desirable to conduct some experiment to make them clear.

iv) Calculate the earthquake response of the building by means of dynamic analysis and examine the results. Improve it to be appropriate design, if necessary.

v) For parapets, interior ornaments, water tanks above the roof, chemneys and others, design them for sufficiently large horizontal forces.

These items are the abstract of the method of antiseismic calculations described in the "Guide for Design of High-Rise Buildings", and further details of them are given in the originals. Abstracting these, the third intem is based on the idea of type ii) design method described in chapter 2.1 and corresponds to the preliminary design parts of this guide. The idea of item iv) recommends the sufficient consideration on safety of the building by means of dynamic analysis. Repetition of the preliminary (trial) design and its dynamic analysis described above will make a building with excessive safety. This routine method is reasonable under the present circumstances where no definite form of seismic force is given. The idea of item v) is already described in the 1.3.3, 2.2.1, and 2.3.2.

"The Guide for Design of High-Rise Buildings" only indicates the fundamental ideas of high-rise building design which is the greatest problem in the present days. There are many problems which are uncertain at present and which have to be solved and clarified in the near future. Not many buildings have been designed according to the guide and further interpretation shall be omitted.

3. FOUNDATIONS AND GROUND CONDITIONS

3.1 Introduction

It is quite clear that foundation structures, which connect buildings and ground, are very important with respect to aseismic design, as the earthquake comes through the ground which supports the buildings.

As far as statical problems are concerned, both soil mechanics which deal with soil problems and foundation engineering which deals with foundation problems have recently been applied frequently to the foundation design of buildings. But concerning the dynamics of soils and foundations which have a deep relation with aseismic design, there are few studies based on both theoretical and experimental works which can be applied directly to the actual design, even though active research in this field can be seen recently in all of the world.

Prototype tests are the most effective way to establish the aseismic design method. But it is almost impossible to make up a reasonable artificial earthquake and to get the response of the building by such an earthquake. Then, the basic design method for the soils and foundations has to be considered by the design engineer himself based mainly upon experience of past earthquake damage and on the results of theoretical and experimental studies.

The following basic principles can be given from past experience :

i) Foundations should fundamentally be supported by rigid ground and not by soft ground.

ii) Parts of the foundation of the same building should not be supported by ground where its properties are significantly different.

iii) Different foundation types have to be avoided for the same building.

iv) Foundation has to be assured of its reliability of construction.

v) Sufficient safety factors are needed to apply to the bearing pressure and pile loads of buildings, and homogeneous distribution is preferable within the same building.

vi) Sufficient rigidity is necessary for the foundation beams, as there are many examples of earthquake damage not by the structural design failure but by an insufficient connection between columns and foundation beams.

The basic principles, mentioned above, are the same as those needed to avoid differential settlement of buildings in the long-term stress condition. In other words, as the basic principle, if the foundation of building is designed to avoid differential settlement for the long-term stress condition, large differential settlement can not be expected to occur during earthquake and then, the stress distribution caused by the earthquake becomes comparatively clear and the seismic design becomes easy to carry out.

The dynamic design method of foundation, as mentioned above, is not established at present and the following problems have to be clarified in order to deal with the ground and foundation as a complete vibration system.

i) Vibrational characteristics of soil : Amplification and damping characteristics of earthquake waves, phase difference of the wave as can be seen in Fig. 3.1 (a).

ii) Input-output characteristics of foundation : As shown in Fig. 3.1 (b), the problems are : what kinds of forces are induced into the foundation, how the forces are transmitted to the building and what kinds of forces are transmitted to the ground through the foundation by the vibration of the building.

iii) Deformation and strength characteristics of the ground : As can be seen in Fig. 3.1 (c), the foundation on the ground has a tendency to slide or rotate, and in the case of pile foundation, the ground is pressed at the top of pile and near to it. The deformation and ultimate bearing capacity of the ground for such an earthquake, (i.e. dynamic deformation and strength characteristics of the ground) have to be clarified. Dynamic strength of the ground is sometimes significantly different from the static one and this difference strongly depends upon the soil properties and on the acceleration of the foundation.

iv) Strength and deformation characteristics of the foundation : As can be seen in Fig. 3.1(d),

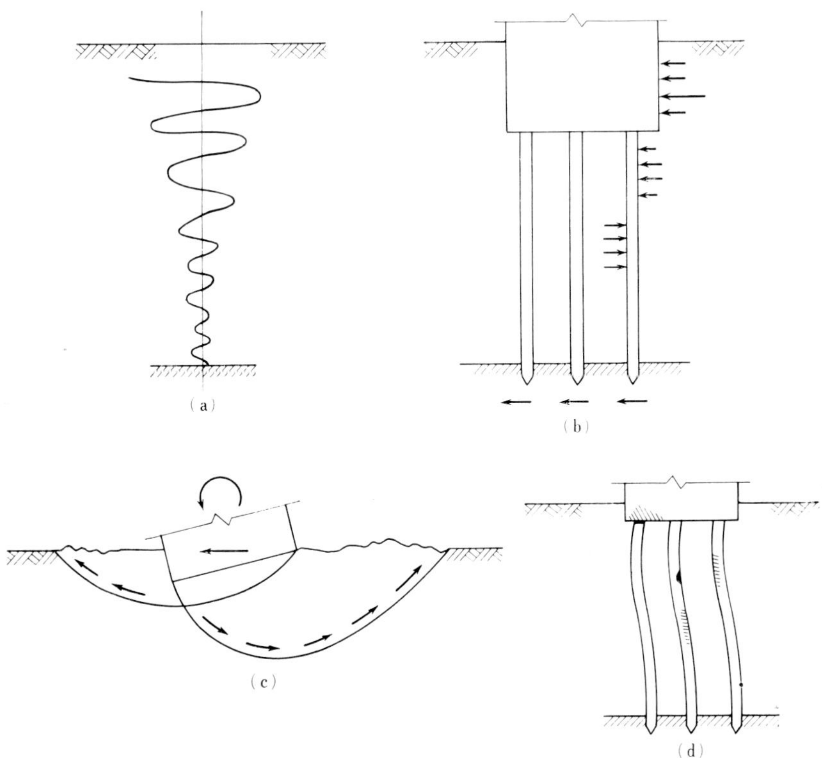

Fig. 3.1

if the foundation itself happens to be destroyed by the force transmitted by itself, then the strength and deformation characteristics of piles or piers become an important factor.

3.2 Earthquake Damage and Ground Condition

It is a well-known fact which has been experienced in every large earthquake, that the soft ground always causes serious earthquake damage. For example, the maximum accelerations of various places were concluded to differ more than ten times depending upon the ground condition, even in the same city area, in the San Francisco earthquake of 1906.

But as far as quantitative information is concerned, with respect to the relation between earthquake damage and ground condition, only little is available inspite of the fairly large amount of experiences. It is due to the fact that the following conditions have to be known to get a quantitative relation, and the latter condition has not been clarified yet until lately except only in the limited cities such as Tokyo, Yokohama and Nagoya.

i) Detailed distribution of earthquake damage in large earthquake.

ii) Ground conditions of the area.

3.2.1 Earthquake Damage of Wooden Buildings

The relation between earthquake damage of wooden buildings and ground conditions has been investigated in every past serious earthquake and has been studied quantitatively by several engineers.[1),2),3),4),5),6),7)]. According to these studies, the following conclusions can be reached in case of heavy damage of wooden buildings.

i) The alluvial soil is deeply deposited. If thickness of the layer is more than 30 m, the damage rate becomes intensely high; when the thickness of alluvium deposit under buildings is 30 m or more, significant damage is observed.

ii) Thick fill. If the thickness of the filled stratum is more than 3 m, the damage increases intensely.

iii) Ground with low rigidity.

iv) Recently constructed fill.

v) Long predominant period of the ground.

vi) Small velocity of surface wave.

Based on the record of damage by the Great Kanto Earthquake, a relation between the damage rate and the thickness of alluvial deposit is given by Fig. 3.2[1)], and the earthquake magnitude ratio for the various kinds of soils are shown in Table 3.1[8)], judging from the results of damage by the Fukui Earthquake.

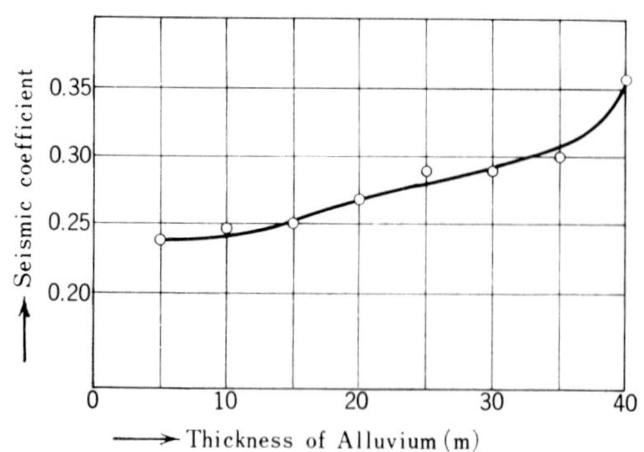

Fig. 3.2

Most of informations mentioned above, are mainly obtained for typical Japanese wooden houses with Japanese tiles on the roof, and with mud plaster bamboo lath wall with bracings. So it should be noticed that for recent buildings constructed under present Building Standard Law, the conclusions may not be applicable directly.

Table 3.1

	Paludal area	Alluvium	Diluvium	Tertiary volcanic rocks
Ratio of seismic intensity	1.5 ?	1.0	0.7	0.4

But it is a well-known fact that softer the sub-soil, the greater the damage rate to buildings. It has been pointed out as the cause of damage of wooden buildings[7),9)] that the differential settlement of buildings on the soft ground should be considered in addition to dynamic phenomena such as magnitude of earthquake and the relation between natural periods of buildings and that of the ground.

3.2.2 Earthquake Damage of Brick Construction and Warehouses

The earthquake damage of comparatively rigid and heavy buildings such as brick construction or warehouses, is different from the case of wooden buildings. The softer the sub-soil, the smaller the earthquake damage of brick construction buildings or warehouses.

That is to say, the damage rate of brick construction in Kanto Earthquake was three times in rigid ground compared with that in soft ground, and in the case of warehouses, as can be

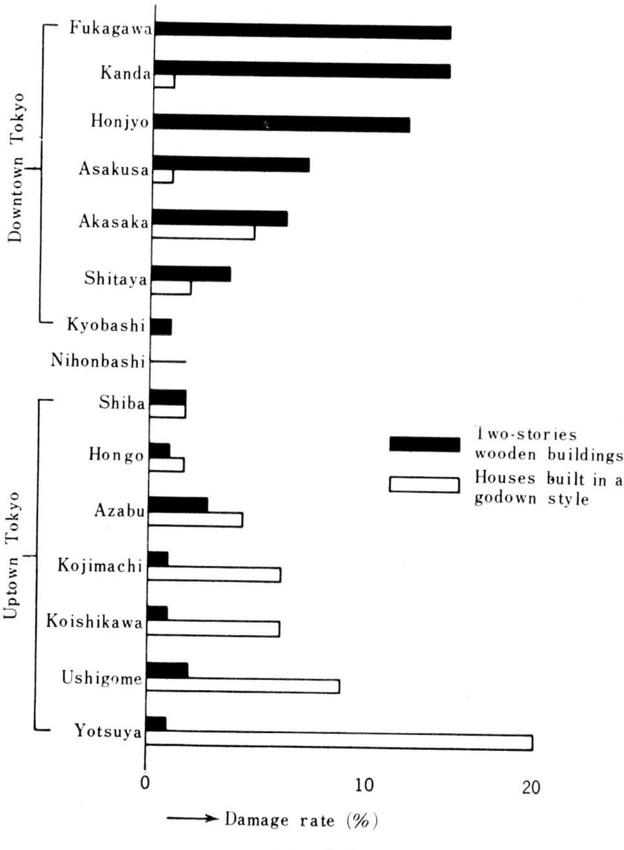

Fig. 3.3

seen in Fig. 3.3[7], the damage rate is much higher in the rigid ground of uptown area compared with that in the downtown soft ground area contrasted to the wooden house case.

These phenomena have been explained[7] by the fact that the natural periods of warehouses is short and close to the predominant period of the rigid uptown ground and that the large damping can be expected by the energy dispersion to the ground especially in soft ground.

3.2.3 Earthquake Damage of Reinforced Concrete Buildings

According to damage statistics for reinforced concrete buildings in the Old City of Tokyo at the Kanto Great Earthquake :

i) The rate of damage is higher in uptown area, where the ground condition is good, if up to "partial damage" is included.

ii) Limited to the "serious damage" only, the rate of damage is higher in downtown where the ground is soft.

For these phenomena, the following explanations have been made[7].

i) The softer the ground compared with the rigidity of buildings, the bigger is the underground dispersion of the vibrational energy and then smaller the damage to be generally expected because of the increased damping of buildings caused by the dispersion.

ii) Serious damage will occur by progressive failure, which is less in buildings with small damping, than in the case of buildings with large damping.

It has also been pointed out that reinforced concrete buildings, as well as wooden buildings, will suffer serious damage of the structure from both stresses by vibration and by additional differential settlement which is caused by the soft ground condition during the earthquake[9].

But as far as the relation between earthquake damage to reinforced concrete buildings and ground conditions are concerned, there is not enough experimental information compared with those of wooden buildings, as the above mentioned statistics are only one material which has become available lately.

It is still within living memory that many reinforced concrete buildings suffered much from the recent Niigata Earthquake. In this particular case, it was an important experience to study the relation between earthquake damage and ground condition, as most of the damage was not by vibrations but by the soil failure during the earthquake. The cause of damage in the Niigata Earthquake can not be considered to be a typical one as in the general ground condition or in Kanto Earthquake, because it was liquefaction of sandy ground……a particular characteristic of the Niigata district.

3.2.4 Ground Conditions and Use of Seismic Coefficient in Design

Seismic coefficient in design can be decreased to some extent when the ground condition is good judging from past records of earthquake damage which, in such a case, can not often be found out with regard to wooden buildings, steel construction and reinforced concrete construction. The seismic coefficient, on the other hand, has to be increased at least for wooden buildings on a newly constructed fill or on a thick alluvial deposit where serious damage has been experienced particularly to the wooden buildings. These proposals have been made for many years.

Article 88, Detailed Regulations for the Building Standard Law and the related Notification No. 1074 (25 July 1952) of the Ministry of Construction are the codes which deal with seismic coefficients for design practice depending upon the kind of ground and upon the type of construction.

The notification mentioned above shows, in the same way as Table 2.2, the definition of the kind of ground and the decreasing factors of horizontal seismic coefficients and also defines the "area where ground is considerably soft" where a standard seismic coefficient $K_0=0.3$ has to be used for wooden buildings. In the application of the said article, the following suggestions should be noticed.

i) The kind of ground should be judged not only by the ground just under the building but also by the soil condition in the surrounding area.

ii) As for the diluvial deposit, shown in Table 2.2, rigidity of the ground may sometimes significantly change depending on the district or location.

iii) The kind of ground is generally defined by the kind of soil which occupies more than half of the layer at a depth of about 10 meters from the surface. But, for example, the response of a structure whose base is supported by rigid piers or basement, has been shown[10] to be influenced not by the surface layer but mainly by the characteristics of the ground under the ground work during earthquake.

iv) As the earthquake damage to the wooden buildings is significant in the following cases, the 2 nd kind of ground is recommended as the kind of ground under consideration even when more than a half of the ground to the 10 meters depth is occupied by the 1 st kind ground ; where there is a soft fill more than 2 meters thickness, independent on the rigidity of natural ground, where there is an alluvial deposit underlain by a strongly inclined base as can be seen in Fig. 3.4 (a) and where there is a fault or some other geological weak points as can be seen in Fig. 3.4 (b).

v) According to Article 88 of the Building Standard Law, the area where the ground is significantly soft is used to be specified by the specific government office, but even when it is not specified, the article should be refered to in the design process.

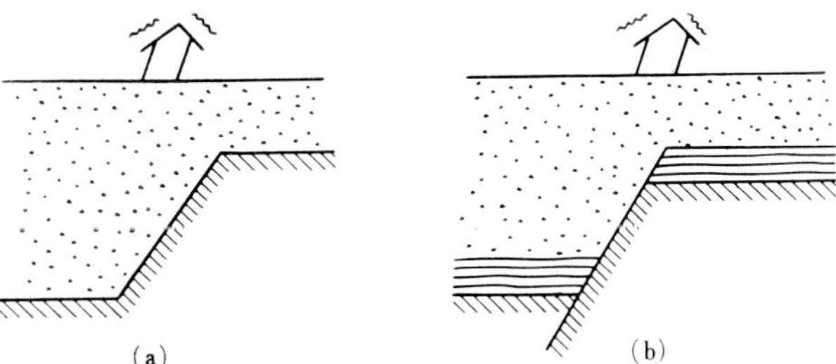

Fig. 3.4

3.3 Soil Investigation

There are two kinds of soil investigations in seismic design. One is a common investigation of the long-term design condition and the other is carried out for special objective to get particular information for the sesmic design. Boring, soil test and vertical load test belong to the former, and horizontal pile loading test and the measurement of K-value and microtremors belong to the latter.

3.3.1 Boring and Soil Test

Several factors, shown in Table 3.2, should be obtained at the site or by the laboratory test depending on the probable foundation pattern, as well as getting ground composition and location of ground water by rotary boring. The static ultimate bearing capacity and the settlement at each part of the building can be computed from these factors. Some information on the dynamic ultimate bearing capacity may be given, to some extent, by N-values and grain size distribution in sandy soil, and by water content and sensitivity ratio in cohesive soils.

Table 3.2 An itemized list of soil exploration

Foundation types	Depth of exploration	Coefficients to be explored
Spread foundation 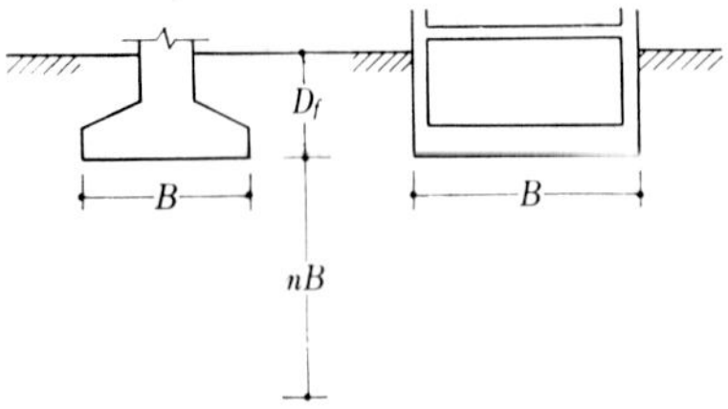	$0 \rightarrow D_f$	r
	$D_f \rightarrow D_f + {}_nB$	$N, r, C, \phi, w, LL, PL, e, Ce, \sigma_0, S_t, GSD$
Pile or pier foundation 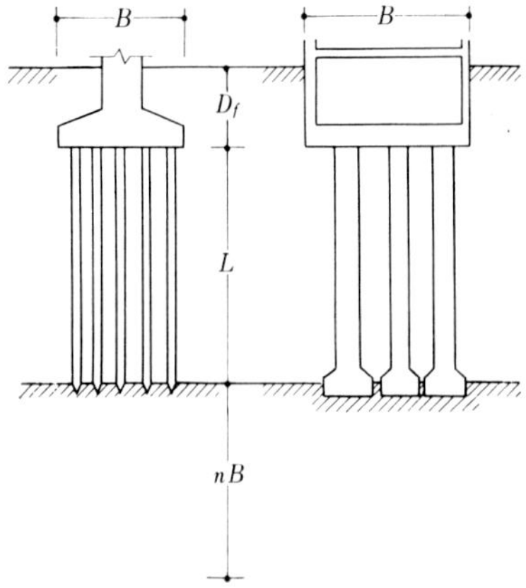	$0 \rightarrow D_f$	r
	$D_f \rightarrow D_f + L$	$N, r, C, \phi, w, LL, PL, S_t, GSD$
	$D_f + L \rightarrow D_f + L + {}_nB$	$N, r, C, \phi, w, LL, PL, e, C_c, \sigma_0, S_t, GSD$

Note :

1. N : N-value, Standard corn penetration test r : Unit weight of soil C : Coheasion ϕ : Angle of internal friction w : Water content LL : Liquid limit PL : Plastic limit e : Void ratio C_c : Coefficient of compressibility σ_0 : Maximum consolidation pressure on soil in field S_t : Degree of sensitibity GSD : Grain size distribution D_f : Depth of foundation B : Shorter dimension of foundation slab L : Length of pile or pier.
2. $n=2\sim4$, increase n when underlain by soft layer.
3. Refer the depth of preliminary boring and exploration items before selection of foundation type to the existing information of the surrounding buildings.

3.3.2 Loading Test

Much information on the strength and deformation characteristics can be obtained by plate loading test for raft foundation and by pile loading test for pile or pier foundation. In both cases, it is necessary to load up to the ultimate bearing capacity or near it at least up to 150% of the load expected to occur during an earthquake. But there are still several uncertainties concerning the size effect of loading plate in the plate loading test and concerning the effect of pile group in the pile loading test.

The horizontal loading test on piles or piers will be very effective to get information on the horizontal resistance of the foundation.

3.3.3 Measurement of Coefficient of Sub-grade Reaction

If a structure is supported by a massless spring which represents the vibrational characteristics of the ground, the constant of the spring is called coefficient of sub-grade reaction or simply K-value of the ground.

There are two kinds of coefficients, namely K_v in the vertical direction, K_h in the horizontal direction, which are employed in the computation of settlement of buildings or earthquake response including the effect of the ground.

The K-value can be obtained by the following methods.

i) Measure of the resonant frequency by the vibration test of model or similar building[11)~13)].

ii) Use the inclination of load-settlement curve by the load test.

iii) Guess from the N-values in the standard cone penetration test.

An example of measurement of dynamic K-values by method i) is shown in Table 3.3. But method i) has several difficulties in that similar buildings can not usually be expected, and that, in the model test, the size effect of the foundation has not been clarified yet, even though many proposals have been made[14)]. On the other hand, a large number of reports can be obtained by the method[15)] ii), but all of them are in statical loading conditions whose application to dynamic case is still not yet clear. Method iii) is also insufficient because of lack of information.

Table 3.3

Soil	K_v (kg/cm^3)
Fill	3～ 5
Loam	3～ 5
Sand	8～10
Gravel	11～13

The coefficient of sub-grade reaction in the horizontal direction will be used for such problems as interaction between the soil and pier or pile in the ground. Pressiometer has been very useful for obtaining such K-values. Pressiometer is a kind of loading test apparatus in which a cylindrical rubber tube is inserted into the boring hole and load applied on the wall of the boring hole by the expansion of the cylinder. The coefficient of sub-grade reaction can then be computed from the given pressure and the volume change of the measuring tube, i. e. displacement of the boring wall.

3.3.4 On Microtremors[16)]

In an ordinary place, a seismograph which has more than about ten thousand times magnification, records the ground motion continuously. The vibrations are the results of transmitted vibrations chiefly caused by artificial disturbances such as traffic, industrial machines, etc. Such ground vibrations are called microtremores and usually the amplitudes of the motions are 0.1～1 microns. It has recently been clarified that the period distribution curve of microtremores shows a definite form for various kinds of subsoils and shows similar characteristics to that of actual earthquakes.

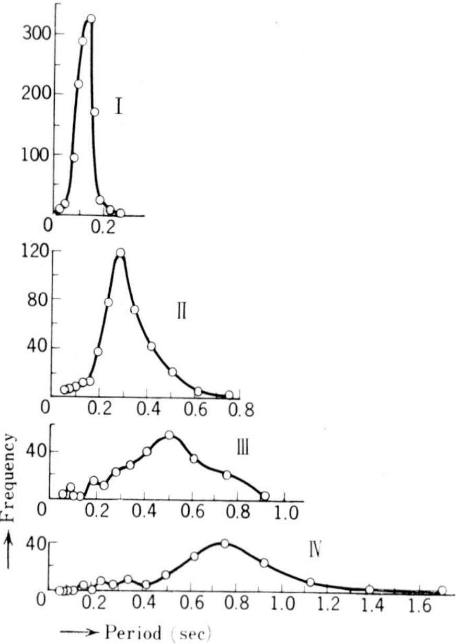

Fig. 3.5

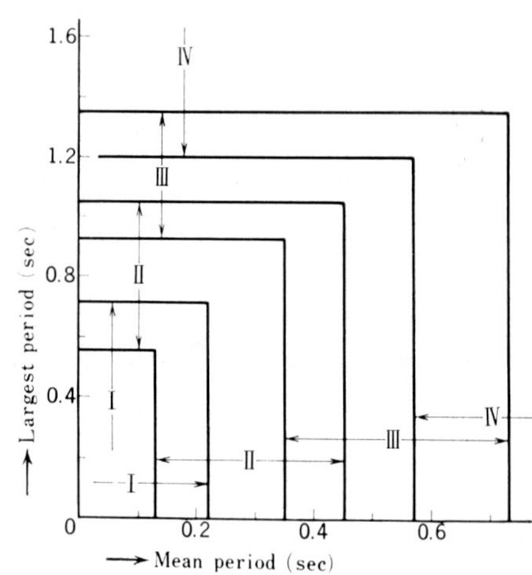

Fig. 3.6

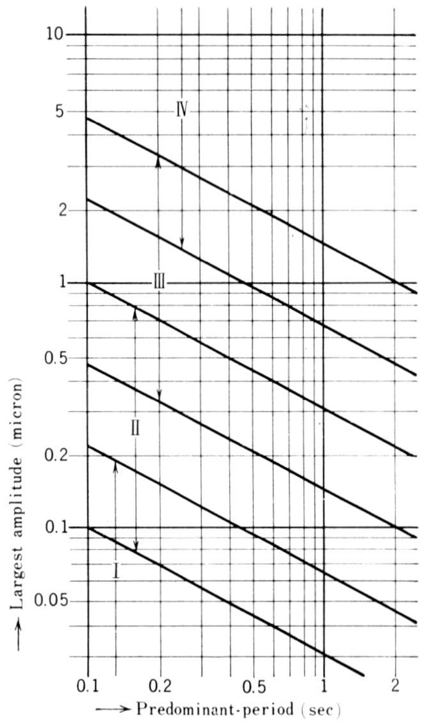

Fig. 3.7

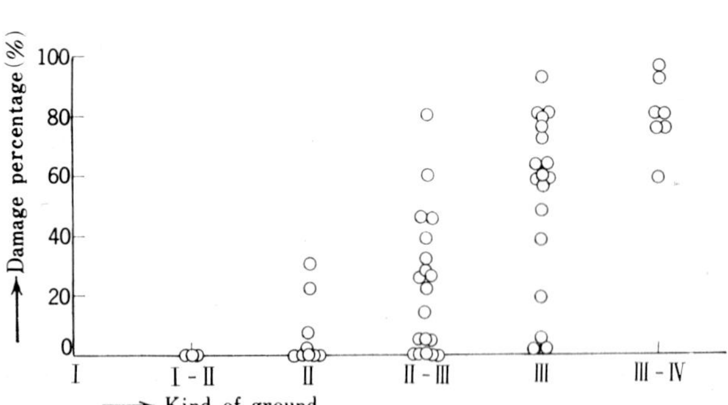

Fig. 3.8

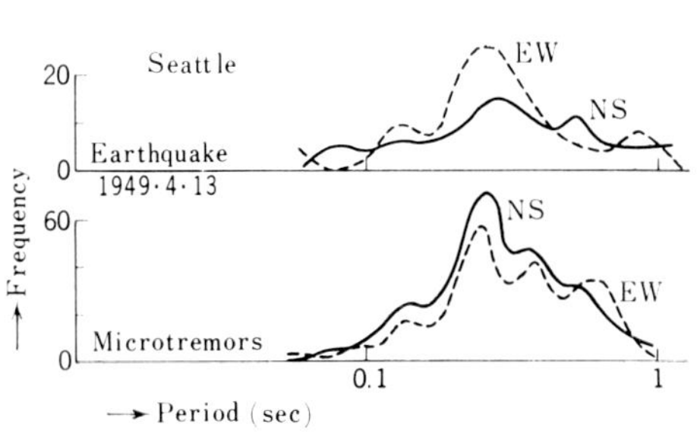

Fig. 3.9

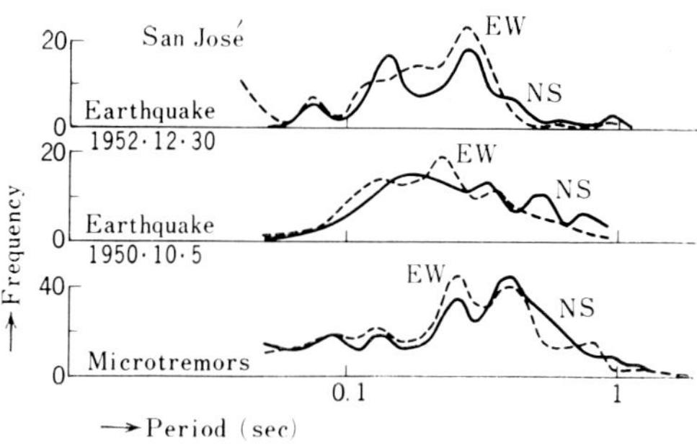

Fig. 3.10

As can be seen in Fig. 3.5, the representative period distribution curves of microtremors at the very stiff to very soft soils show different pattern of curves I ~ IV, where frequency means the number of waves belonging to each period in a specified time.

Using these characteristics, the classification of the ground by microtremors has been proposed with the predominant period, the largest amplitude and the largest period as shown in Fig. 3.6 and Fig. 3.7. In these figures, the kinds of ground I ~ IV correspond to the 1 st kind to the 4 th kind in Table 2.2 (The standards for the designation of four kinds of ground, the Building Standard Law), and III belongs to typical ground and IV to very soft ground.

An example of the relation between the percentage of earthquake damage to Japanese wooden houses and the kind of ground classification at each place is shown in Fig. 3.8. The figure, in the case of the 1944 Tokai earthquake at the Kinugawa River district, Shizuoka prefecture, shows a satisfactory correlation between them.

Other examples showing the comparison between the period distribution curves of strong motion seismographs and those of microtremores, also show good agreement in both San Jose, California and Seattle, Washington.

One of the proposals for an input data to be used in the computation of earthquake response of buildings is to get the vibrational characteristics by microtremor measurement first and then find a record of strong motion seismographs with similar period distribution curve as the input.

3.4 Seismic Design of Foundations

The seismic design of building foundations, as mentioned above, can only be called pseudo-dynamic since the earthquake force is replaced with static horizontal forces acting on the upper structure. The design method for the stress transfer from the structure, by the statically replaced horizontal force, to ground or to pile or pier through foundation is carried out as follows.

i) The vertical stress on the column base is carried directly to the ground or pile or pier.

ii) The bending moment on the column base is carried to the foundation beam and not to the foundation slab.

iii) The bending moment on the column base is distributed to the foundation beam and to the equivalent members which are composed of foundation slab and ground, or foundation slab and pile or pier.

3.4.1 Raft Foundation

If the method i)+ii) is applied to the seismic design of foundations, the equally distributed bearing pressure, computed from the vertical load on the foundation during earthquake, must be smaller than short-term allowable bearing capacity of the ground.

Twice the long-term allowable bearing capacity is allowed for the short-term allowable bearing capacity of the ground by the standard of Architectural Institute of Japan[17]. As the long-term allowable bearing capacity has a safety factor of more than 2.0 for yielding point and 3.0 for ultimate bearing capacity by the code, the safety factor during earthquake is more than 1.0 for yielding point, and more than 1.5 for ultimate bearing capacity. Incidentally, it is not based on theoretical or experimental judgement to estimate the short-term allowable capacity twice that of the long-term bearing capacity.

Though economy and a fairly low probability of suffering strong earthquake during entire life of the buildings have to be considered, the safety factor more than 1.5, above mentioned, may not be always assured during earthquakes, as the dynamic ultimate bearing capacity of the ground is not always larger than the static case, differing from typical structural materials which are not so influenced by dynamics or statics. It is a well-known fact that some type of sand looses its bearing capacity by the liquefaction during cyclic loading, and serious damage of reinforced concrete buildings has actually been experienced during the Niigata Earthquake caused by liquefaction.

Table 3.4[18] shows the consideration of short-term allowable bearing capacity outside of Japan with respect to the ratio of short-term case to long-term case.

Table 3.4

Country	$\frac{\text{Short term allowable bearing capacity}}{\text{Long term allowable bearing capacity}}$	
Algeria	Rocks	3
	General soil conditions	2
	Saturated loose soils	1
Argentina	1.25	
Germany (West)	1.50	
Greece	1.50	
India	Having a bearing pressure greater than 45 t/m^2	1.5
	Having a bearing pressure greater than 20 t/m^2 and equal to or less than 45 t/m^2	1.3
	Having a bearing pressure greater than 10 t/m^2 and equal to or less than 20 t/m^2	1.0~1.3
Portugal	2.0	

If the method i)+iii) is applied to the seismic design of foundation, a bending moment will be transfered to the footing or, in raft foundation, to the foundation slab. In this case, the

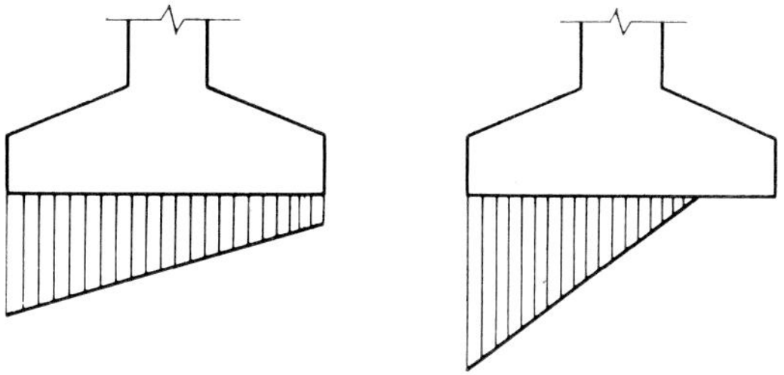

Fig. 3.11

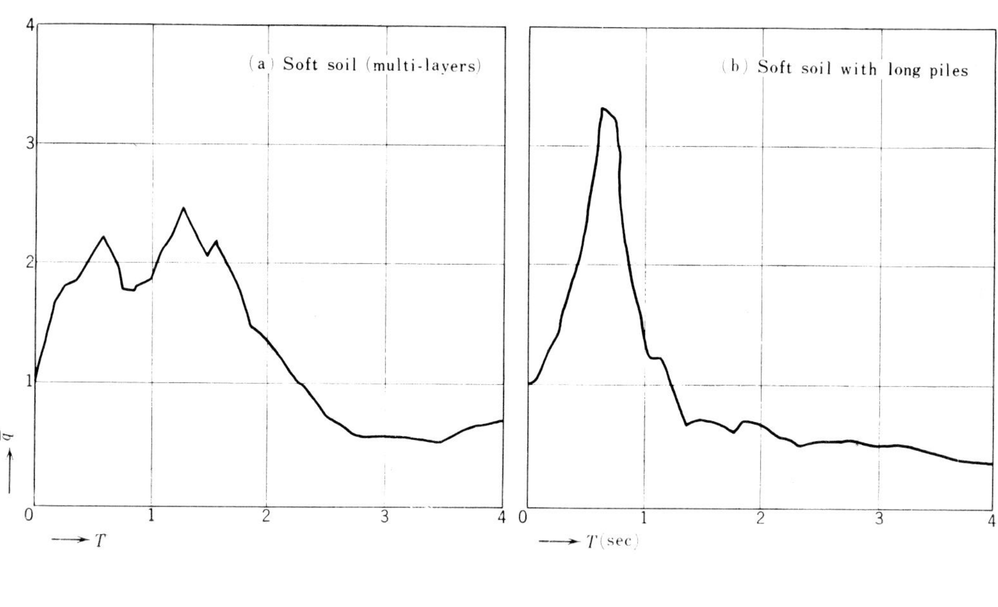

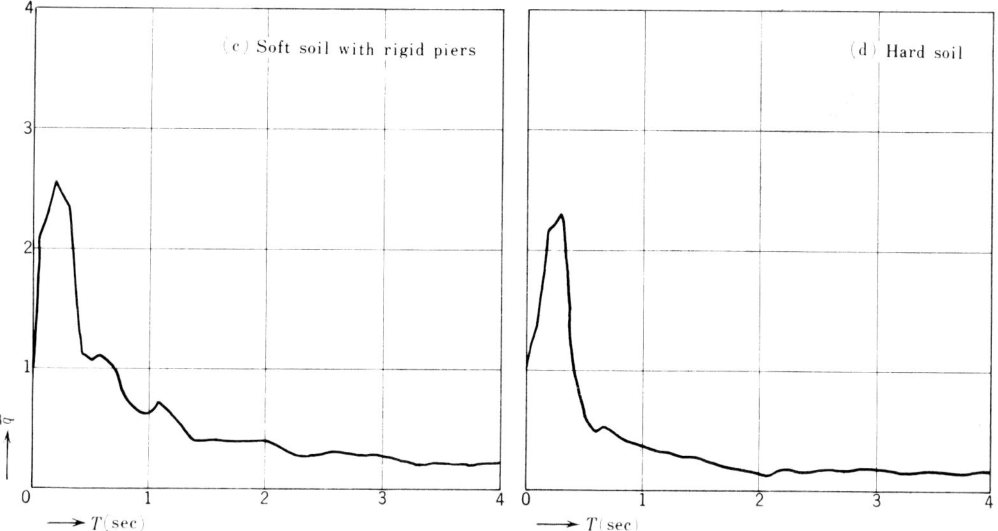

Fig. 3.12

rotational rigidity of the footing, used in the distribution of moment of column base to foundation beam and footing, is generally expressed by αKI (where K : coefficient of sub-grade reaction of soil, I : secondary modulus of footing base, α : coefficient, function of coefficient e.). Then it should be noted that the rotational rigidity of the footing depends upon the eccentricity coefficient e, in other words, the amount of vertical load (i.e. if $e \leqq 1/6$, $\alpha=1$, if $e>1/6$, $\alpha<1$). The earth pressure distribution under the footing or raft foundation by the bending moment can be assumed as a triangular or trapezoid shape, as can be seen in Fig. 3.11 and the maximum bearing pressure may not exceed the short-term allowable bearing capacity. In this case, it may be desirable to avoid expecting large amount of bending moment of the footing with the limitation of $e \leqq 1/3$, i.e. so as to get more than 50 percent of footing base to work effectively.

In the raft foundation, as far as a safety factor is assumed to be satisfactory for long-term vertical load, failure or sliding of the foundation by the base shear during an earthquake may not usually occur, which is different from pile foundations mentioned below.

3.4.2 Pile or Pier Foundations

Let us consider the effect of pile or pier foundations with respect to earthquake resistance. It is needless to say that pile or pier should resist, with no significant settlement the vertical load, caused by the horizontal earthquake force during earthquake.

The so called normalized acceleration spectrum is used, in which the maximum earthquake force induced in one-mass system is shown in the unit ($\bar{q}$) of the corresponding maximum ground acceleration of the earthquake ($T=0$), with damping coefficient $h=0.05$ as shown in Fig. 3.12[10].

The following can be described concerning earthquake response of the structure.

i) As can be seen in Fig. 3.12(a), when the building has a raft foundation resting on soft soil, several peaks are observed in the spectrum over the range of short T through long T ($T=1.6$ sec). The value of T_{cr} which corresponds to $\bar{q}=1$, is about 2.5 sec.

ii) When the building on soft soil is supported by long piles, the spectrum has a peak at a larger value of T ($T=0.7$ sec) and T_{cr} is more than 1 sec as shown in Fig. 3.12 (b).

iii) An example is shown in Fig. 3.12 (c), where a building is supported by rigid piers. In spite of the existence of soft soil on the rigid layer, the characteristics of the spectrum are the same as case iv).

iv) The spectrum which, as can be seen in Fig. 3.12 (d), is obtained on the firm ground with raft foundation, has a peak close to 0.2 sec, and the value of $\bar{q}$ decreased rapidly with increasing T. The value of T_{cr} is about 0.5 sec.

It can be concluded from these spectra that, even on soft ground, the response of the structure with piles or piers shows as if the rigidity of the ground changes in the order of i) to iv). This is an important effect of pile or pier foundations and the effects on the seismic resistance of buildings. The more rigid the foundation, the more significant is the effect that can be expected and, on the other hand, the effect on the spectrum may not be so significant if long slim piles are used. Then, if a basement is on a rigid layer, which is generally more rigid than short piers, great seismic resistance can be expected and can be shown by the comparison of the normalized acceleration spectrum as mentioned above.

It will not be so serious a problem in raft foundation, but the treatment of base shear, in the case of pile or pier foundation with soft soil near the ground surface, becomes important.

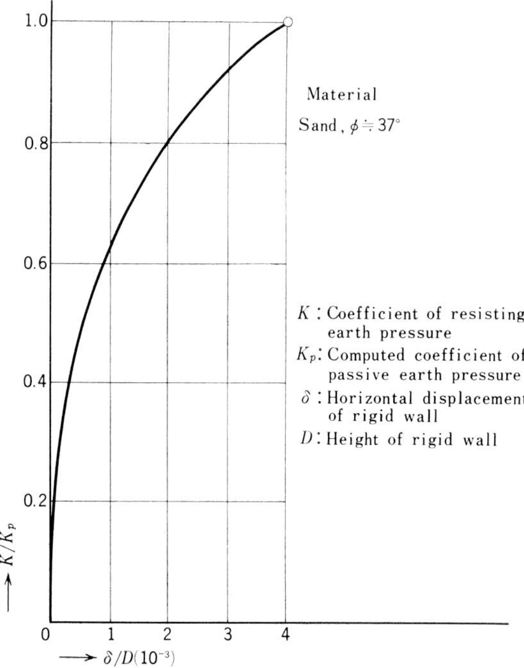

Fig. 3.13

In the following expression, that follows

W : Weight of building

R_v : Bearing capacity of pile foundation

R_h : Horizontal bearing capacity of pile foundation

C : Base shear coefficient

F : Safety factor,

it is generally easy to get the following formula for the vertical load.

$$R_v \geqq FW$$

But as a general characteristics of ground, where pile foundation is applied, these bearing capacities are ;

$$R_h \ll R_v$$

Then the following condition may often occur.

$$R_h < CW$$

In such a case, horizontal resistance of the pile is insuficient to resist the base shear from the semi-static viewpoint.

In pile foundations, the friction between the foundation slab and the ground may not be relied upon and no frictional resistance can be expected in likely to settle areas. In this case, the base shear has to be resisted by both horizontal resistance of pile $R_h(\delta)$ and passive earth pressure $P_p(\delta)$ on the basement wall or the side wall of the basement beam. The relation between R_h and δ, where is the horizontal displacement, can be obtained by the results of horizontal loading test or with reference to Chang's formula. The relation between P_p and δ is shown in Fig. 3.13[20], as one of the few experimental results available at present.

Judging from the result, each ratio of two elements in the right hand side of the following formula can be obtained, (i.e. the resisting ratio of pile and wall, and displacement).

$$CW = R_h(\delta) + P_p(\delta)$$

In pile foundation design, it can not be simply decided whether pile head should be fixed in the foundation slab with reinforcement or should be placed so as to work as a pin connection. The advantage of fixed pile top has been described in that the increasing degree of redundancy can prevent serious structural damage, even if suffering from partial damage.

In high-rise buildings, tensile force sometimes occures on piles or piers during earthquakes. In such a case, each structural part, including pile top and joints of pile, has to resist seismic forces satisfactorily.

The allowable tensile bearing capacity with the assured structural strength of each part, is computed from frictional resistance of soil around the pile or pier, except when distance between each pile is less than a certain amount where decreasing allowable tensile strength has to be considered from the pile group viewpoint[17].

3.5 Dynamic Properties of Soil

Several characteristics of ground and foundation described related to Fig. 3.1 have to be clarified in order to employ dynamic analysis in the design of building foundations. But there are some attempts to move into dynamic design in Japan influenced by the recent high-rise buildings. Dynamic strength of soil, one of the above mentioned characteristics, will be investigated in what follows.

3.5.1 Dry Sands

A fairly large amount of laboratory tests[21] have been carried out concerning shear strength or bearing capacity of dry sands under vibration. The results of experiments are influenced by the method of experiments and by researcher, but it can be approximately described that the decrease of shear strength or bearing capacity is not a predominant influence on the earthquake resistance but only to a small degree. Damage of houses on dry sands during earthquake is mainly caused by the slide of the slope under vibration.

3.5.2 Saturated Sand

It is a well known fact that saturated sand looses its strength or bearing capacity, under vibration with fairly small acceleration compared with dry sand, by the so-called liquefaction phenomena.

There are examples of serious damage by the liquefaction in the United States[22] and in India[23] and, in addition to that, many reinforced concrete buildings were damaged by the world-wide know Niigata Earthquake June 1964. The particular ground conditions for the liquefaction in sand layer under the ground water can not be defined yet, but the ground which satisfies the following five conditions, may have the possibility of suffering serious damage of buildings from liquefaction during earthquakes[24].

i) Almost pure sand with less than ten percent of silt and clay is distributed from the surface to the depth of 15~20 m.

ii) Uniform sand (coefficient of uniformity less than 5), and medium grain size (mean grain size=0.2∼1.0 mm).

iii) The sand layer is under the ground water.

iv) Loose sand, where the N-value is smaller than a critical N-value[25] with more than a certain thickness.

v) Period distribution curve of microtremors is in wide range including long periods.

The ground which satisfies the above conditions can be found in a filled area or around a new river which is less than 300 years old.

3.5.3 Cohesive Soil

It may be described from past experiments that there is no significant influence on the strength, as in the case of dry sands, under vibrations up to 200 gals. But it may not be denied that the cohesive soil with very high sensitivity looses strength under vibration and, with relation to that, one proposal has been made with respect to the correction of safety factor for long-term load depending on sensitivity as shown in Table 3.5[26].

Seed[27] has proposed to decide the allowable limit composed of long-term stress σ_L and earthquake stress σ_E with the function of number of earthquake waves as can be seen in Fig. 3.15. In the figure q_u means static unconfined compression strength.

3.5.4 Modal Analysis of Ground-Foundation System

In the analysis of dynamic response of buildings during earthquakes, the effect of the ground

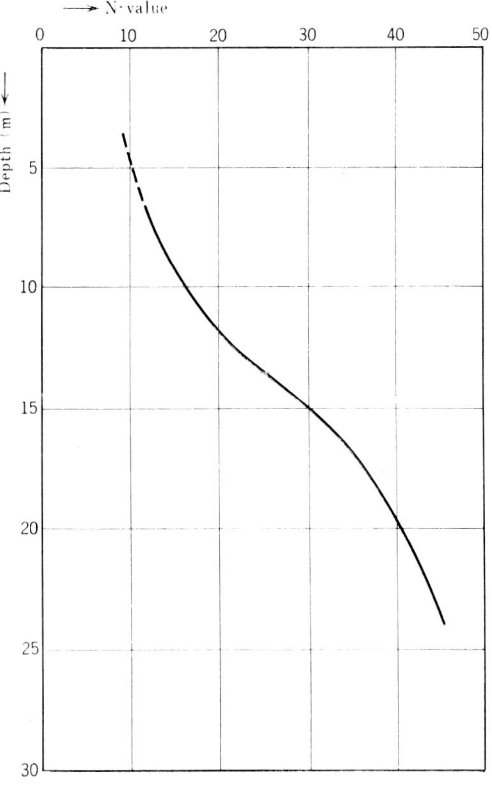

Fig. 3.14

Table 3.5

Degree of sensitivity, S_t	Safety factor
less than 10	3
10～15	$1+\frac{S_t}{5}$
more than 15	4

and foundation should be added to the modal analysis of buildings. It may not be correct to assume a fixed base in a mass-spring structural system, as is generally carried out at present. It is attempted, as the next stage of analysis, to introduce dynamic coefficient of sub-grade reaction K, or to employ a model with horizontal displacement and rotation to the foundation based on the deformation of the ground by half-space theory. But this treatment still has such problems as an analysis of layered ground or the characteristics of in-put on the basement wall.

Recently, a mass-spring system has been applied to each divided slice of ground on the rigid base, or in case of pile foundation, there is an interesting study[28] which considers a soil-pile interaction system to get an earthquake response of structure.

The following definitions are used in this chapter.

Foundation, Foundation structure : Foundation or foundation structure is composed of foundation slab and ground work.

Foundation slab : Structural part which transport the stress of structure to the ground or to

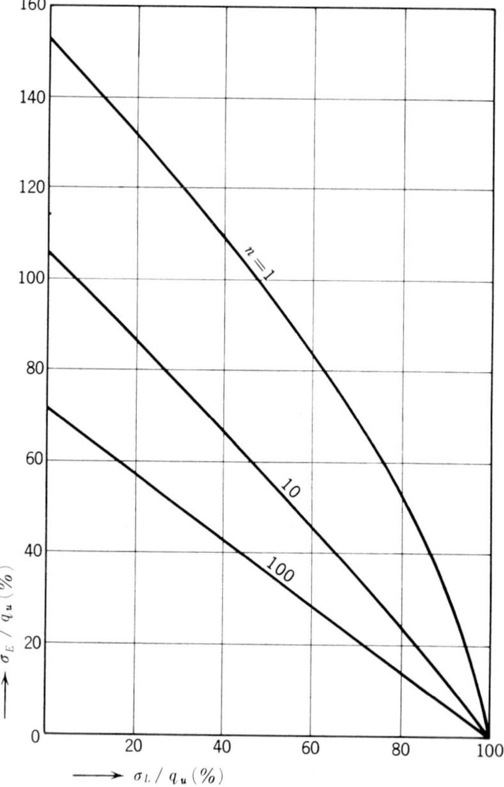

Fig. 3.15

the ground work ; footing in footing foundation, slab in raft foundation.

Ground work : Generally piles or piers or rubbles which are prepared to support foundation slab.

Pile : Ground work, with comparatively small diameter to its length, constructed by driving.

Pier : Ground work, with comparatively large diameter to its length, constructed by excavation.

References

1) Kawasumi, H.; Distribution of Earthquake Damage and Ground Condition in Tokyo and Osaka, Distribution of Danger Extent in Disaster of Japan. Building Disaster Handbook, Metropolitan Disaster Prevention Committee, Society of Natural Resources, Feb. 1947 (in Japanese).
2) Kitasawa, G.; On the Ground of Tokyo, Trans., A.I.J. No. 40, Feb. 1950. (in Japanese).
3) Omote, S.; The Relation between the Earthquake Damage and Structure of the Ground in Yokohama, Bull. of the Earthquake Research Inst., Univ. of Tokyo, Vol. 27, 1949 (in Japanese).
4) Omote, S. and Miyamura, S.; Relation between the Earthquake Damage and the Structure of Ground in Nagoya City, Bull. of the Earthquake Research Inst., Univ. of Tokyo, Vol. 29, 1951. (in Japanese).
5) Takeyama, K., Hisada, H. and Ohsaki, Y.; Behavior and Design of Wooden Buildings Subjected to Earthquake, Proc. 2 nd World Conference on Earthquake Engineering Vol. III, 1960.
6) Ohsaki, Y.; Earthquake Damage of Wooden Buildings and Depth of Alluvial Deposit, Trans. A.I.J. No. 72, March 1962.
7) Otsuki, Y. and Kanai, K.; Aseismic Design of Structures, Corona Pub. Co., Ltd. March. 1961. (in Japanese).
8) Special Committee on the Hokuriku Earthquake Damage Investigation; Report on the Earthquake Damage in Fukui Earthquake 1948. (in Japanese).
9) Takeyama, K. and Ohsaki, Y.; Some Problems on the Structural Design Method and the Safety of Buildings, Report, A.I.J. Vol. 31, May 1955. (in Japanese).
10) Hisada, T., Nakagawa, K. and Izumi, M.; Normalized Acceleration Spectra for Earthquakes Recorded by Strong Motion Accelerographs and Their Characteristics Related with Sub-Soil Conditions, Building Research Institute, Occasional Report No. 23, June 1965.
11) Nakagawa, K.; Coefficient of Sub-grade Reaction Obtained from the Vibration Test of Buildings and Model Footing, A.I.J. Vol. 67, Oct. 1961. (in Japanese).
12) Hisada, T. and Nakagawa, K.; Coefficient of Sub-Grade Reaction Obtained from the Proto-type Loading Test on the Building, Report A.I.J. Vol. 22, 1953. (in Japanese).
13) Hisada, T., Nakagawa, K. and Kimura, E.; Study on the Dynamic Properties of Ground, Report, A.I.J. Vol. 22, 1953. (in Japanese).
14) Yamamoto,; Dynamic *K*-value, Dynamic Properties of Soil and its Application, J.S.S.F.E. Jan. 1965. (in Japanese).
15) Ohsaki, Y.; Measured Values of Coefficient of Subgrade Reaction and its Application to Footing Design, Report of Building Research Institute No. 21, March 1957.
16) Kanai, K. and Tanaka, T.; On Microtremors VIII, Bulletin of the Earthquake Research Institute Vol. 39, 1961.
17) A.I.J.; Design Standard and Explanation on the Foundation Structure of Buildings, Nov. 1952. (in Japanese).
18) International Association for Earthquake Engineering; Earthquake Resistant Regulations A World List, 1963.
19) Chang, Y.L.; Discussion on "Lateral Pile-Loading Tests" by Feagin, Trans. A.S.C.E. Vol. 102, 1937.
20) Matsushita, K. et. al; An Experimental Study on the Displacement of a Rigid Wall and the Earth Pressure (Part. 2, The 1 st result of experiment), Trans. A.I.J. Sept. 1965. (in Japanese).
21) J.S.S.F.E.; Dynamic Properties of Soil and its Application, Nov. 1964. (in Japanese).
22) Fuller, M.L.; The New Madrid Earthquake, United States Geological Survey, Bulletin 494, 1912.
23) Officers of the Geological Survey; The Bihar-Nepal Earthquake of 1934, Memoires of the Geolo-

gical Survey of India Vol. 73, 1939.

24) Ohsaki, Y.; Sandy Ground Susceptible to Earthquake Damage, Technical Report Submitted to Japan-U.S. Conference on Soil Dynamics, July 1965.

25) Koizumi, Y.; Changes in Ground Condition of Sand and Building Damage by Niigata Earthquake, Lecture Material, B.R.I. Nov. 1964. (in Japanese).

26) Ohsaki, Y.; Foundation Engineering (theory and practice), Building Structure Series 11, Corona Pub. Co., Ltd. July, 1961. (in Japanese).

27) Seed, H.B.; Soil Strength during Earthquakes, Proc. of 2nd World Conference on Earthquake Engineering Vol. 1, July 1960.

28) Parmelee, R.A., Penzien, J., Scheffey, C.F., Seed, H.B. and Thiers, G.R.; Seismic Effects on Structures Supported on Piles Extending through Deep Sensitive Clays, Institute of Engineering Research, University of California, SESM 64-2, August 1964.

4. WOODEN STRUCTURES

4.1 General

Wooden structure is a general term refering to a structure composed of beams, columns or their equivalent of framed timber. Besides traditional Japanese and western wooden houses, it includes glue-laminated timber structure, a brick building framed by wood, a stone building with wood framing and a concrete block building with wood framing.

Timber framed brick construction, timber framed stone construction and timber framed concrete block construction were proved unacceptable structures by past examples of earthquake damages, so they should not be built except for special purposes. When they are built of necessity, the curtain walls of a brick, a stone and a concrete block that suffers much damage by itself shall be the self-supported structure, and the effect of bearing wall upon it is not usually expected. Therefore, as the proceeding of structural planning it is right to adopt the following common method for general wooden structures.

Putting aside timber framed brick construction, timber framed stone construction and timber framed concrete block construction, to think of general wooden buildings, the light weight is the advantage of wooden structure and yet structural technique of wooden construction has no disadvantage against earthquakes. But there has been great damage to wooden buildings in past earthquakes. The main reason is that almost every building is much concerned with only small things such as how to compose joints instead of considering rational structural planning.

According to past experiences, a wooden building suffers damage when the building is in the following conditions.

i) Inadequate site or ground.
ii) Imperfect foundation.
iii) Inadequate plan and solid style.
iv) An outworn building, or a new building that was poorly executed.
v) Insufficient bracings or walls.
vi) Inadequate way of putting bracings and knee braces.
vii) Small columns or other thin members compared with the scale of the building.
viii) Inadequate joints and connections.
ix) Heavy roofs.

The key point of wooden structure is how to make it resist horizontal forces. The methods of resistance are classified into the type of unit frames, the type of bearing wall and their combination type.

The unit frame type is the type that resists horizontal force by each unit frame or rahmen, an arch, shell, etc.…. The bearing wall type is the type that resists horizontal force by adequately set bracings, shores, walls which are called a bearing wall system. The former typical buildings are a gymnasium, a hall and a factory. The latter typical examples are a school building, a house and a shop.

In the type of bearing walls, the horizontal planes of a roof and a floor shall be strengthened by horizontal trusses in order to transmit horizontal forces to bearing walls. The consideration that the whole building may resist horizontal force in one body is necessary. Careful consideration shall be given to the form and arrangement of foundation in order to make the frame and the bearing walls efficiently against the earthquake forces. The flow of stresses, ground conditions and the method of execution shall be examined, and the pertinent design shall be adopted. The continuous footing constructed by reinforced concrete or by plain concrete shall be arranged especially at the bottom of bearing walls without fail.

In wooden structure, as stated above, it is important to consider the composition of frame of the whole building. At the same time it is necessary to plan each part of the structure and conversely plan the whole structure considering the details of each part of the structure concretely. It can be said not only in wooden structure but also in other structures. Above all in wooden structure the whole frame is apt to deform greatly because of the construction of joints and the material properties of wood. As it is difficult to preserve rigidity, in planning the structure, first of all this reversible relation shall be considered seriously.

Joints shall be constructed so as to have sufficient strength as well as the required rigidity and ductility. Careful consideration shall be given to the design and the execution in order to avoid the looseness of joints.

The wood constituting members in wooden structure has the following peculialities that are not seen in other structural materials.

i) Contraction is relatively large. Particularly the joints slack easily by the contraction in the direction perpendicular to fibers. Therefore, dry wood shall be used, and the average value of percentage of moisture content of members subjected to loads immediately after the execution must be less than 20%.

ii) The elastic modulus is small. Consequently, members are apt to show large deformations.

iii) The amount of deformation from the yield point to destruction is relatively small. Therefore, from the points of this property and the problem of stability against earthquake the design of wooden structure shall be based on elastic supposition.

iv) A notable creep phenomenon is seen due to vertical loads. Considering strength and deformation at earthquake, this factor must not be ignored. This is important especially in snowy area considering the earthquake in deep snows.

v) Sinking occurs by compressive force in the direction perpendicular or oblique to fibers. It has a great influence to the amount of deformation of horizontal members and chord members of built-up members.

vi) The defects and the notches of wood influence greatly the strength and rigidity. In consequence, it is necessary to select members carefully and arrange them with mechanical consideration.

When calculating stresses and deformation in wooden structure, usually the following are assumed.

i) The panel point of a truss, the panel points of a column and a beam, and a column base shall be assumed hinged. When they cannot be so supposed, the influence of secondary stresses shall be considered.

ii) All members and joints shall show elastic behaviors. But in sinking of horizontal members and chord members, and in calculation of the section of joints, sometimes plasticity may have to be considered.

Considering the peculiarities of wood itself and joints, the influences of dryness and contraction, and the suppositions in the calculation of stress and deformation, the exactness of the calculation of wooden structure, especially the exactness of calculation of deformation cannot be expected.

Therefore, in order to raise the calculation exactness as high as possible, the structure composed of a plain and simple frame type shall be planned so that the stream of stresses may be easily grasped.

In a long-span building of unit frame type, it is desirable to select the structure of relatively low degree of redundancy so that the stream of stresses may be easily grasped. The structure of supporting points and panel points shall be as close as possible to the conditions assumed in the stress calculation. This can be said in glued laminated timber construction of long-span that have become popular recently. In glue-laminated timber construction, the laminated lumbers shall be laminated in a factory in order to raise the reliance of production and they shall never be laminated by adhesion at the site of execution.

In buildings of bearing wall type, it is safe to think that a group of frames including bearing walls bear the whole horizontal forces. The rate of its bearing will be stated later at the required amount of walls. When the arrangement of bearing walls is extremely unbalanced the influence of twist shall be considered.

In buildings of the mixed type of unit frames and bearing walls, rigidity and the arrangement conditions of the two shall be compared and examined as carefully as possible, and the rate of bearing of horizontal force shall be decided.

Wooden structure has different properties from reinforced concrete structure and steel structure in strength and rigidity. Therefore, when wooden structure is combined with those other structures in one building, generally wooden structure shall be designed separately and independently from other structures. When they must be combined inevitably, rigidity of each frame shall be considered so that the occurrence of unfavorable stresses may be prevented.

Most firewalls constructed of reinforced concrete are not separated, however when the horizontal forces are not transmitted to them, the connection as the structure shall be clearly separated. This can be applied in timber framed brick structure and timber framed concrete block structure.

Recently beams have often been constructed of steel. Careful attention shall be paid to the joints of the steel beams and wooden columns, notching of timber columns and sinking of timber.

As many outworn wooden buildings suffer earthquake damage, the construction method of preservation and protection (from ants) shall be considered in the planning for the whole building. It is important to construct free from danger of fire caused by earthquake. Preservation, protection and fire prevention will be treated later in detail.

4.2 Structural Design

4.2.1 Methods

In designing, it is important to consider the form and the dimension of a building from the point of planning and space to increase the resistance against earthquake. The building whose principal structural parts are constructed by wooden construction shall be less than 13 m in height and the height of the eaves shall be less than 9 m under the provisions in Japan.

The plan which is simple and is easily united is desirable. The building which is narrow and long yields large deflection in the direction of beams by earthquake forces, so especially the structural precautions are necessary. The heights of buildings shall be rendered uniform as much as possible. When there is much difference in height, each one shows different vibration properties. Therefore, the connecting parts shall be connected lightly. The connecting parts of wooden structure and other structures shall be treated in the same way as the above.

Wooden buildings shall be constructed so that horizontal deformation may be as small as possible, for if the deformation makes the inclination of building great the stability of frame loses due to the influence of vertical loads, and there is fear of destruction. Accordingly, first of all, the weight of roof finish shall be made as light as possible. Secondly, the horizontal planes of structure such as roof plane and floor plane shall be strengthened so that the whole building may resist as one body. At the same time, unit frame and bearing walls that resist earthquake forces shall be arranged proportionately and pertinently so that the required strength and rigidity may be preserved. Particularly the balanced arrangement of the bearing walls from the point of planning, shall be considered so that the building may not twist in an earthquake. From the space point of view, bearing walls of upstairs and downstairs shall overlap each other.

When materials, structure and arrangement of fire wall is appropriate, it is effectively worthy of a bearing wall. Therefore, it should be used sufficiently. But as a simple fire wall constructed by wooden frame easily yields, cracks and exfoliation by earthquake and sometimes it cannot fulfill the function as a bearing wall, careful consideration shall be given to it.

Stairs can resist earthquake forces as a great bracing when arrangement and structural system of stairs are good. Therefore in designing, it is desirable to consider not only the traffic convenience within the building but also the position that yields the effect from the earthquake-proof structural point of view. In a wooden school house, the stair hall constructed by reinforced concrete has a great deal of effect in resisting earthquakes.

In comparison with other structures, as a wooden building has small dead weight and a small number of stories, the small axial force of a column and relatively slight foundation will do. Even though the slight foundation is good enough, it should never be placed on filled-up ground or the soil of insufficient bearing capacity. It shall be set directly on the bearing layer that has required bearing power of soil. As stated before, the lower part of a bearing wall must have a continuous footing constructed of reinforced concrete or plain concrete by all means. Otherwise, the effect as a bearing wall cannot be expected. Further, according to the conditions of loads and ground, piling foundation is necessary in wooden structure. The attention for the work is the same as the caution of the general piling foundation.

4.2.2 Design for Vertical Loads

a) General

The frame arrangement shall be decided so that vertical loads may be distributed as evenly as possible to each frame. Columns shall be equally disposed and the columns of each floor shall take the same position in the upper and lower floors. The form and the arrangement of foundation shall be designed so that the foundation reaction caused by vertical loads may be distributed equally. In case of considerably unequal distribution, the influence on the upper structure due to uneven settlement shall be considered. When there is partially unequal vertical loads (for example unequal snow distribution in snowy areas) its influence should be considered in the whole construction.

In the frame with angle braces, attention shall be paid to the bending deformation and the destruction of columns, as there are many examples of earthquake disaster that the joints of columns and angle braces were broken.

b) Unit Frame

The unit frames (independently stable frames like a rahmen and an arch) against vertical loads shall be disposed so that loads may be distributed as equally as possible. Speaking of the frame type, it is desirable to make the stream of stresses natural and make the structure simple so that the stress distribution may be grasped clearly.

Tie members and diagonal members shall be provided in each framework to secure lateral rigidity of unit frame sufficiently. The footings of unit frames shall be tied rigidly to each other to avoid uneven settlement and movement, as far as possible.

4.2.3 Design for Horizontal Loads

a) Structural Type

Unit frames or bearing walls shall be installed so as to make them resist seismic forces, wind pressure forces and other horizontal loads. The method of transmitting lateral forces to them will be described in section d).

b) Unit Frame Structure

Each frame to resist lateral forces shall be arranged so as to bear equal loads. In case of unequal arrangement, horizontal trusses or other similar construction shall be installed so as to form one rigid body, considering the influence of twist. The frame type shall be the type which makes stress distribution reasonable. Even when each framework has equal rigidity and symmetrically arranged so as to bear equal lateral force, the whole building shall form one rigid body by installing horizontal trusses or other similar construction.

Further, in order to increase the lateral rigidity of unit frames, each shall be connected to others by tie members perpendicular to the plane of the frames.

c) Bearing Wall Construction

Bearing walls shall be arranged as symmetrically as possible both in plan and in elevation; otherwise, the effect of torsional deformation shall be considered and the bearing rate of horizontal force shall be examined.

Distance between two adjacent parallel bearing walls shall be less than 8 m (less than 12 m when bearing walls are tied by horizontal trusses so as to form one rigid body).

When bracings or column buttresses subjected either to tension or compression are provided, bracings or column buttresses shall be installed so as to resist the approximately same amount of lateral force in the opposite direction.

In case of a design not based on structural calculations, the arrangement of bearing walls in a building of bearing wall construction shall be in accordance with the followings (cf. Building Standard Law, Enforcement Ordinance Article 46, Paragraph 3).

For earthquakes, the effective length of the bearing wall shall be not less than the values, per square meter of floor area of each story, given in Table 4.1, both in the direction of beam span and in the direction of wall girders, respectively. When the soil is unusually soft, however, the values shown in Table 4.1 shall be increased by 50 percent.

Table 4.1 Required Effective Length of Bearing Walls for Earthquake
(unit : cm/m^2 (floor area))

Kind of buildings	One story	Two stories	
		2 nd floor	1 st floor
Buildings in which the roof is covered with light materials such as metal, slate, asbestos board, shingle, and others of similar type	12	12	21
Buildings in which the roof is covered with comparatively heavy materials such as roof tiles or others, and not included in the above, or buildings with heavy walls as in wooden framed brick construction, wooden framed stone construction, Dozo or others of similar type	15	15	24

Further, in case of wooden buildings with light roofs or two-story buildings, wind pressure may be more dangerous than earthquake force. Therefore for wind pressure as well as earthquake force, the effective length of bearing wall shall be calculated and examined both in the direction of beam span and in the direction of wall girders of each floor. In this case, instead of the effective length per square meter of floor area, the effective length of bearing wall per square meter of the projected area of the elevation perpendicular to the direction of wind in the building shall be considered (required values are omitted here). In this case, the projected area of the elevation to be computed shall be the sum of that of the 1st and 2 nd stories for 1st story of two-story buildings; for one-story buildings and for the 2 nd story of two story buildings, the sum of each part shall be taken.

The effective length of bearing walls shall be computed by multiplying the actual length of each by the multiplier given in Table 4.3, however joints for bracings and column buttresses shall be in accordance with the methods that will be stated later at Section 4.3 Structural Parts. Further, more than half of the required effective length of bearing walls shall be those provided with bracings or column buttresses.

The frames resisting lateral forces with diagonal members such as bracings or column buttresses shall be so constructed as to form trussed structures, and columns, beams, diagonal members and footings shall be firmly tied to one another for safety.

The joints of bracing, column or column buttress acting as tension member with the footing, shall be carefully designed so as not to cause pulling out.

Table 4.2 Multiplier for Calculating the Effective Length of Bearing Walls

	Kind of bearing wall		Multiplier
(1)	With bracing members, 1.5 cm or more thick and 9 cm or more wide, or steel bars larger than 9 mm in diameter, or other steel members having equal or higher strength		1.0
(2)	With bracings larger than one-third of columns used in the frame, or steel bars larger than 13 mm in diameter, or other steel members having equal or higher strength and connected to horizontal members		1.5
(3)	With bracings larger than one-half of columns used in the frame, under which continuous footing of reinforced concrete or of plain concrete with depth of 45 cm or more provided.		3.0
(4)	With bracings equal in size to columns used in the frame, under which continuous footing of reinforced concrete or of plain concrete with depth of 45 cm or more is provided		4.5
(5)	With X bracings in accordance with the provisions for bracings given in items (1) to (3) inclusive		double the value of each item involved
(6)	With X bracings in accordance with the provision for bracing given in item (4)		6.0
(7)	Earth-plastered wall with no back coating, or wall plastered on wooden laths, with studs smaller than one-third of columns in the framework		0.5
(8)	Earth-plastered wall with back coating or other similar wall		1.0
(9)	Walls having wooden laths nailed to the surface of columns, with studs larger than one-third of columns used in the framework	single wooden laths double wooden laths	3.0
(10)	Combined bracings of items (1) to (6) inclusive with wall of items (7) to (8)		the sum of the value of each item concerned
(11)	Wooden column buttresses whose dimensions are larger than those of columns used in the framework		3.0

d) Construction to Transfer Lateral Forces

Construction having large rigidity in the horizontal plane such as horizontal trusses shall be employed in order to transmit lateral forces to the bearing walls and the frames.

Horizontal trusses and other similar structures shall be tightly joined to walls or frameworks to secure transmission of forces and to obtain rigidity of joints.

Lateral forces acting in the direction of wall girders in long span costruction shall be resisted first by the studs in the gable wall, next by the horizontal trusses provided along the gable wall, and finally transmitted to walls or frames in the direction of wall girders.

Horizontal trusses and other similar construction to resist lateral forces shall be made sufficiently rigid and strong. In calculating the rigidity and strength of horizontal trusses or other similar structure, the effect of finishing materials may be considered.

Horizontal trusses shall be made free from lateral deformation by adoption of appropriate construction.

e) Joints

The details and precautions of connection and joints will be stated at Section 4.3 Structural Parts. The method to use metal fasteners or the method to use adhesive materials is desirable. For the joints that require rigidity, joint by adhesion or nails is pertinent while for the joints that require ductility, joints by bolts, doweled joint and abutting joint are proper. As the large deformation of each member, and subsequently cause large deformation of the whole framework to occur due to the looseness and the large deformation of joints, attention shall be paid to this matter.

The timber being almost free from defects such as wane and knots shall be used near the joints. The position subjected to large stresses shall be avoided for the joints.

When the type of joint subjected to eccentric loads is adopted, the influences due to eccentricity shall be considered. When the splice plates are used for the joints of the members subjected to bending, attention shall be paid especially to their detail.

When spliced joint is used for tension members, the sum of the sectional area of the splice plate shall be increased by 50 percent of required sectional area. At the tension joint, the extra length of tension members shall be taken sufficiently. For joints that become slack due to the dryness of timber, it is important to design so that the joints may be fightened at any time.

4.3 Structural Parts

4.3.1 Foundations

Foundations shall be constructed on solid ground after removing surface soil. The base of foundation shall be more than 45 cm below the ground surface. In the districts with the fear of freezing, the base of foundations shall be set below the freezing line. In case of thick surface soil and soft ground, broken stone foundation or other foundations shall be made. Thus foundations shall be constructed on firmly pounded ground.

When the soil is especialy soft and the loads transmitted to the foundation is great, piling foundation shall be adopted. When wooden piles are used, they shall be unseasoned wood. After the completion of the building, they shall be permanently below ground water level.

Footings shall be constructed of concrete or reinforced concrete. The footings arranged beneath important parts from the point of structural bearing strength such as the exterior walls

Photo 4.1 Collapse due to Unevensettlement of Footing

and principal interior walls shall be the continuous or raft construction. Independent footings shall be used only in unavoidable cases such as for the footings of independent columns. Particularly in case of soft ground, uneven hardness of ground and great differences in loads to the footings, and in case of bending stresses that are likely to occur at the footings, the footings shall be constructed with reinforced concrete.

When pulling out of footings in long-span construction, in frames with bracing, or in column buttresses, is likely to occur, the footings shall be constructed so that the pulling out may be resisted by their own weight or by the resistance of continuous footings of reinforced concrete construction.

4.3.2 Frameworks

a) General

Frameworks constitute wall frames to support the loads of roofs and floors, and at the same time they are the principal structural parts that resist the earthquake force and other horizontal forces. The propriety of this part of the structure affects directly the resistance of the building against earthquakes.

Pertinent framework construction must have the following conditions :

i) Each member shall be arranged appropriately so that there is no big opening and diagonal timbers having required sectional size are disposed appropriately.

ii) Sectional size of members shall be the required size, and there is no big notch by connection and others.

iii) The positions of connections and joints are pertinent and they will bear stresses expected at those positions.

iv) Constructed frameworks do not cause large displacements and deformations under design external forces.

It is needless to say that the sections of braces, columns and horizontal members must have required sizes so that the frameworks with braces may resist the horizontal external forces sufficiently. It is necessary that end connection of brace, the connection of foot of column that yield tensile forces because of putting braces shall transmit the forces completely to sills, footings and soil by anchor bolts, and yet hinder deformaion, sinking, splitting, settling and lifting.

b) Sills

The base of principal structural columns in the lowest story shall rest on sills. In other words, not only the base of external columns but also the base of internal columns shall rest on sills so that the foot-parts of buildings shall be unified and fixed to increase resistance against earthquake forces. But no sills are allowed when the column is firmly tied to the footing or when employed in one story houses (cf. Architectural Standards, Ashigatame (horizontal

Photo 4.2 Slide due to Insufficient Tightness between Sill and Footing

Photo 4.3 Collapse due to Decay of Corner Sill

members connecting the base of columns) Enforcement Ordinace Article 42).

Sills shall be firmly tied to footings. But there is no need of firm tie in one story houses whose total floor area is less than 50 m^2 (cf. Building Standard Law, Enforcement Ordinance Article 42, Paragraph 2).

Anchor bolts shall be disposed near the lower ends of tension braces, just below the upper ends of compression braces, near the joints in sills and ends of sills and at intervals of about 3 m in general in other cases. The strength of anchor bolts, the size of washers, the length of anchoring shall be determined to make it possible to transmit the forces to sills by columns and braces to footings.

Shoulder area shall be determined so that sills may have ample bearing strength against compression forces from columns and braces, while the joint method that shall transmit completely tensile forces from columns and braces to sills in combination with metal fasteners shall be applied to tie firmly.

Horizontal brace shall be arranged in every corner of sills and in other required places, except when the building is surrounded by continuous concrete footings. Durable lumbers such as Japanes cypress, Japanese cedars, etc. shall be used for sills. The bottom of sills shall be set generally more than 20 cm above the ground level and in addition wood preservation treatment shall be applied.

c) Columns

Columns shall be arranged not only at external perimeter of the building but also at important internal places as far as there is no interference. The extreme unequal load burden shall be avoided and the rigidity and bearing strength against horizontal forces shall be increased by the arrangement of bracings.

Columns shall be safe not only against compressive force and bending moment but also against sinking at their ends. In case of the long columns, the safety against buckling shall be examined.

When not based on structural calculations, the width of principal structural column shall be not less than the values shown in Table 4.3 in the direction of beams and ridge, respectively. For columns of buildings built in heavy snow districts, however, the values shown in Table 4.3 shall be increase by coefficients not less than 1.1.

Long columns having particularly great vertical and horizontal loads shall be built-up columns or shall have column buttresses. According to the conditions, it is sometimes better to use steel members jointly.

At the joints of columns and horizontal members or knee braces, pertinent metal fasteners shall be used to tie firmly, giving no big notching to the column.

Notching at or near the middle length of columns shall be avoided and, when more than one-third of the required cross-sectional area is notched, that part shall be properly reinforced (cf. Building Standard Law, Enforcement Ordinance Article 43, Paragraph 3).

At the column ends, the connection that transmits sufficiently the force of that part and preserve the required rigidity shall be installed so that pulling, sinking, lateral movement and others may not occur. Especially when compression braces are installed, attention shall be paid so that tensile forces occuring in columns may be transmitted completely through the connection.

Table 4.3 Width of Column Sections

Column		Story / Type of buildings	1	2	3
			Ordinary buildings		Buildings with heavy walls as in wooden framed brick construction, wooden framed stone construction, Dozo or others of similar type
			Buildings in which the roof is covered with light materials such as metal, slate, asbestos board, shingle, and others of similar type	Buildings which the roof is covered with comparatively heavy materials such as roof tiles or others, and not included in the left item 1	
A	Columns in such buildings as schools, public baths, theaters, cinemas, show-halls, auditoriums, stadiums, public assembly halls, or columns having spaces in the directions of beams or wall girders exceeding 10 m	2 nd story of two-story buildings or one-story buildings	$3.3\times\frac{l}{100}$	$4.0\times\frac{l}{100}$	$4.5\times\frac{l}{100}$
		1 st story of two-story buildings	$4.0\times\frac{l}{100}$	$4.5\times\frac{l}{100}$	$5.0\times\frac{l}{100}$
B	Columns other than those mentioned above	2 nd story of two-story buildings or one-story buildings	$3.0\times\frac{l}{100}$	$3.3\times\frac{l}{100}$	$4.0\times\frac{l}{100}$
		1 st story of two-story buildings	$3.3\times\frac{l}{100}$	$3.5\times\frac{l}{100}$	$4.5\times\frac{l}{100}$

Note :
l is vertical distance between horizontal members of principal structural part such as sills, girts, beams, girders and others, connected to the column. This distance shall be considered in the directions of beams and ridges, respectively.

The bottom of columns set directly on the footing shall be more than 20 cm above the ground level, and the foot of columns shall be firmly tied to the footing by metal.

Corner columns of two or more story buildings, and similar columns shall be continuous pieces, or joints shall be reinforced so as to give the same strength as continuous columns (cf. Building Standard Law, Enforcement Ordinance Article 43, Paragraph 5).

Joints shall, as a rule, be avoided in single columns. When unavoidable, the joint must be made with special care of connection method avoiding the position at or near the middle of the timber. The column shall be properly reinforced if the column at the base is spliced for the purpose of repair.

It is important that separated columns of upstairs and downstairs shall be connected. Especially when compression braces are used, separated columns shall be firmly tied by metal fasteners so that tensile force occurring to the column shall be transmitted completely.

d) Girths, Girders and Beams

Girths connect middle part of the building at the part of connection of upper and lower stories of two-story buildings, and they have the role to tie and fasten the middle part. Therefore, girths shall be used between continuous columns and joints if any shall transmit tensile forces adequately.

Wall girders shall be the joints that can transmit the axial force in order to stiffen the upper external parts of the building in combination with sills and girths. Especially when compression bracings are installed within the framework, girths and girders shall be connected by straps and strap bolts so as to transmit completely the horizontal component of forces occuring in the bracings.

Bending members of beams and others shall have sufficient bending and shearing strength against the loads existing in these members. At the same time they shall have the rigidity in accordance with the purpose of use and shall have the section that will not produce excessive deflection, vibration and other obstacles. But unnecessarily big timbers shall not be used.

Notching of beams shall be avoided as much as possible, especially on the lower side of middle part (cf. Building Standard Law, Enforcement Ordinance Article 44).

e) Diagonal Bracings

Diagonal bracings are the most important members because they stiffen the frame of wooden buildings and increase the resistance against earthquakes.

It is desirable for the diagonal bracings to be put with the slope as gentle as possible (generally less than 3 to 1). One diagonal bracing in a unit rectangular frame, shall be arranged in the diagonal position, e.g. the ends of diagonal bracings shall be firmly tied to near the joints of columns and beams or other principal horizontal members (cf. Building Standard Law, Enforcement Ordinance Article 45, Paragraph 3).

When the space between columns is narrow, diagonal bracings are often arranged in K-form as shown at Fig. 4.1 (b), (c) and (d) to avoid the steep slope of diagonal bracings. But according to this structural method, as the joints of two diagonal bracings yield so excessive stresses that the column is damaged, the structural method of Fig. 4.1 (a) is desirable even if the diagonal bracing slope is steep and stresses become great.

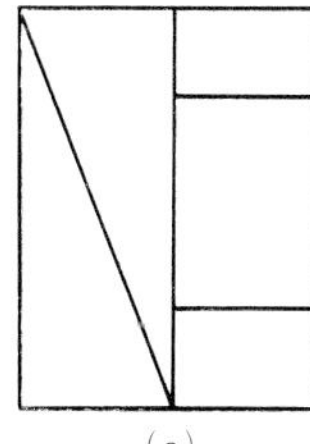
(a)

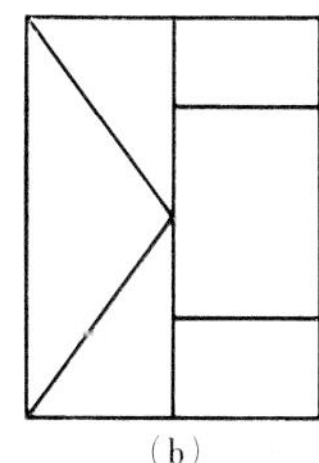
(b)

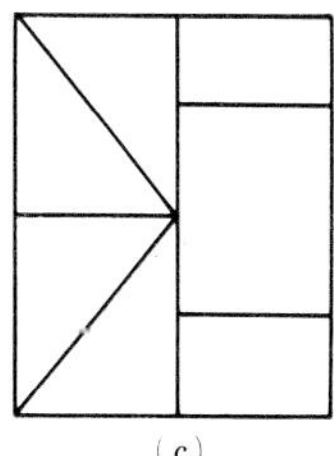
(c)

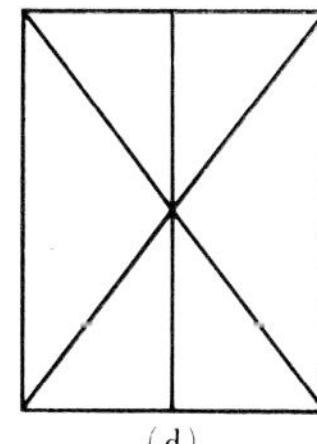
(d)

Fig. 4.1

For diagonal bracings under compressive forces, the timber whose cross-section is not less than 3.5 cm in thickness and not less than 1/3 of the timber used in the framework shall be used. Careful consideration shall be given to buckling and bearing failure at the ends. The buckling shall be considered not only in the structural plane of the wall but also in the direction perpendicular to it.

The joints of the diagonal bracings under compressive forces shall be arranged as shown at Fig. 4.2 in accordance with the diagonal bracing slope. In either case, compressive stresses perpendicular to the axis of horizontal members, due to the axial force of diagonal bracing shall

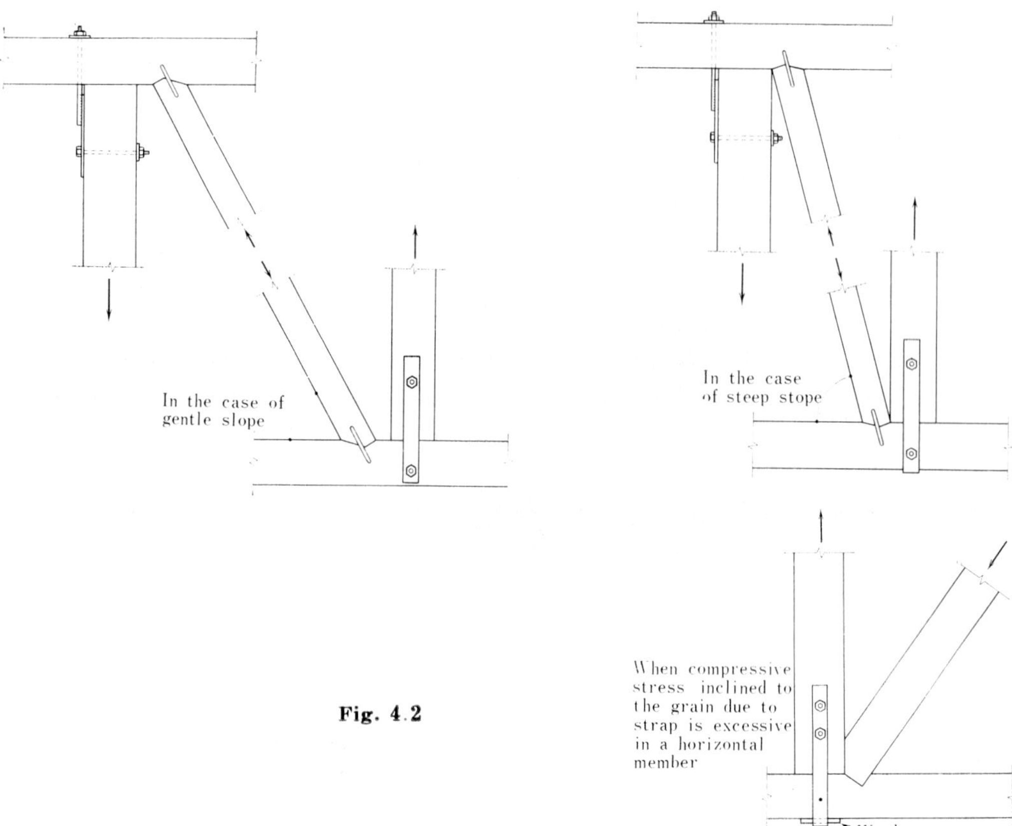

Fig. 4.2

not be excessive.

Columns and horizontal members shall be firmly connected by metal fasteners so that they will not be separated by vertical component of forces in the bracings.

The diagonal bracings that are under tensile force shall not be as shown in Fig. 4.3 (a) but as shown in Fig. 4.3 (b) as far as possible so that the buckling of bracings may be restrained by studs.

For diagonal bracings under tensile force, the timber with the cross-section not less than

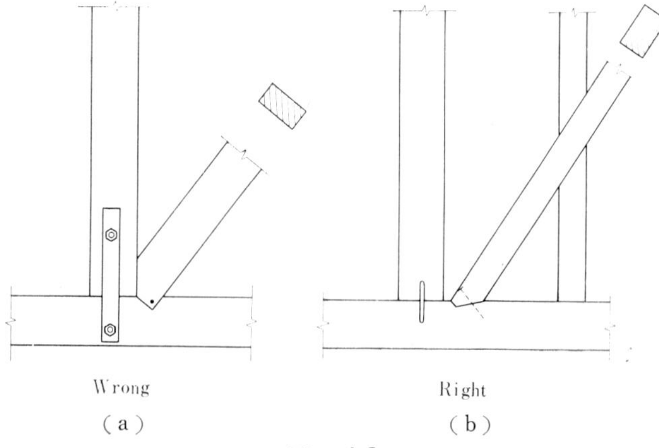

Fig. 4.3

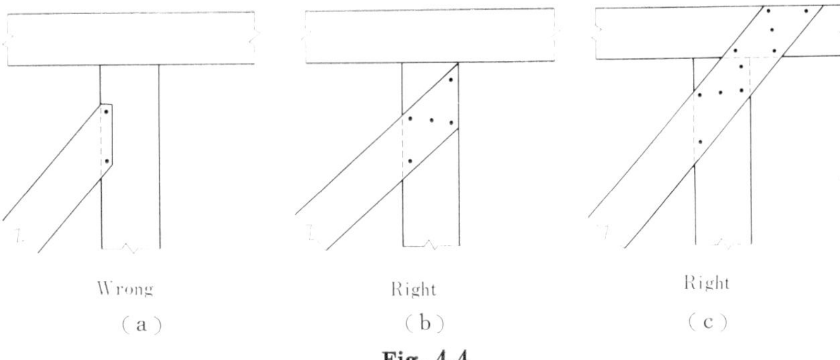

Fig. 4.4

1.5 cm in depth and 9 cm in width (in the external wall in case of shinkabe-style wall in which columns are not covered by wall), steel bars not less than 9 cm in diameter, or other steel members having equal or higher strength shall be used.

The joints of diagonal bracings under tensile force shall have sufficient annexation area by notching in the column or the girder as shown in Fig. 4.4 (b) or (c), and nails whose length is 2.5 to 3 times in thickness of diagonal bracing member shall be used as many as possible to the extent that the member shall not be splitted (cf. [Structural Design Standards of Timber, Section 603.3], the smallest intervals of nails shown in the Table 6.3 of precautions of nail joints). Great bearing strength for the case of the column shown in Fig. 4.4 (a) can hardly be expected. When the head of the nail has a washer of zinc iron plate, the bearing strength will be increased.

No objectionable notching in diagonal bracings shall be allowed for the bearing strength (cf. Building Standard Law, Enforcement Ordinance Article 45, Paragraph 4).

When the wooden diagonal bracings cross studs, battens and others, the member of diagonal bracing shall not be notched, but studs, battens and others shall be notched. On the

Photo 4.4 Damage of a Building having no Diagonal Bracing

Photo 4.5 Inclination of Second Story due to Imperfect Composition of End Connections of Diagonal Bracings Arranged in Second Story

occasion of earthquake, diagonal bracings often fail at the notched part.

When diagonal bracings are put in X-form, both shall not be notched at the intersecting point, but one bracing shall be cut without notching the other and it shall be an abutting joint.

There is no question about this when steel bar bracings are crossed.

Diagonal bracings yield their effect only when they operate together with the surrounding frames. Therefore, when the bracings are used, the actual stresses shall be transmitted completely not only for the bracings and the joints of their ends but also for the joints of the columns and the horizontal members related directly to these members and they shall be connected in the equilibrium of proportionate forces.

f) Knee-Braces

Knee-braces tighten rigidly the corner of horizontal timbers and columns, and they have the effect to reduce the deformation of frames against horizontal forces. The bending resistance of column will be depended on so as to produce this effect sufficiently. Therefore, the column shall have sufficient cross-section against bending failure.

According to the examples of earthquake damage, there are many cases that the column was broken at the junction with knee-braces and subsequently the building was destroyed. So very careful consideration should be given to the use of knee-braces.

Large notching for the connection with knee-braces shall not be made in columns, and the columns reinforced by reinforcing posts and struts shall be connected to knee-braces. At or near the place where the column section is weakened (for example, at the joints of lintel), knee-braces shall not be installed.

In fastening knee-braces, bolts generally shall be inserted normally to columns and horizontal members (cf. Fig. 4.5).

When coupled knee-braces are fastened to columns by bolts, splitting sometimes occurs at

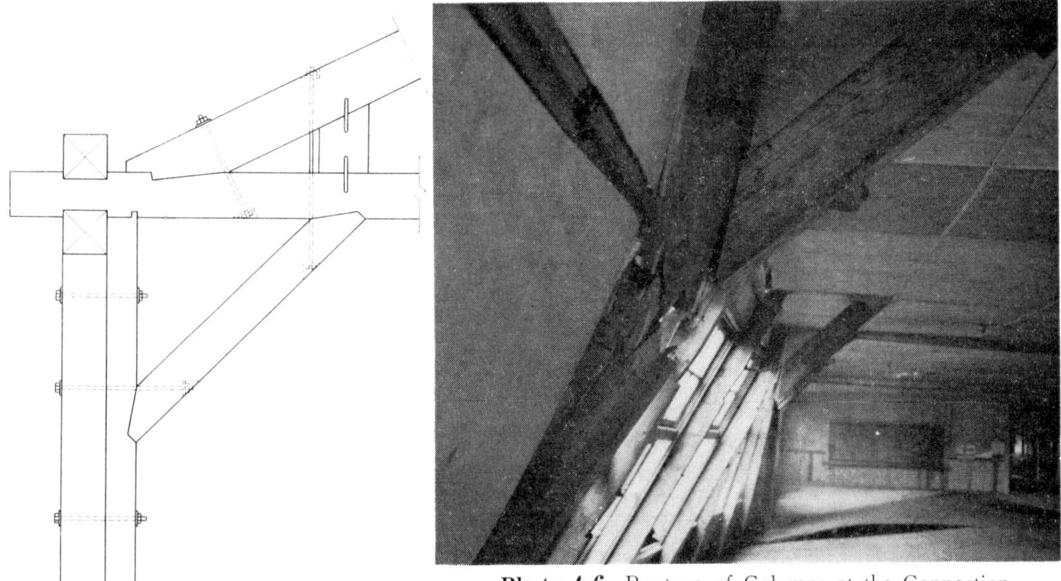

Fig. 4.5

Photo 4.6 Rupture of Columns at the Connection of Knee Brace and Column

the ends of columns and knee-braces so that this shall be prevented by stop-splitting bolts and others. When columns and beams to which the knee-braces are connected do not have sufficient strength to permit the knee-braces to act effectively, the stiffening knee-braces shall be connected with bolts of about 9 mm in diameter and rigid joint method shall be avoided.

g) Column Buttresses

When bearing walls are not constructed within the building, especially in the case of a narrow and long building in a straight line in plan, column butresses are simple and effective way to resist lateral forces.

Column buttresses are set inside the building and outside the building. Generally, however column buttress is placed on the outside of the building. The column buttresses outside the building do not present a pleasant external appearance and if the aseptic care is neglected, the expected bearing strength will not be gained.

Column buttresses should be arranged generally at the symmetrical position on both sides of the building. The slope of diagonal members of column buttresses shall be less than 3 to 1, except in the second floor. It is better that column buttresses be disposed generally to bear compressive force.

Chief members of column buttresses should be timber with no joints, and in case of two-story buildings, it is desirable to be supported by horizontal brace members on the second floor face (Fig. 4.6 (a)). Even if lattice members are closely installed, they are ineffective to lateral buckling (Fig. 4.6 (b)).

Therefore, column buttresses should be simplified and be supported separately around the eaves and the second floor beams from the footing as shown in Fig. 4.6 (a).

The sectional size of column buttress shall be of the same size as that of the column adjacent to it or of bigger size. When column buttress acts as a tension member or when its

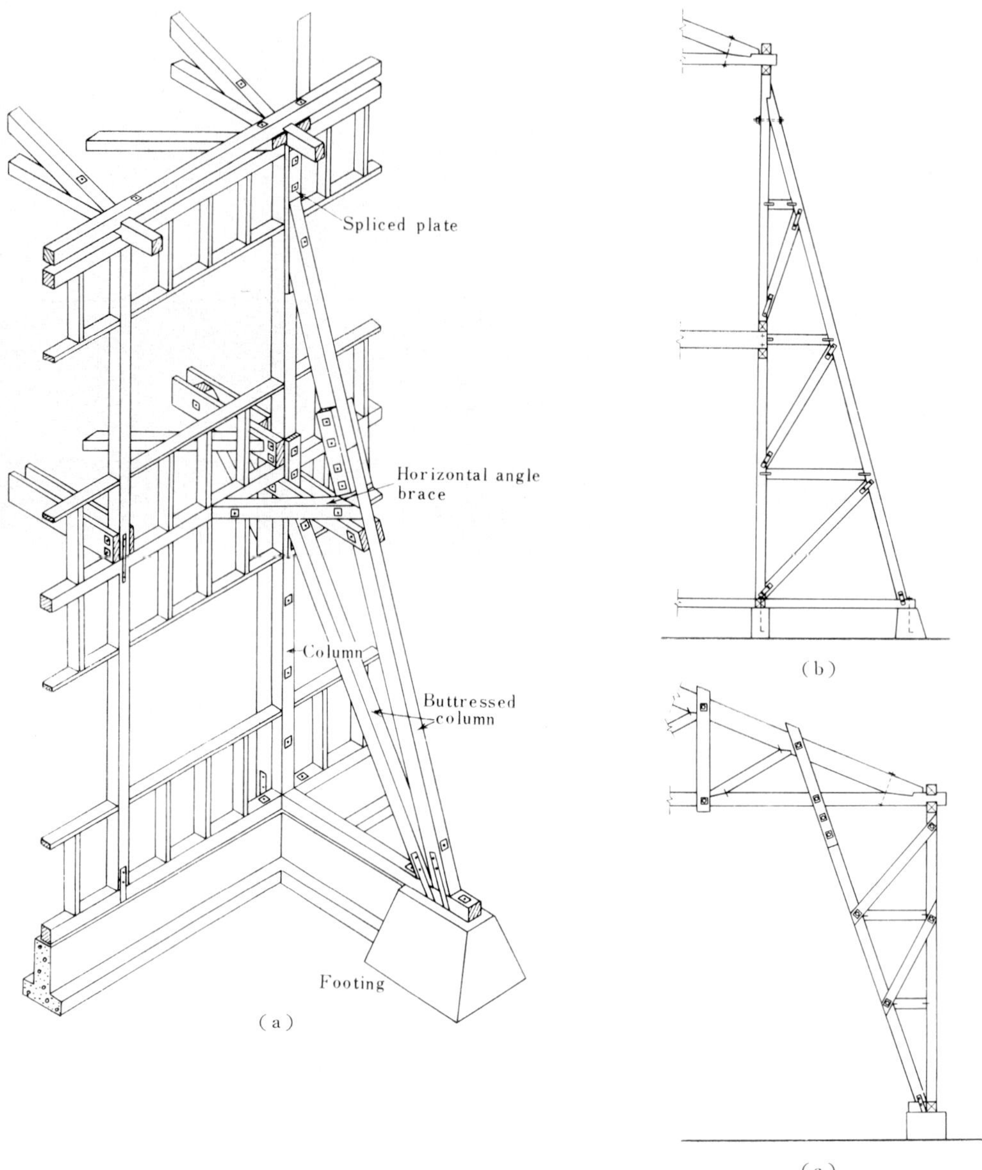

Fig. 4.6 Shore

height exceeds 4 m, the strength of the column buttress shall be verified to be safe by structural calculations.

Column buttresses shall not be placed at the middle of column to yield bending moment to the column, and shall support the positions of wall girder and girth, and further shall fasten those parts with reinforcing members by bolts. When column buttresses are installed within the building, they shall be reinforced not to yield great stresses to roof truss (cf. Fig. 4.6 (c)).

Antiseptic treatment or other wood preservative treatment with the same effect shall be applied to the column buttresses exposed outdoors. In this case, durable lumber shall be used for column buttress, and effective preservative construction should be adopted. The structure shown in Fig. 4.6 (b) is improper because rainwater on the diagonal members flows to the building.

It is desirable that the footings of column buttresses shall be made in one body, connected with the continuous footings. In the case of independent footings, they should be constructed not to settle or be displaced.

4.3.3 Walls

a) General

Walls, when their position and structure are pertinent, restrain the deformation of buildings and act very effectively to increase the resistance to earthquakes.

Walls effective against earthquakes are the walls whose frameworks are constructed in triangular form by bracings. The cautions for installation of bracings are described in the article on bracings.

The arrangement design of walls is stated in Article of 4.2.3, c) on Bearing Wall Structure.

Photo 4.7 Falling of Mortar in Exterior Wall

b) Walls

Generally, bracings act effectively in Ohkabe-style wall (stud wall framing finished on both sides) than in Shinkabe-style wall.

When bracings as thin as narrow-plank are used for Ohkabe-style wall, they shall be arranged in form on both sides of wall and they shall be nailed at the intersecting point. When compression bracings of square timbers and sawn square timbers are installed in the Ohkabe-style wall, finishing planks shall be arranged on both sides of wall diagonally in the reverse direction to bracings or horizontaly if the relation with finishing material makes it impossible, or thick plywood board shall be arranged, serving as finishing and bed materials so that the resistance to earthquakes may be increased. The walls finished with thick wood siding boards, which are nailed by nails of proper length, or which are nailed to compression bracings, are effective from earthquake proof point of view.

In case of Shinkabe-style wall, bracing having a cross-section not less than 1/2 of the column section shall be arranged in places where external appearance is not regarded as important, such as the inside of closets.

4.3.4 Floor Framing

a) General

In wooden buildings, bracings, knee-braces, column buttresses, etc. consolidate the vertical plane, and at the same time horizontal angle braces and horizontal bracings shall be arranged on the planes of floor and roof tie beams to increase rigidity of horizontal plane. It is important to make the whole building in one rigid body in three dimensions from the point of the resistances not only against earthquake but also wind. When bearing walls are arranged eccentrically or at long intervals in the plan, careful consideration shall be given to secure rigidity within the horizontal plane, and to make the bearing walls effective.

Floor frames shall be structured with sufficient strength and rigidity not only to be safe against vertical loads, but also to transmit safely the shearing force due to lateral loads to the bearing walls.

b) Beams, Sleepers and Floor Joists

The joints in floor framing members such as beams, sleepers and floor joists shall be joints that will not be loosened easily under earthquake forces. Without making large notch to members to be connected, sufficient connection, as far as possible, shall be taken. Sleepers shall restrain the movement of floor joint, and the floor joint shall restrain the movement of beams. The principal joints and splices of beams shall be connected firmly by bolts and other metals.

When horizontal angle bracings are arranged, or when serving as one of horizontal truss members, joints in beams, sleepers and floor joists shall have the cross section to resist sufficiently against the stresses due vertical loads and other additional stresses.

Too great a notch shall not be made in beam members in the connection of diagonal members to them.

On the floor of the first floor, battens of floor posts shall be put on the floor posts, and especially in high floor building, bracings shall be installed.

Further, it is economically advantageous that long-span beams subjected to heavy loads shall be beams of coupled timbers, or steel beams.

Photo 4.8 Rupture of Column due to Large Notching at the Connection of Girder and Column

c) Horizontal Angle Braces

In the floor framing, horizontal angle braces shall be installed to strengthen the connection of the principal bearing walls and horizontal members. The horizontal members shall be firmly connected by bolts, nails and other metals (cf. Building Standard Law, Enforcement Ordinance, Article 46, Paragraph 2).

However, horizontal angle braces shall not be installed in the part where horizontal trusses are placed on the floor framing.

d) Horizontal Bracings and Horizontal Trusses

It is very effective for resistance against earthquakes to connect the bearing walls adjacent and parallel to each other by horizontal trusses or by horizontal bracings. Particularly when the space between bearing walls adjacent to each other is large, horizontal trusses are necessary. According to the conditions, column buttresses shall be set between bearing walls so that the span of horizontal trusses may not be too long.

Not only each joint of horizontal trusses and of horizontal braces but also their joints with floor framing and bearing walls of both sides shall be constructed to have sufficient strength and rigidity. In the above joints, large notching and eccentricity shall be avoided.

The structure whose flooring boards are placed diagonally can be substituted for horizontal trusses. But in this case, it shall be constructed so that the joints of floor joists and beams have little slack and that it transmit the horizontal forces adequately to the bearing walls.

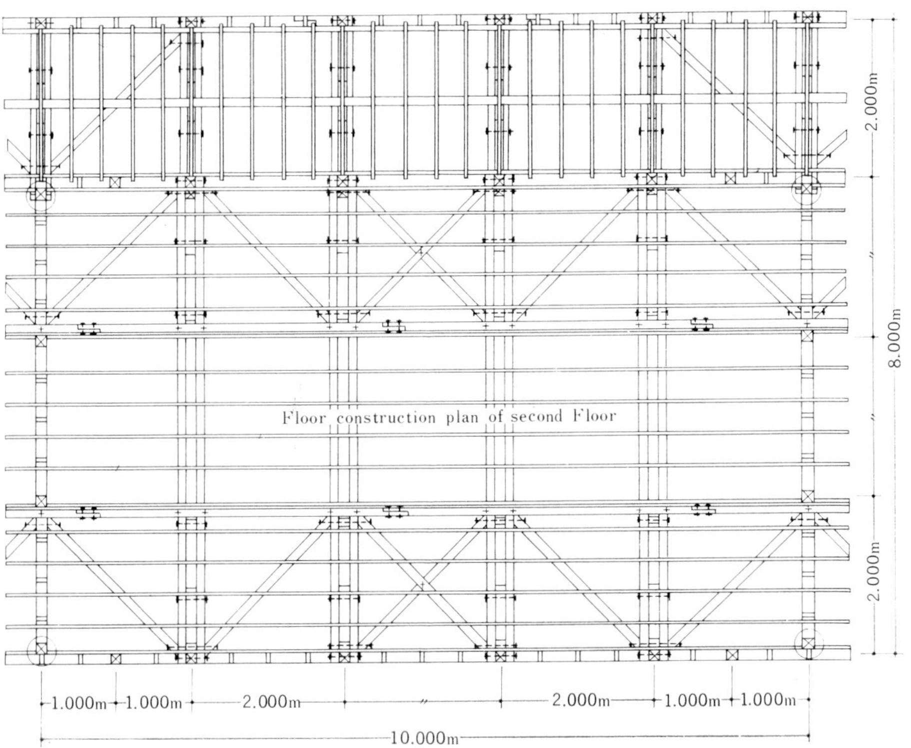

Fig. 4.7 Horizontal Truss (School Building)

4.3.5 Roof Trusses

a) General

A roof truss itself is the part that suffers relatively small earthquake damage. However, the joints with frameworks are the weak points. As stated in Article on Floor Framing, it is important to strengthen the whole building and to increase the resistance against earthquakes by the floor framing and the rigidity in the horizontal (tie beam) plane.

b) Roof Trusses

Roof trusses shall have sufficient strength and rigidity not only to be safe against vertical loads acting on the roof, wind pressure, ceiling weight and others, but also to transmit safely the shearing force due to lateral loads to the bearing walls.

Roof trusses shall be Western roof trusses except for short spans. In the case of a Japanese roof truss, the deformation due to earthquake shall be restrained properly by bracings. Especially when the inside of Japanese roof truss is used as a storage, this precausion is necessary.

Unnecessarily big timber shall not be used for tie beams.

Roof trusses in both cases, Japanese and Western, shall be firmly connected to the framework members by metal fasteners and shall be constructed to be safe against movement and collapse resulting from earthquakes and wind pressure.

c) Bracings

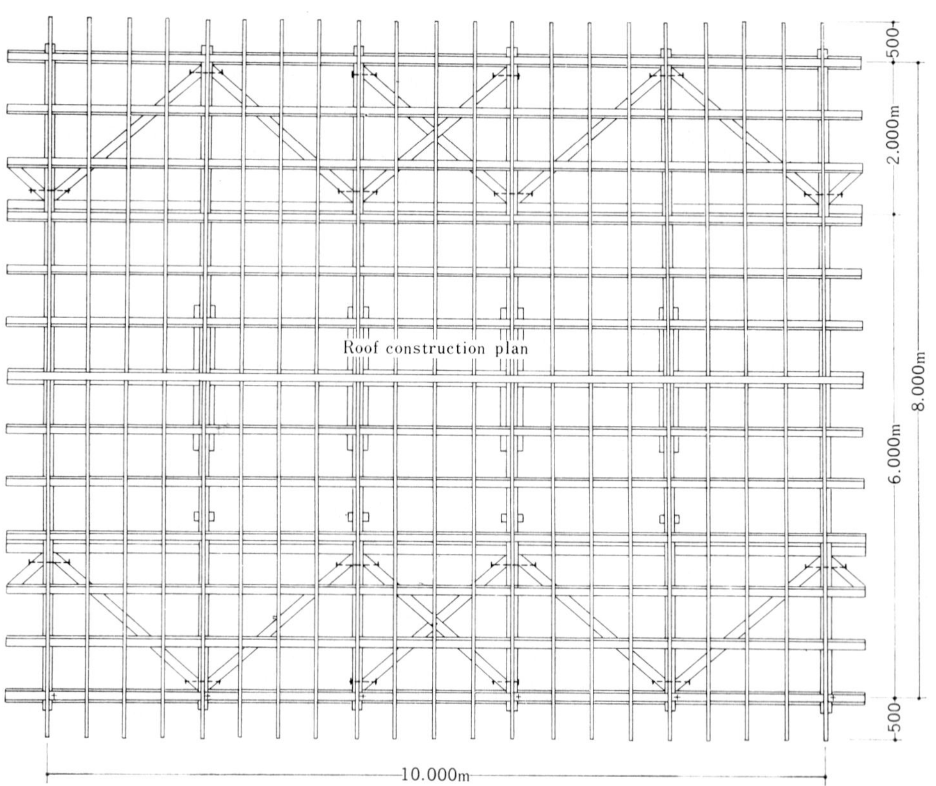

Fig. 4.8 Horizontal Truss

For roof truss whose span is not less than 6 m, bracings and roof truss braces shall be installed between the roof trusses (cf. Building Standard Law, Enforcement Ordinance Article 46, Paragraph 2).

d) Horizontal Angle Braces

In the tie beam plane, horizontal angle braces shall be installed to strengthen the intersecting point of the principal bearing walls and horizontal members crossing at right angles to each other. Besides the crossing mentioned above, horizontal angle braces shall be installed at 4 to 6 m intervals in the ridge direction. Generally they shall be placed at the external circumference of the building (cf. Building Standard Law, Enforcement Ordinance Article 46, Paragraph 2).

e) Horizontal Bracings and Horizontal Trusses

With the same principle as stated in Article on Floor Framing, horizontal braces and horizontal trusses shall be installed.

When the slope of the plane of principal rafter is gentle, according to the conditions, the plane of principal rafter may be substituted for the plane of tie beam and the rigidity in the plane may be increased by horizontal braces, bracings and trusses. Sheathing boards may be placed diagonally. However, the boards and roof truss members shall be constructed to transmit the forces sufficiently.

4.3.6 Roofs

The most effective method to lessen the earthquake force is to reduce the weight of upper parts of buildings, the weight of roof in particular. Therefore, roofing members as light as possible shall be selected for resistance against earthquakes.

Next, roofing materials shall be firmly tied so as not to be separated and fall in earthquakes. Very careful consideration shall be given to use of clay tile roofing since it is heavy and its slope is steep.

4.3.7 Ceilings

From the view point of reduction of weight in the same way as roofs, light ceilings are desirable because they diminish the earthquake force directly. Careful consideration shall be given to bed structure, fixing, and finish, not to be peeled off or separated and fall in earthquakes. It is to be noted that bathrooms and kitchens easily decay. Therefore, there are many cases that they collapse suddenly.

4.3.8 Stairs

When the position and the structure of stairs are proper the stairs have the effect of bracing members, and they are effective for resistance against earthquakes. In order that stairs work as bracing members, the connection of upper and lower side-girders shall have sufficient joint bearing strength and joints that can transmit the stresses of vertical members and floor framing members that are in equilibrium with diagonal members are necessary.

4.3.9 Fire Walls

Fire walls of reinforced concrete are effective as quake-resisting walls, so that they should be considered in structural planning. Care shall be taken so that bolts do not penetrate the fire walls. Fire walls shall be fireproof and self-supporting construction, and shall not be of plain concrete or masonry construction (cf. Building Standard Law, Enforcement Ordinance

Article 113, Paragraph 1,2).

4.4 Preservation

4.4.1 General

To make buildings strong against earthquakes, careful consideration shall be given not only to design but also to maintenance after completion, and buildings shall be examined periodically. When decay and failure of timber, looseness of bolts and other defects are found, they shall be repaired and reinforced immediately, and the condition of buildings shall always be safe against earthquakes.

4.4.2 Preservation from Decay

For principal structural members such as sills, which touch hydrated materials such as concrete, brick, stone, earth and others, and for column buttresses exposed outdoors and for other principal structural members in places apt to be easily decayed, the timbers with much heart wood selected from durable species such as Japanese cypress, and Japanese cedar shall be used, and wood preservatives shall also be applied (cf. Building Standard Law, Enforcement Ordinance Article 37, Paragraph 1).

For parts under the floor, the base of framework inside the external walls with mortar finish on metal lathing, internal walls with mortal or tile finish, places such as bathrooms and kitchens where water is always used, etc., the air shall be well ventilated and dry to prevent decay (cf. Building Standard Law, Enforcement Ordinance Article 22, Paragraph 2 and Article 49).

When decay of principal structural members is discovered, the rotted parts shall be removed, or shall be exchanged with new members and they shall be reinforced as required so that the original strength and rigidity may be maintained.

4.4.3 Protection against Ants

The damages caused by Leucotermes Speratus are frequent in damp places. Generally the wood preservation method can be applied to the prevention method for them. The contrivances from the structural point of view combined with of wood preservation treatment shall be applied to Coptotermes Formosanus.

When ant damage is discovered, the ants shall be immediately exterminated. The parts where the damage is severe shall be exchanged with new timbers and preventives shall be applied to both new and old timbers and to nearby parts which are likely to be damaged.

Ant damage is often discovered by winged ants flying together and from detection and tracing of ants' paths. The mixed chemicals such as P.C.P. and γ B.H.C. are recommended for preventives.

4.5 Fire Prevention

4.5.1 General

Wooden buildings are susceptible to fire damage that occurs accompanying earthquakes. To prevent this, the buildings first of all, shall be so strong that they will not collapse. Secondly, the outside of buildings and places where fire is used and its vicinity, shall be made fire proof so that the spread of fire from adjacent buildings may easily be prevented. Further, if possible, internal partition walls and floors shall be made fire proof.

If wooden buildings are made fireproof, fire damage accompanying earthquakes may be lessened and earthquake damage diminished. So it is desirable that the precautions already described be followed not only in the case of new buildings but also in the repair of buildings.

4.5.2 Fire Prevention

The methods to make the outside of wooden buildings and floors fireproof are the following :

i) External Walls, Floors and Underside of Eaves

External walls, floors and lower side of eaves shall be mortar finish on wire mesh lathing or plaster on wooden lath finish whose thickness shall be not less than 2 cm.

Mortar or plaster not less than 1.5 cm in thickness shall be applied on cemented excelsior board finishing or gypsum board finishing. Tiles shall be placed on mortar finish and the sum of their thickness shall be not less than 2.5 cm.

Mortar shall be plastered on cement board finishing, magnesia cement board finishing or clay tile finishing and the sum of their thickness shall be not less than 2.5 cm.

Earth-plastered wall with back coating shall be adopted. Zinc iron plates or asbestos cement slates shall be lined on gypsum board finishing (not less than 1.2 cm in thickness) or rock wool heat retaining board finishing (not less than 2.5 cm in thickness). Asbestos cement slates not less than 0.6 cm in thickness shall be lined on cemented excelsior board finishing (not less than 2.5 cm in thickness). More than two asbestos cement slates or asbestos pearlite board shall be lined, and the sum of their thickness shall not be less than 1.5 cm (cf. Building Standard Law, Enforcement Ordinance Article 108, Paragraph 2).

ii) Roofs

Fire-proof materials for roofs shall be selected. Such as clay tile roofing, asbestos cement slate roofing, or metal roofing on cemented excelsior board. However, it is desirable that the lower side of eave satisfies the conditions of i) (cf. Building Standard Law, Enforcement Ordinance Article 108, Paragraph 3).

iii) Windows and Entrances

At the openings of windows and entrances, fire doors, even simple ones, shall be installed.

The part where a fire door touches a frame or a fire door, shall be shiplap or shall have an astragal or a door stop, so that there shall be no open space when the fire door is shut (cf. Building Standard Law, Enforcement Ordinance Article 110).

iv) Fire Walls

When there is but little space between the building and adjacent buildings, fire walls, fire proof wing walls, fences and others, even simple ones, shall be installed between the building and adjacent buildings.

The fire walls installed inside the same building shall be in accordance with the requirements of Section 4.3.9, and they shall project over 50 cm from the external wall surface of the building and the roof surface. When an opening is made, its width and height shall be less than 2.5 m, and A class fire door shall be installed (cf. Building Standard Law, Enforcement Ordinance Article 113, Paragraph 3,4, and Article 110).

v) Partition Walls in Roof Trusses

When the roof trusses of a building, whose floor area is not less than 300 m^2 are of wood construction, partition walls of fireproof construction or fire protection construction on both sides must be installed in roof trusses spaced at less than 12 m intervals in the ridge direction.

When roof trusses of a connecting corridor which connects buildings whose total area is over 200 m^2, are of wood construction and the length in the ridge direction is more than 4 m, partition walls of fireproof construction or of fire protection construction on both sides shall be installed (The above is according to Building Standard Law, Enforcement Ordinance Article 114, Paragraph 3,4).

4.6 Emergency Repair

4.6.1 General

The emergency repair mentioned here is the temporary repair immediately after a wooden building has inclined and joints have been damaged. In case of such great damage that the building has inclined, it shall be rebuilt after the emergency repair, and the raising up of the inclined building is not desirable.

When joints of the principal members such as columns, beams and others, are damaged, it is desirable to exchange these with new members after the emergency repair.

4.6.2 Repair of Each Part

In order to prevent progressive horizontal displacement of the whole inclined building, it shall be propped up as a temporary measure so that it may not fall down by aftershock. Next, the building shall be straightened. The building which is supported in the inclined position is not desirable in many respects. The moved sills should be pushed back by a jack. Further, it is simple and effective to arrange properly bracings, column buttresses, kneebraces and horizontal braces at places that will not get in the way.

When this cannot be done, or when the number is not sufficient, walls should be destroyed and bracings be put in.

The damaged parts of joints of the pushed up building shall be reinforced with reinforcing spliced plates, spacers and metals.

5. STEEL STRUCTURES

5.1 General

5.1.1 Characteristics of Steel Structures

Compared with other structural members such as wooden members, reinforced concrete members and masonry members, steel members are prominent in their strength and in their ductility, and they have the advantage that huge members can easily be constructed, so they are the most distinguished structural members that make large-scale buildings resistant to earthquakes.

As for the past earthquakes, especially Kanto Earthquake of 1923, steel construction buildings were damaged relatively little. That is, a factory or similar one-story building was hardly damaged, partly because the structure is light. But several multi-storied buildings in Marunouchi suffered great damage, for the foreign design method was applied to them. As it was, the consideration for lateral forces of frames was lacking and brick masonry was used for walls. On the other hand, the buildings which were very carefully designed regarding lateral forces according to original Japanese design method, were damaged as slightly as surface tiles were separated, and displayed fully the earthquake resistance of steel structures.

Further, as the earthquake resisting design method based on dynamic vibration phenomenon of a building caused by seismic movement has progressed recently, it has become possible to estimate, in direct quantitative way, earthquake resisting effects resulting from the great unit strength and high ductility of steel members. As the result, their earthquake resisting superiority has been made clearer.

It is safe to say that the most pertinent structural material for multi-storied buildings is steel.

The important problems in the design of steel structures are connections and buckling.

Structural shapes such as H-shape steels, steel pipes, light gauge steels have recently been increased in kind and their sizes have become large. Many of them can be used for structural members such as a column, a beam, etc. Thus, steel structure has the character to be considered that each member is half industrialized. Therefore in the design of steel structures, the joints have extremely large role in the structure, and their strength and deformation should be examined carefully.

At important joints, no matter what the actual stresses may be, they should be designed so that the full strength of the members may be developed.

Especially, in dynamic design, when the design allows yield and destruction of joints first, then it becomes very difficult to estimate the load-deformation properties of structures and generally ductility is lowered also.

Another problem is buckling. Compared with other structural members, steel has slender cross section due to high unit strength, and buckling is necessarily possible. In the buckling phenomena that appears in steel structures, there are various types of bucklings such as flexural buckling of the whole member, torsional-flexural buckling (lateral buckling), local buckling of plate elements constituting a member, etc. In multi-storied buildings, they are often possible in the plastic range.

Generally the occurrence of buckling lowers the bearing strength and brings sudden change of the load-deformation curve of a frame in many cases. Accordingly, even when the occurrence of buckling does not cause destruction of a structure immediately, careful consideration should be given to the prevention of buckling because the response to earthquake often becomes unstable due to the buckling of some members.

Steel construction is weak in resistance to fire and corrosion.

As the construction methods are being modernized, the increasing use of pure steel and light fire-proof coated structures is expected. In this case, careful attention should be paid to fire-proof efficiency.

Photo. 5.1 shows the state of damage of steel structure caused by fire accompanying the Kanto Earthquake.

Further, the consideration from the point of construction methods not to rust easily and the prevention of the decline of bearing strength due to rust by means of applying rust-proof paints

(a)

(b)

Photo 5.1

are important for earthquake resistance.

5.1.2 Steel Materials

For structural materials of steel structures, mainly SS 41, SM 41 and SM 50 have been used.

As the buildings become colossal such as high-rise buildings and the buildings of long-span, the steel members of higher unit strength have been required, and the steel members of unit strength of 55 kg/mm^2 and 60 kg/mm^2 class have already been in the stage of practical use.

Of course it is necessary to examine welding and ductility sufficiently when high tensile strength steels are used. The full understanding about its mechanical character is important when considering bearing strength of frame and the peculiarity of dynamical stability.

Fig. 5.1 shows the stress-strain relation of each steel member in the standard way (the uniform elongation by JIS Z 2201, No. 4 test piece). From this figure, the followings are pointed.

i) As yield point of steel is increased, the yield ratio (the ratio of yield point to tensile strength) is increased.

ii) As unit strength of steel is higher, elongation is less.

Table 5.1 shows the relation of Fig. 5.1 regarding these two points in figures.

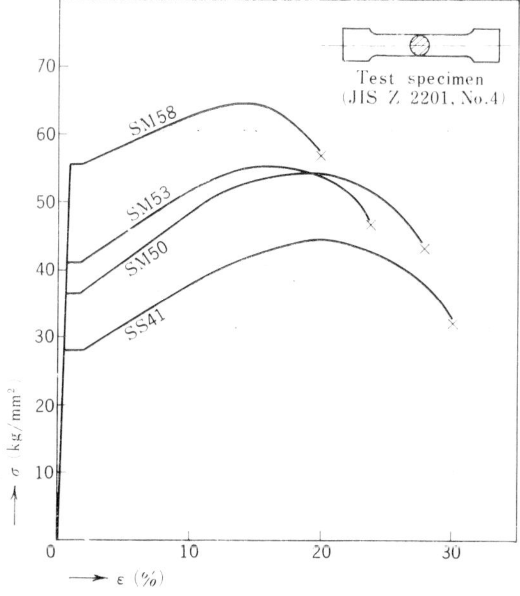

Fig. 5.1

Table 5.1

Classification of steel materials	SS 41	SM 50	SM 53	SM 58
Yield ratio σ_y/σ_{max}	0.6	0.7	0.72	0.86
Elongation (%)	30	28	24	20

Allowable unit stress of steel has been considered on the basis of yield point. But as the steel with higher yield point has higher yield ratio, safety percentage to maximum strength is lower, as yield point of steel is higher when allowable unit stress on the basis of yield point is equally given to these steels. In this sense, as for high tensile steel, it is necessary to decide allowable unit stress considering both tensile strength and yield point of materials. In the estimation of the pecuriality of dynamical stability, it shall be carefully considered that frame performance, and load-deformation relation of members in plastic range depends largely on yield ratio of materials.

Further, as seen in Fig. 5.1, the values of elongation of materials shows uniform decrease.

In actual structures, the places where the conditions of uniform elongation are formed are quite rare. Generally the stress in a member changes along the length of the member due to change of sectional area of a member and change of moment distribution. Under these conditions elongation concentrates partially and such big elongation values as shown in Fig. 5.1 cannot be expected in the whole member. This partial elongation ability depends on yield ratios of materials. For example, when the member whose sectional area changes continuously as shown in Fig. 5.2 (a), is tensioned, it yields at the point A–A of minimum section. However, in case of the materials with 1.0 yield ratio without strain hardening, the member cannot yield except at A–A line. As the A–A is the line without width in geometry, plastic elongation at this part is 0. That is, when the A–A reaches yield point the member breaks without plastic elongation.

In the strain hardening materials (less than 1 yield ratio, in Fig. 5.2 (b) below), as the A–A line comes into strain hardening area, A's neighborhood yields. Its yielding area relates to yield ratio, and the relation between elongation ability of the whole member and yield ratio is shown in Fig. 5.2 (c). That is, the materials with higher yield ratio have less elongation ability.

This influence is much greater than the difference of elongation ability by the classification of steel materials in the case of uniform elongation seen in Fig. 5.1.

In actual structures, the decrease of elongation ability by the stress inclination will probably be the greatest problem of at the part of riveted and bolted connections.

Phot 5.2 is an example of the break of truss lower chord member that occurred at the time

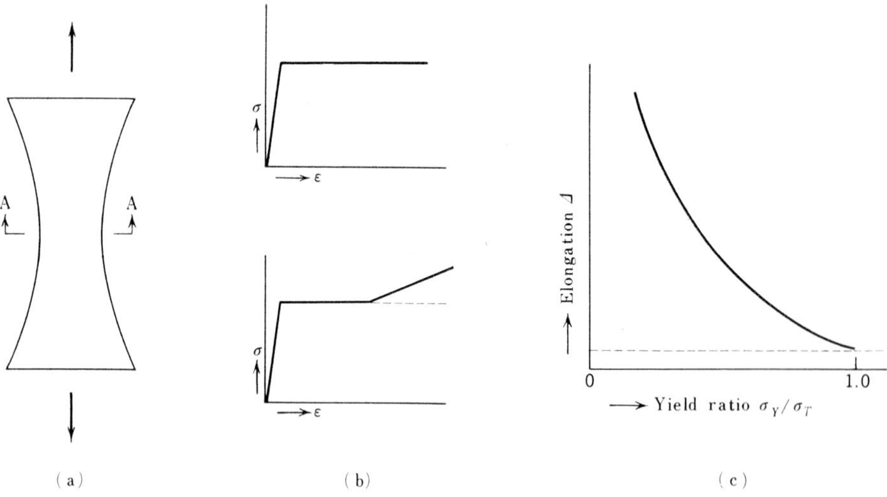

Fig. 5.2

of the Niigata Earthquake, and shows the state of extremely easy destruction which is considered to be due to the reasons stated above.

Photo 5.2

5.2 Structural Design

5.2.1 Use of Buildings and Classification by Scales

a) Multi-storied Buildings

Many multi-storied buildings will use pure steel structure. Regarding dynamic analysis of multi-storied buildings of this sort, its guide is given in Chapters 1 and 2. When the restoring force characteristic of the frame for response analysis is to be decided, the following must be carefully considered.

i) Additional bending moment caused by horizontal deflection of frame and axial force.

As the axial force at a lower part is great, this value shall not be ignored (Fig. 5.3 (a)).

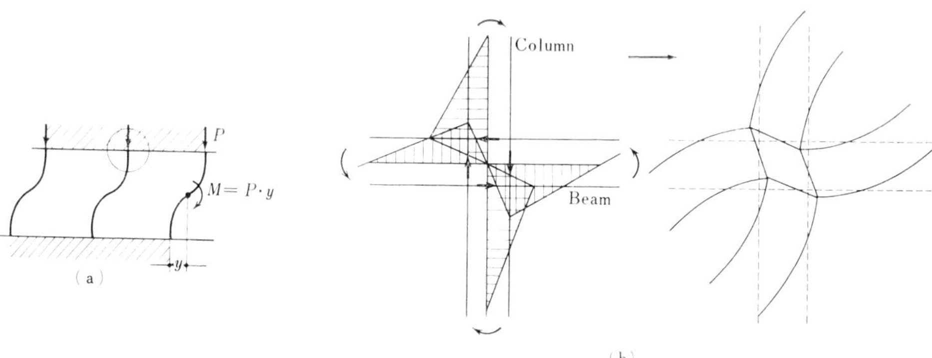

Fig. 5.3

ii) Rotation of panel point caused by shearing deformation of rahmen corner surrounded by columns and beams, and the bearing strength of corner panels (Fig. 5.3 (b)).

As this part receives large shearing force, its bearing strength and deformation must be examined.

The elements stated above lower the rigidity of frame and the maximum bearing strength.

Next, local buckling of plate elements (flange, web) that constitute columns and beams, should be carefully examined. Especially when plastic deformation of frame is expected, a flange and a web shall have the width/thickness ratio so that the expected plastic deformation may occur without causing local buckling.

When the resistance against quake and the resistance against wind by frame bracings are planned, the vibration analysis shall be made considering bending deformation.

It is desirable to decide bearing strength of members on the basis of the actual condition of use of steel materials.

The structural method that can secure plane rigidity shall be adopted when floors made of steel deck and precast concrete slabs are applied, in order to make the building light.

To be safe against deformation of the frame by earthquake, internal walls and external walls may be constructed in the following ways:

i) the structural method that wall materials will not receive deformation, so that they will not give resistance against earthquake force, and ii) the structural method that wall materi-als deform with frame.

For case i), wall panels shall be loosely connected requiring only mutual rotation and slip-ping of wall panels. However, it is necessary that wall panels will not fall by connectors being unfastened.

For case ii), wall materials shall be ductile materials that can deform with frame. Fire-proofing of floors, walls and frame cover shall be considered fully.

b) Low and Small Scale Buildings (such as houses, shops and small offices)

Buildings such as houses, shops and small offices, many of which were formerly built with wood, are constructed by steel structure, possess generally higher reliability for resistance against earthquake and it is recommended.

In this type of small scale steel structures, comparing building area, the amount of internal walls and external walls is relatively large, so that there will be greater percentage of resistance of these wall materials in earthquakes. That is, there is greater resistance than the calculated.

But it is not expected in this sort of ordinary buildings that the consideration for the resistance against earthquakes is especially paid in the placement of internal walls and external walls. Therefore, the resistance of external walls and internal walls against earthquake means the dangers of the destruction of wall materials of this sort in earthquakes and of separation of joints at the same time.

From the statements mentioned above, in small scale steel buildings, it is desirable that frame bracings shall bear earthquake force, and that steel frame itself shall have enough rigidity to prevent excessive deformation without adding force to wall materials.

In case of small scale buildings of this type, assembling and production of steels are liable to be done in an easy manner without enough supervision. The damages from earthquake, wind and snow that small scale steel buildings have recieved are mostly due to defective execution.

c) Factory Buildings

Generally in factory buildings, there are many conditions that will not satisfy the assumptions made in structural calculations. For example, when the frame with travelling crane as shown in Fig. 5.4 (a) receives horizontal force like an earthquake, supposing the column base fixed, the inflection point appears near the position of the crane girder where the section changes suddenly and column has the properties similar to a long cantilever.

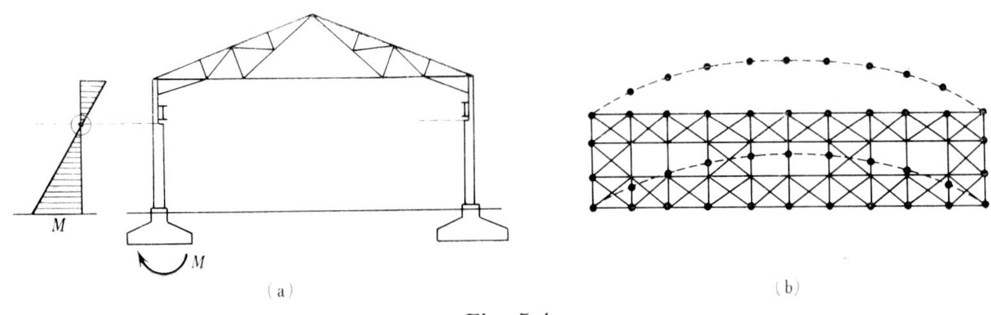

Fig. 5.4

In this case, the column base can be designed to have required moment. However, it is difficult generally to install tie beams in the direction of span in a long span factory, and as result rigidity is apt to be lacking. That is, stresses more than supposed occur at the column head by rotation of footings. Further, there are many cases that the behaviour of roof truss as a rahmen member is not always treated clearly, so the parts near the joints of columns and roof trusses must be designed very deliberately considering these points.

Factory buildings are constructed mostly on reclaimed land and places near seashore, and generally poor soil is one of the bad conditions.

Many factories are the buildings that are long in the direction of the girder, and there are many cases in which partition walls cannot be installed from considerations of function. In this case, planning to raise the rigidity of roof by horizontal braces, careful considerations shall be given to heightening of the rigidity of each frame paying attention to the points stated above.

In case of a long building, even when the roof is made solid, it is dangerous to make the frame bracings of double gable bear earthquake force unconditionally. As shown in Fig. 5.4 (b) the differences in deformation of the middle frame and gable frame should be considered.

In factory buildings, they sometimes take very complicated form according to function and it is difficult to do clear structural planning. In this case, the plan shall proceed considering that the members shall be arranged so that the stresses may be transmitted smoothly to the ground and places where excessive stress concentration may occur shall be avoided.

When there is a factory which deals with chemicals that make iron corrode, or when there is such a factory in the neighborhood, it is desirable to examine periodically the decrease of earthquake resistance ability of frames due to corrosion.

d) Long-Span Structures

The suggestions for earthquake resistance regarding long-span structures have many points in common with the case of factory buildings of (c) stated before. The general hints on long-span structures are the following:

i) In case of small scale buildings, there is a limit to the smallest size of rolled steel shapes that can be used. The absolute amount of required steel materials is small, and some increase of size is required in the design of joints. For these reasons, member section is ample for the design stress. On the other hand, in long-span structures as the economy in amount of steel materials becomes an important condition, the section is often selected to utilize full allowable unit stress. In long-span structure, this is advantageous not only for the purpose of the economy of amount of steel materials but also for transportation and building method, and this is often a factor in the settlement of design details. This design with full allowable unit stress influences the safety of buildings in connection with the following:

ii) For the long span frames, the percentage that walls occupy is small. Accordingly so-called extra-resistance cannot be expected as much as in the case of small scale buildings.

iii) Even a slight mistake in the calculations of dead load and live load gives a great deal of influence to frames as roof (floor) area that one frame bears is large. This is the cause of giving excessive stress to the member in relation to i).

iv) Enough attention shall be paid to buckling as columns and beams will have large distance between supporting points.

v) Rigidity of roof (floor) shall be considered when frame bracings resist earthquake force (cf. 5.2.1 Article c)).

vi) As footing beams in the direction of span are often not installed, attention shall be paid to the stiffness of the column base. Even when a footing beam is installed the stiffness decreases due to the long span.

vii) In case of space frame construction, the transmission of stresses between it and supporting parts of a column, a beam and a footing shall be examined carefully. Further, as there are many panel points in case of space frame construction, bearing strength and slack of joints shall be very carefully examined.

viii) In case of small rise arches and curved structures, the perimeter of roof shall be reinforced sufficiently because the roof has the danger of being inverted by lateral movement of supporting point and up-and-down movement of roof an earthquake.

5.2.2 Use of Steel Members Sorted by Types

a) Hot Rolled Steel Shapes and Members that Use Steel Plates

Angle shapes, channel shapes, I-shape steels and steel structures made of many combinations of these steels and steel plates have been used for a long time and the formulation of highly reliable design methods has been established to a high degree. This construction may be applied to every field from low and small buildings to multi-storied buildings from the point of use and scale. However, it is not advantageous for too small scale buildings from the view point of economy and design.

H-shape steel can be used as column and beam members as it is, for the efficiency of section is high and local stress and local buckling have been examined sufficiently from the point of elasticity and plasticity studies.

Multi-storied buildings are often designed as a rahmen structure using mainly the rolled steels and steel plates. In this case, especially important factor from the point of design are the column and beam connections.

As stated in 5.2.1, Article a), ii), the parts of panel plates surrounded by columns and beams receive extremely great shearing force in an earthquake and they shall be designed to resist it sufficiently.

Further, there is partial deformation of flanges at the connection and unequal distribution of stresses accompanying it.

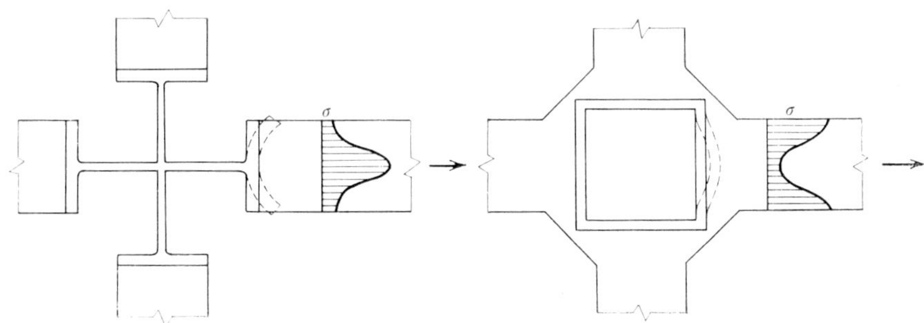

Fig. 5.5

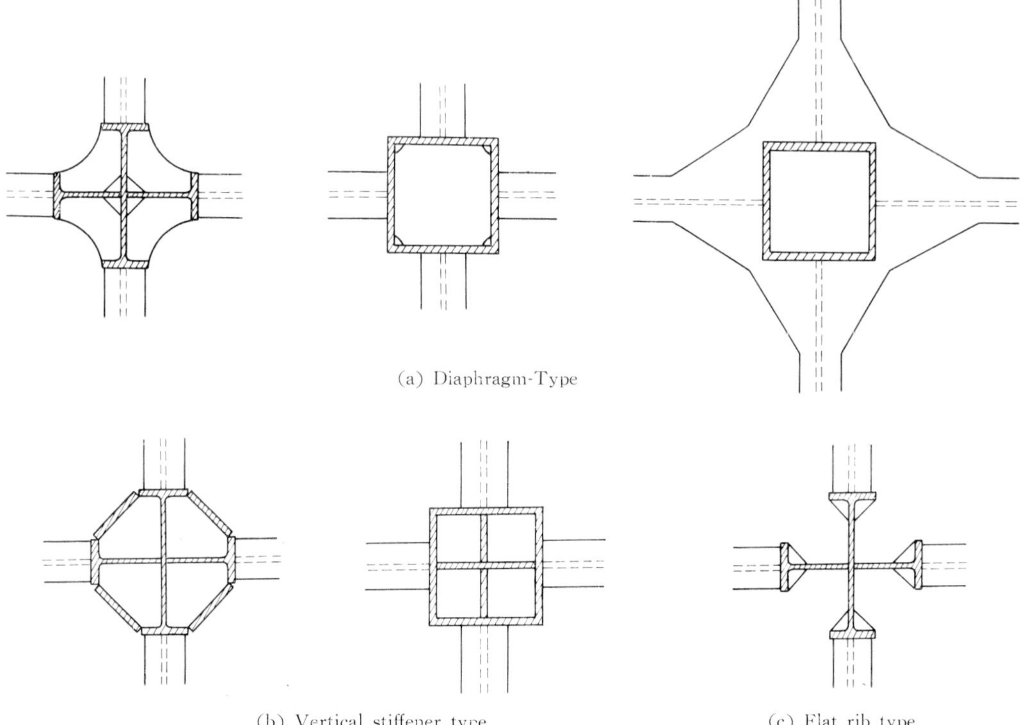

Fig. 5.6

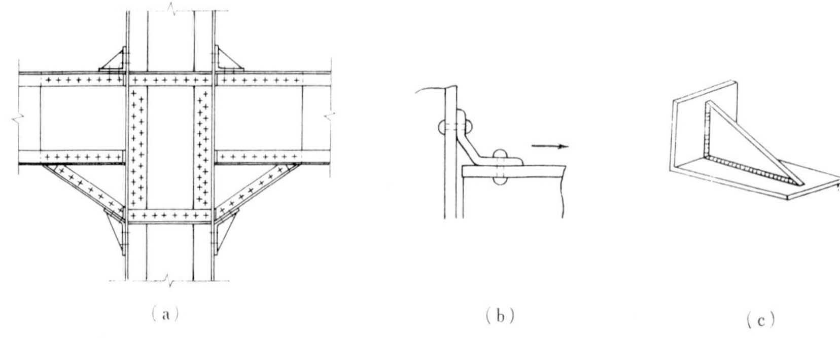

Fig. 5.7

As shown in Fig. 5.5, column flange partially deforms due to the forces from the beam flange. This causes the rigidity of joints to decrease and at the same time makes stress distribution within the flange unequal. When it passes the potential limit of plastic stress redistribution, it sometimes makes maximum bearing strength at joints to decrease.

Therefore, it is necessary to arrange reinforcing members that prevent the partial deformation as shown in Fig. 5.6.

In case of assembling joints by rivets, joint angle steels are often used as connectors (Fig. 5.7). In this case attention shall be paid to the tensile rivets. As force operates on rivets through angle steels, eccentric tensile force necessarily occur and bearing strength of rivets decreases. Further, joint angle steels receive bending deformation due to the eccentric tension (Fig. 5.7 (b)).

To prevent this, it is desirable to fix reinforcing rib on angle steels (Fig. 5.7 (c)).

In welded structures, it is necessary to pay attention to the steel plate which at joints receives great force in the direction of the thickness of plate. Mechanical properties of steel plate are not completely isotropic and comparing the strength in the direction of rolling, the strength in the direction of thickness is somewhat low. Further, where there is a defect such as lamination in the plate, the strength in the direction of thickness decreases still more. Elongation in the direction of thickness is much lower than in the direction of rolling.

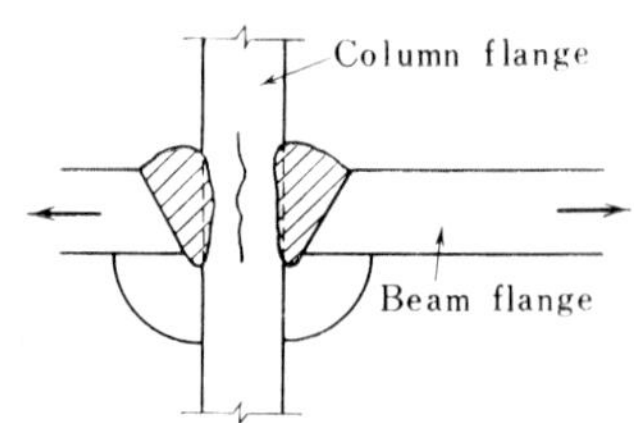

Fig. 5.8

When a column flange plate receives the influence of heat by welding, the strength is further reduced and this is true regardless of thickness. Larger the plate thickness, the greater the amount of weld heat and steeper the temperature slope, so that the quality of the metal is easily changed. Considering the possibility of the defect to influence the quality of the metal such as lamination is large, a thick plate not less than 40 mm shall be checked when it is used.

b) Members that Use Structural Steel Pipes (Steel Pipe Structure)

A steel pipe as a structural member is advantageous against buckling because it has a closed section and complete isotropy.

When it is used as a truss member, the strength against lateral buckling is much larger than in ordinary steel structures. It is because, as shown in Fig. 5.9, in the rolled shape steel structure, panel points are usually composed of gusset plates so that the in out-of-plane rigidity is weak and they are deformed as shown in the figure. On the other hand, in steel pipe structure, chord members and diagonal members are usually connected by welding directly, so that the twisting rigidity of a steel pipe member itself is high and the rigidity in the lateral direction also becomes very high.

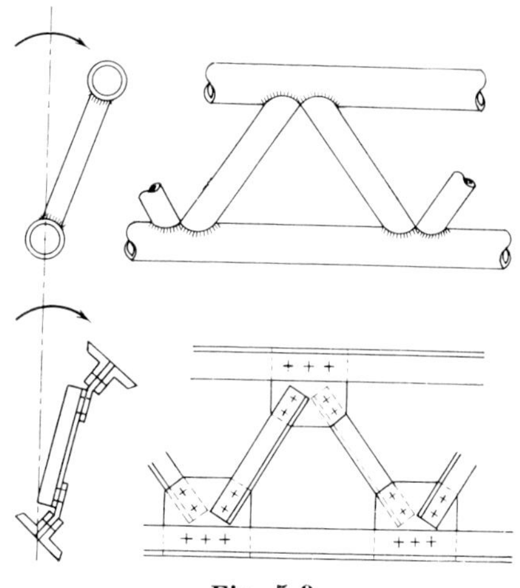

Fig. 5.9

Steel pipe structure is suitable for a long-span structure because of very high potential bearing strength in reserve against buckling.

It can be used as columns of a multistoried building and in structures such as a tower. The precaution in steel pipe structure is the local deformation at joints. As the inside is hollow, it is weak against local deformation in spite of its appearance (Fig. 5.10).

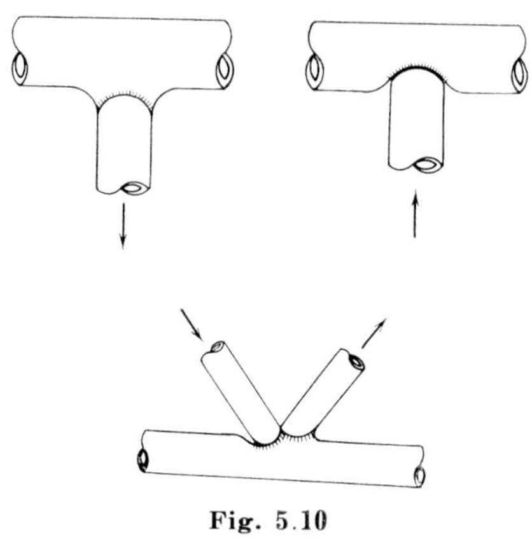

Fig. 5.10

It is better to use a reinforcing rib unless sufficient safety against local deformation is confirmed by theoretical analysis or by experiments at the important parts in the structural (Fig. 5.11).

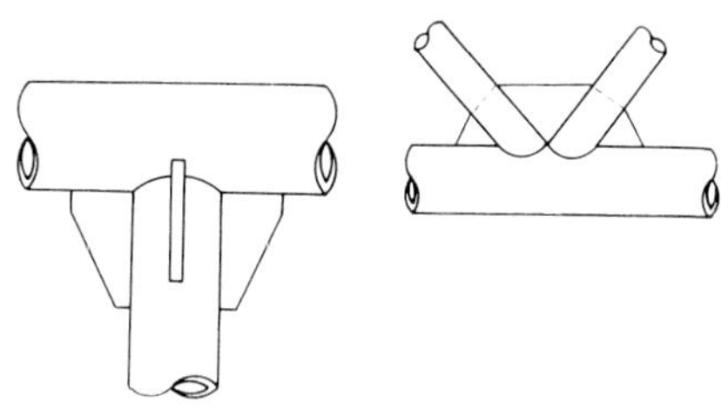

Fig. 5.11

c) Members that Use Light Gauge Steels (Thin Walled Steel Structure)

Light gauge steels made by cold rolling of thin steel plates have the open section of one axis symmetry.

When they are used as beam members and column members, attention shall always be paid to twist since in such a section the center of shear S and the center of gravity G do not coincide (Fig. 5.12).

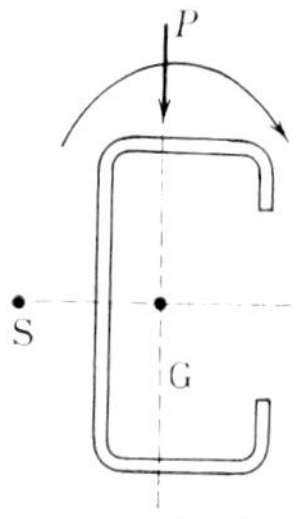

Fig. 5.12

Further, the thinness of the plate makes the resistance against twist of the section weaker.

Light gauge steels have small local buckling bearing strength because plate thickness is small. To this, the design method taking advantage of bearing strength after local buckling can be applied. However, as the occurrence of local buckling sometimes causes twisting deformation of a member, the whole structure is destroyed at last due to the occurrence of local buckling and the resulting twist giving great stresses work at the places that are unexpected in the design.

Photo. 5.3 shows the damage of a column of light steel structure at Niigata Earthquake. The state of coincidence of local buckling and twist can be seen. Provided that enough attention shall be paid to these points, member can be used for a purlin and furring strips advantageously because the section has high efficiency.

Generally, light gauge steels are suitable for relatively small scale buildings, and when the light gauge steels are used, it is impossible to expect plastic design or great plastic deformation.

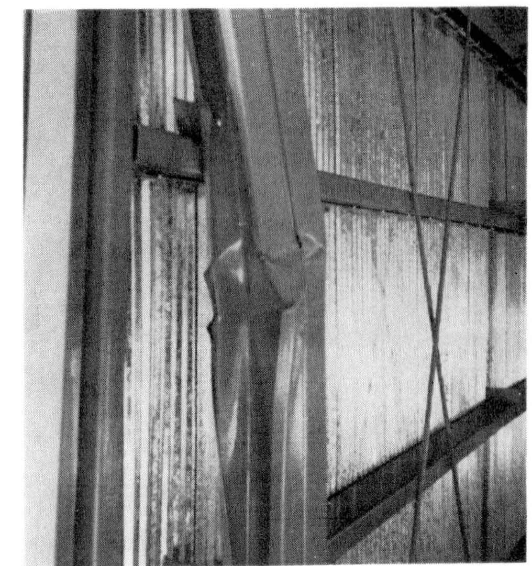

Photo 5.3

5.3 Structural Precautions

5.3.1 General

It is needless to say that even in steel structures, the form of a building as simple as possible is desirable. When it receives vibration, attention shall be paid to it not to cause extremely great twist. Accordingly, the effects of twist shall be considered when the features of a building are irregular and quake resisting walls and braces are not arranged proportionately. In case of the building made of some blocks of different rigidity, stress concentration occurs generally at the joints, due to different rigidity of blocks. To deal with it, sufficient unit strength is necessary, or an expansion joint shall be used to avoid the occurrence of excess stresses.

Frames are generally designed by elastic calculations with some assumptions, supposing a static force or by considering the dynamic effects. These suppositions are sometimes quite far from the actual circumstances, so the suppositions shall be selected carefully so that the structural load carrying capacity may not change remarkably as the suppositions change slightly.

Supposing that the supporting points in a long-span structure of an arch or a shell do not move, there is little stress and a slender frame can be designed. However, it is impossible to realize this unless the distance between supporting points is connected by the direct rigid tie beam to ensure that the supporting points may not move in fact. Especially in case of small rise the damage from inversion occurs easily due to the movement of the supporting points. Therefore, it is necessary to consider beforehand so that there may be no obstacles even when the supporting points move. The examination of deformation is important since there are examples of vibration resistance due to the rigidity of frame which is apt to become low for small support movements.

Besides, in producing actually a column foot calculated as a hinge, as it is made in the form with some unit fixation, excess stresses occur due to bending accompanying partial buckling, and the whole structure is sometimes damaged due to twist.

5.3.2 Foundation

In steel structures except of multi-storied buildings, less attention is paid to its foundation structure than expected in many cases, because the loads added to its foundation are relatively small. But in most such steel structure, there is not a basement and the footings are often on the surface of the ground so that sometimes it receives unexpected great damage due to sliding of the surface of the ground and differential settlement in earthquakes.

As such a destructive power is difficult to calculate, attention shall be paid to the foundation in the steel structure built on soft soil, trying to avoid the horizontal and vertical relative movements of a column foot by making mat foundation, or by installing as many tie beams as possible.

5.3.3 Joint

a) Joint by Rivets and Bolts

As the joint by rivets and bolts has been used for a long time and its standard method has already been established, it can be applied with almost no special design in general case if it is fastened faithfully to the specifications. In the past earthquake disasters, the shear of rivets is rare while the destruction of members is frequent.

There are two kinds, SV 34 and SV 41, of the quality of materials for rivets used in structures, and they have the same allowable unit stress according to the standards in our Institute AIJ. In the case of members with high tensile strength, better if the unit strength of rivets is high, careful execution with adequate temperature is necessary since ductility becomes less. Generally SV 34, which has greater ductility and is easy for execution, is recommended.

Advantages of rivet structure are the following:

i) When loads concentrate at one part, stress concentration will be relieved because the rivets

Photo 5.4

at that part slacken and the forces are transmitted to other parts.

ii) There is no question of the remaining stresses and the distortion in fabrication. They complement the shortcomings of welded connection that will be mentioned later.

In bolts there are ordinary bolts and high strength bolts for friction grip joints. First of all, in case of ordinary bolts, they are not permitted by the law at the principal joints of a large scale building. Even when the use is allowed, they shall be fastened in such a way as by double nuts that such nuts may not slacken or turn back. Especially the connection of the members that receive vibration and stresses repeatedly shall be friction grip joint as ordinary bolts are inadequate. Further, it is prescribed that the hole diameter for ordinary bolts shall not be over 0.5 mm larger than the shank diameter. To achieve this, it is desirable to pay attention to the execution of a different fabricating method from the rivet construction, making the hole diameter a little small in member construction in a factory and making it reamed to the required size in fabricating at the site.

Friction joint by high strength bolts is widely used in buildings, due to high rigidity of joints and little noise, instead of rivets. The characteristics that stress concentration is little, due to the forces widely transmitted on friction surface and that fatigue strength is great comparing rivet joints, are not used effectively in building.

In case of friction joint, as unit strength that yields large sliding at joints is the basis that decides design unit strength, the consideration of friction surface and the adjustment of tightening force of bolts are the most important, and the conditions of execution which are closely related to design shall not be forgotten.

Even though it is made a rule to clean thoroughly the surface of friction joint and to remove black skin, floating rust, dust, paint and varnish before fabricating, it is inadequate if it is immediately after polished brightly. After it is cleaned sufficiently in a factory, it is desirable to be rust properly when carried into the site. After brought in, only floating rust from the rusts that have been gathered till that time is cleaned by a wire brush before assembling. Therefore the efficiency of joints is sometimes worse than expected because such splice plates do not rust properly when they are piled, processed and preserved. When it is reamed in place it is important to care for oil not to soak. Paying attention to these points, it is easy to get 0.35 coefficient value of sliding, and it is possible to adopt maximum 0.46 if confirmed by experimentations. These coefficient values of sliding should be obtained at the actual joints and established, considering the conditions of execution when it is designed.

High strength bolts for friction joints, as the combination of bolts, nuts and washers as a set is decided, shall not be set apart and used changing its combination. As the combined use of the friction joints and rivets or weld is not finalized yet, the experimental examination is necessary when the combined use is required.

Further, in case of friction joint, as the difference between the hole diameter and bolt shank diameter is considerably large, the deformation is great when sliding occurs. Accordingly when there is the fear that the deformation of joints gives bad influences on the construction itself, attention is necessary. But the bearing strength after the occurrence of sliding, is advantageous for earthquake resistance due to large margin of safety, for the unit strength of steel materials for bolts is considerably high.

The defects of the connection by rivets and bolts are the decrease of effective section area at the joints and the occurrence of stress concentration due to notching of the rivet hole. Stress concentration shall be avoided to prevent the danger of the brittle failure with no increase in the extension.

In case of mild steel, there is almost no fear of brittle fracture until unit stress reaches the yield point. However, considerable attention is necessary for execution when sufficient extension must be expected. That is, when the member that receives tensile stress is holed, the construction method that shall not yield the effects of cold treatment around the hole, either holing by a drill at the required position or holing smaller than the prescribed size when a punch is used, and making the size as prescribed by removing the damaged zone near the hole, is recommended. In the similar meaning, it is recommended that plates shall not be sheared but cut smoothly by automatic gas cutting.

Recently, as the joint form that is not loose, tensile bolt connection with high strength bolts has been applied. In this case high stress concentration often occurs also below the head of a bolt and at the threads. Such destructive property of joints is inclined to yield the decrease of bearing strength suddenly when bolts are cut. It shall be considered for earthquake resistance if sufficient extension can be expected. As a rule, it is considered that such a joint form shall be in a structural form that shall not give plastic hinge at the joint position.

b) Welded Joint

The joint by weld has an advantage that base metal area does not suffer a loss. Recently as the execution techniques have been developed, welded connection is going to be used more and more. In light gauge cold-formed steel structure, though the use of gas welding and electric resistance welding is admitted, the basis is arc welding. In arc welding there are each kind of automatic arc welding and manual arc welding. The former yields great power to the long continuous weld such as of built-up collective section, however, the manual arc welding is generally done at the places where short irregular welding occurs at the joints of columns and beams. As for the key points of the connection by weld, it is needless to say that not only unit strength and rigidity shall be satisfied, but also the connection form that secures ductility shall be selected. Therefore it is important to consider that proper execution shall be made by the selection of the most pertinent steel materials, and by the decision of the form of joints, welding method and electrodes according to use.

The ordinary mild steels such as SS 41, whose weldability is good, have no trouble to be welded as usual. Especially where it is exposed to cold air or when it is a thick steel plate SM 41 more pertinent to welding is recommended. When it is necessary to use the steel materials with higher unit strength than those steels for welded structure, SM 50 with good weldability and great absorptive energy against shocks is proper. Generally SS 50 with much carbon equivalent and STK 51 of steel pipes for scaffold shall be avoided, and the allowable unit stress of welds using them is not prescribed.

An electrode is an important element as well as the base metal. As every kind of electrode is produced according to use purpose the most suitable one shall be selected. The coated electrode of low hydrogen type is recommended in case of a thick plate or high tensile strength steel whose mechanical characteristics are important.

The last important element is proper execution. The factory welding that makes horizontal position common due to the equipment of revolving jigs and positioning devices is more reliable than the field welding that is executed under bad field working conditions. Accordingly high allowable unit stress is given to shop welds, and it is desirable to do construction execution making shop welding as frequently as possible. The field welding shall be done by very skilled welders as they are restricted by the welding posture and working at high positions.

The weak point of welding connection is that some occurrence of welding distortion cannot be avoided. In order to keep it at a minimum, careful attention is necessary for the joint form and the welding procedure, otherwise excessive strain is left and excessive stress concentration occurs.

There are butt joints and fillet joints in the welded joint forms. The butt weld is the structure with little tolerances that generally requires accurate size considering lost length; however its transmission of stresses is smooth. On the other hand, the fillet weld is easy to take an error in length and to be executed; however it is inadequate at the place where large stresses are transmitted and it is better not to use it where stresses change repeatedly.

Attentions for the design and execution of welding are the following:

i) A short bead shall be avoided. Generally fillet weld shall be over 10 times the minimum throat width and over 40 mm.

ii) The welding under the conditions where strain is restricted, shall be avoided because it gives considerable contraction causing damage.

iii) To take full advantage of bearing strength of members, at the joints of a column and a beam, a beam flange is connected directly to a column flange by bevel groove weld: however, bevel groove weld at such joints is generally difficult to execute, so that it is desirable to deal with it carefully because the defects are easily left.

iv) A coated electrode shall be used in well dried state. Especially in low hydrogen electrodes, it is important to be dried beforehand.

v) When the steel materials or high tensile strength steels, which are over 40 mm in thickness are used, it is essential that proper research and study shall be done before they are used.

5.3.4 Tensile Members

The typical tensile members are a truss member and a brace member. As the consideration for the complicated phenomenon such as buckling is almost needless in tensile members, generally one with large slenderness ratio is likely to be used. Especially in a bracing member, the member which is very slender or irregular compared with other principal members, is used and its damage is often seen in earthquakes (Photo 5.5~5.8, Fig. 5.13).

Under even tension, tensile members that absorb energy very well are able to bear large deformation; however, they are easily fractured especially at a rivet hole where the stresses concentrate. Thus most of them cannot bear such large deformation as usually considered.

However, if it is used at the part of the structure with high degree of redundancy such as rigid frame structure, even though a brace is broken, it has still bearing strength. In a truss form building, however, the fracture of a member is directly related to destruction. In this case, except in a small building, it is necessary to use a member with considerable rigidity

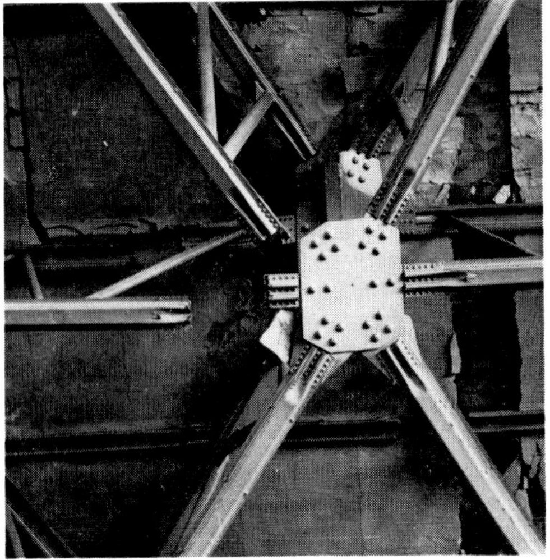

Photo 5.5

Photo 5.6

Photo 5.7

Photo 5.8

and bearing strength, even though it is a tensile member.

When a single angle is used for a tensile member, it is calculated by the effective section ignoring 1/2 of the attached side by considering the conditions of connection of ends. As eccentricity occurs at the rivets of joints, it is sometimes necessary to consider the use of lug angles and an increase in the number of rivets (Fig. 5.14).

Another attention to a tensile member is the buckling caused by unexpected compressive force at earthquake. There is the way of use that has an effect only in tension, putting braces diagonally. In this case, the buckling at the compression side must be expected to occur. If there is a wall, secondary damage of breaking the wall surface may occur.

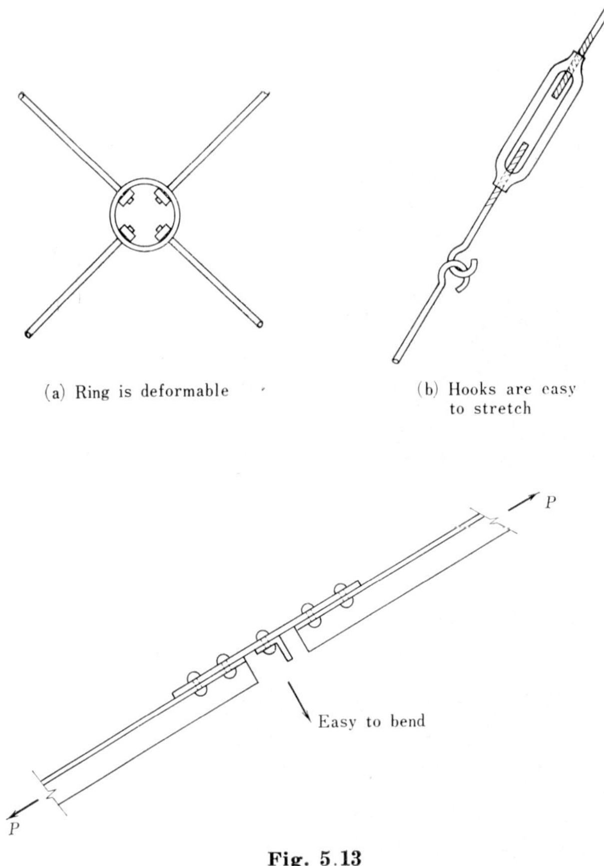

Fig. 5.13

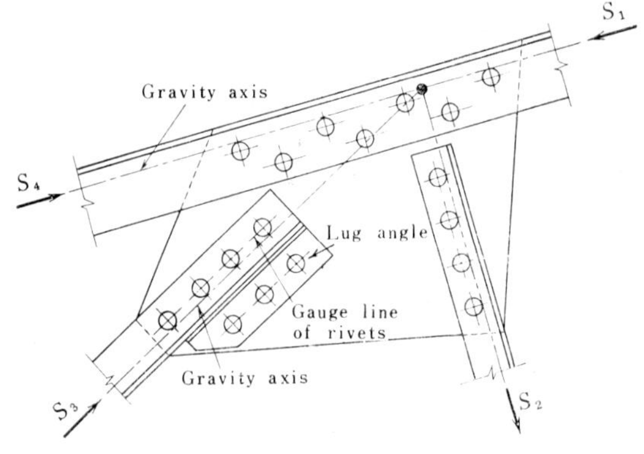

Fig. 5.14

5.3.5 Compression Members

The special characteristic of compressive members is in the buckling phenomena. There are bending buckling and bending twisting buckling as the forms of buckling yielding of compressive members. Generally, design is made on the basis of the bearing strength fixed by bending buckling. In the case of thin walled open section, especially the members of L–

form, T-form and cross form with little bending twisting rigidity, it is necessary to pay attention to bending twisting and buckling also. The use of a single angle steel shall be limited to minor members due to the bad influence of eccentricity generally that occurs at the connection.

Buckling may occur to the plate such as web of compressive members. But in case of the plate whose edges are reinforced by a flange and a stiffener, the bearing strength after buckling is available in the form of increased effective width in AIJ Standard for Structural Calculation of Light Guage Steel Structures, as bearing strength does not decrease even after the buckling of the plate. The important thing is that the rigidity and the unit strength of reinforcing members around the plate such as flanges and stiffeners shall be sufficient. As

Photo 5.9

Photo 5.10

such required rigidity is not prescribed in AIJ Standard for Structural Calculation of Steel Structures, a design that will not produce buckling shall be adopted where possible.

Built-up compression members shall be designed as a single member. In such a design, it is important to observe the detailed structural rules to avoid early buckling of individual members under a low load.

Further, as there is no change of Young's modulus in the elastic buckling limits even when high tensile strength steels are used, it is needless to say that the development of bearing strength cannot be expected except in the plastic buckling range with small slenderness ratio.

The reinforcement of supporting points shall not be forgotten in order that compressive members fulfill the expected purpose. When the ends of member move easily the stability of the whole building shall be considered. When the length of a member shall be considered the length of buckling, the reinforcement is necessary so that the ends of members may not move. As many members are collected in a space truss, it is hard to attach them without shortening some of them, and they are often connected by a big gusset plate. A gusset plate is weak in flexual rigidity and the bending of a gusset plate occurs. In some cases, the reinforcement of a gusset plate is necessary (Photo 5.9~5.10).

5.3.6 Bending Members

The typical bending member is a beam; however, the columns of a low building are often designed only for bending. Full web bending members have the character that they can resist great deformation without a remarkable decrease of bearing strength when there are proper lateral supports. As the members to secure bearing strength resisting such great deformations, a full web of stumpy form is desirable.

When strength and rigidity are required, a trussed girder is used when the member is tall and the web thin. Many beams of a building are supported enough in lateral direction generally in a body with a roof and a floor. When there is a large span between supports in the lateral direction such as crane girders in a factory, it is necessary to pay attention to lateral buckling that is the whole buckling of beams. In bending members, there are local bucklings of web and flange besides the lateral buckling. When a flange buckles, beams are twisted and the bearing strength is lost suddenly. Therefore, generally, the outstanding parts of a flange and the width of a cover plate are restricted, so that the section may not buckle before yielding. According to AIJ Standard for Structural Calculation of Steel Structures, as a rule web does not buckle when it is used; however, when a flange and a stiffener are strong the load carring capacity still increases even after buckling. This bearing strength after buckling is the basis of design in AIJ Standard for Structural Calculation of Light Gauge Steel Structures.

Further, there is no effect unless supports in lateral direction are installed so that compressive flange may not move. Where supports in lateral direction are not provided properly, the use of a box section is better. In the case of truss beams, there is much effect when steel pipes with high twisting rigidity are used as flange members. Especially, in steel pipe structure as stated before, direct welded connection of steel pipes is advantageous for high rigidity.

5.3.7 Bending Compressive Members

Typical bending compressive member is a column. Not only in the factories of one-storied buildings but also in multi-storied buildings, the section is often prescribed mainly by bending

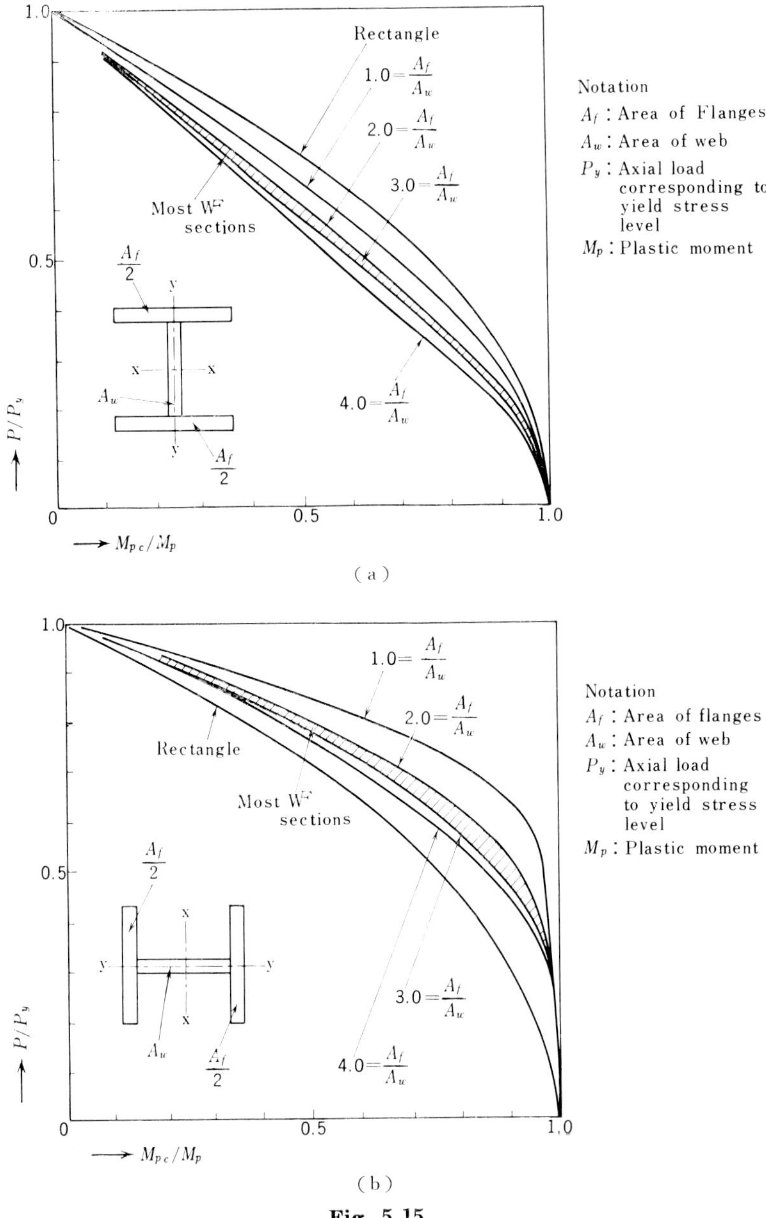

Fig. 5.15

moment. Generally, while existent axial force is small, the resisting bending moment is almost unaltered, but as existent axial force becomes larger, resisting moment decreases. Taking the final resistance to bending as the object, the relation between the two is shown in Fig. 5.15 where so-called Interaction Curve is drawn. As slenderness ratio becomes larger, more influence of buckling is given to bending compressive members. Therefore, even if axial force increases slightly, bearing strength decreases considerably in some cases. The example of I-section, which receive axial and lateral loads at the same time, shows the relation such as Fig. 5.16, and when slenderness ratio becomes large, it is necessary to pay attention to them.

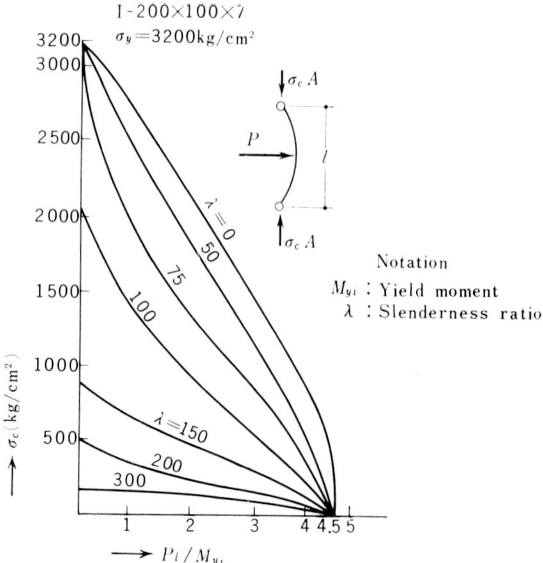

Fig. 5.16

Steel pipes are advantageous against buckling and twisting and recently they have been used with a single pipe as columns of a building. Especially when its inside is filled with concrete with a high unit strength, it is effective for resistance against earthquakes, as bending compressive bearing strength increases considerably because the local buckling of the pipe wall is prevented.

5.3.8 Connections and Supports

In trusses except slight ones, the axes of each member (central axes) shall meet at one panel point. When gusset plates are used, considering the extension of stress their unit stress shall be examined not to be excessively small. The pipe structure without gusset shall be designed so that cross angle of members may not be excessively small. When reinforcing bars are used as lattices, considering the length of weld, the joint form with large deformation ability such as side fillet weld is desirable.

Beam to column connections of rigid frames shall take the form that satisfies both rigidity and strength requirements. Otherwise, the design may be uneconomical as the deformation of the structure increases and its load carring capacity is lost before sufficient use of bearing strength of the member. Therefore, the use of continuous gusset has been recommended in rivet structures. In this case, reinforcing stiffeners shall be put around the parts of web of a column and of haunch to prevent the buckling of gusset plates without fail (Fig. 5.17).

As mentioned before, sufficiently strong stiffening rib must be used at the connection with clip angles not to yield eccentricity to tensile rivets (Fig. 5.18).

When a single web beam is used in the box section column, the stiffeners to make the stresses transmit to the flange of a column smoothly are necessary as stated before. Further it is better to change the section of a gusset plate gently without changing it suddenly (Fig. 5.19).

In welded structures, the use of strong butt joints is coming into vogue recently. In

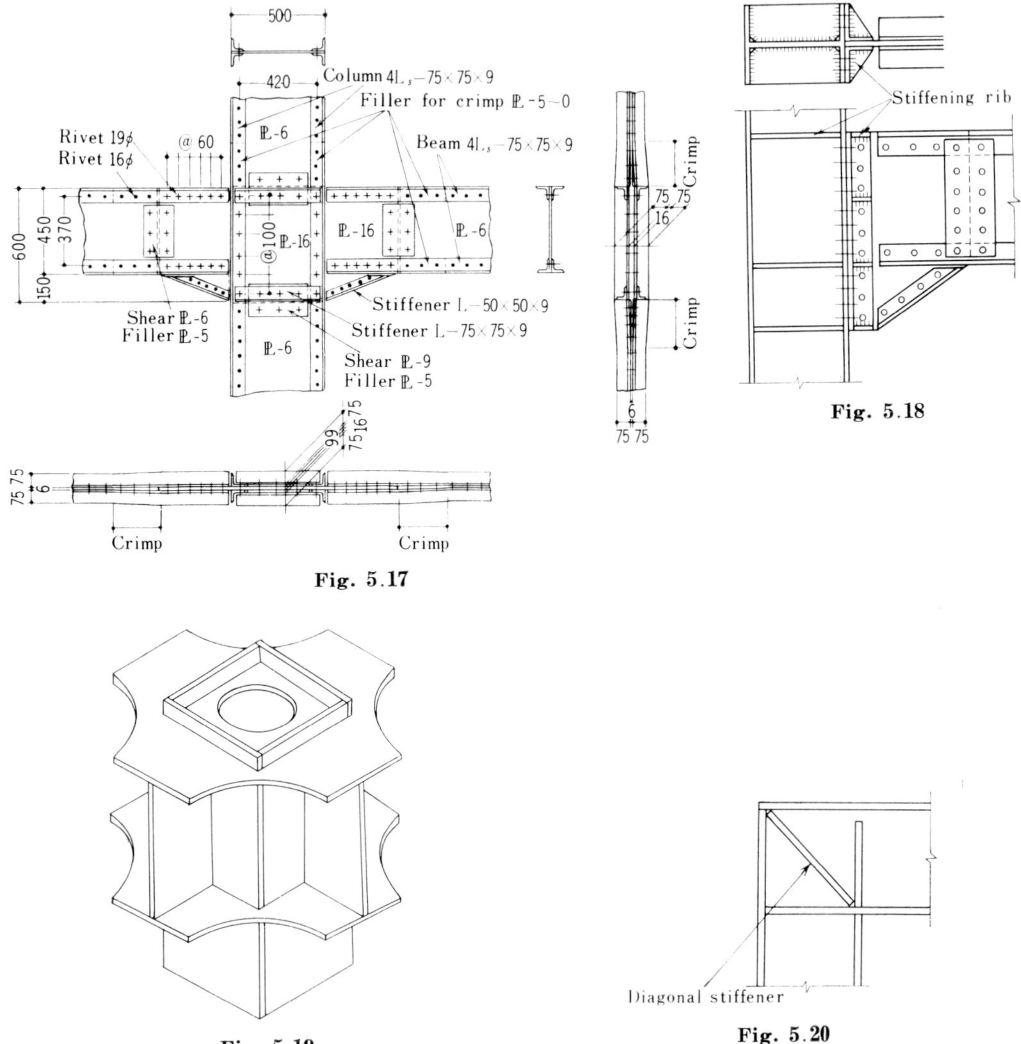

Fig. 5.17

Fig. 5.18

Fig. 5.19

Fig. 5.20

this case, attention shall be paid to the deformation of the flange of the column to which beams attached. Generally a stiffener is necessary for the web of a column. Further, as it is common that the large shearing force occurs at the corner panel of rigid frames, it shall be stiffened by a stiffener or a cover plate, as stated before, in order to use the bearing strength of member sufficiently (Fig. 5.20).

In the case of repeated stresses, the corner connections of rigid frames are sometimes made curved and it is necessary to pay attention to the stresses at the curved part.

In order to assume the column foot as fixed, it is necessary to use sufficient anchor bolts or encasement with stiff reinforced concrete. Except in the case of column foot encased with stiff reinforced concrete, shearing force operating at the column foot must be considered to be transmitted to the bottom of the baseplate. That is, the direct shearing force of anchor bolts shall not be expected. Therefore, in case of great shearing force, it is better to make

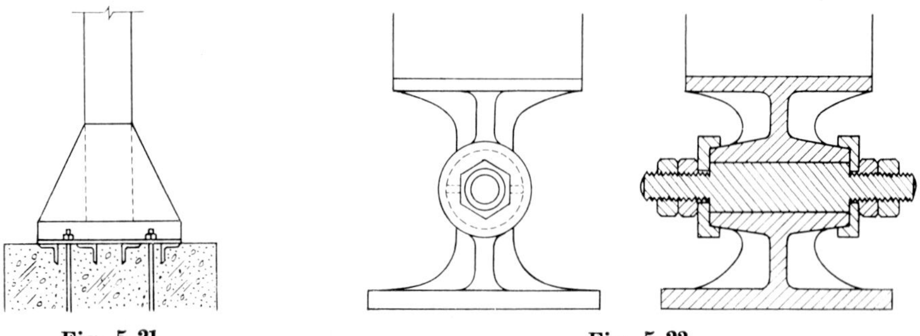

Fig. 5.21

Fig. 5.22

keys at the base of the baseplate such as by steel angles (Fig. 5.21). In this case, it is necessary to pay attention so that concrete goes around sufficiently.

When the column foot or the ends of other members are designed as a hinge, except minor ones, they shall be considered and designed to constitute a hinge. For example, in the case of a joint that has rounded plates and bolted to be a hinge, when loaded without provision for sufficient rotation of the hinge, extraordinarily high stress occurs in the bolts and there is fear that it may cause breaking of the bolts and the destruction of the whole structure. Therefore, much attention shall be paid especially to a long-span structure with large deformations. Further, in constructing a hinge or a roller, it is insufficient to only put parts together, but also necessary to hold the ends against separation in the event of earthquakes (Fig. 5.22).

6. REINFORCED CONCRETE STRUCTURES

6.1 Introduction

6.1.1 Earthquake Resisting Characteristics of Reinforced Concrete Structures

Reinforced concrete structure is a general name of structures in which the reinforcing steel bars and concrete are used, they being cast monolithically. In a wide sense under the term "reinforced concrete structure" are included not only monolithic reinforced concrete structures but also precast reinforced concrete structures and prestressed reinforced concrete structures. The monolithic reinforced concrete structure is further classified into the ordinary framed structure, the bearing wall structure and the shell structure. This chapter describes only the ordinary framed structures.

One of the strong points of reinforced concrete framed structures is that the foundations, columns, walls, girders and slabs are constructed monolithically so that the joints of members are rigid. It is capable of resisting as a monolithic body against not only vertical loads but also the forces due to earthquakes and other forces in all directions.

In monolithic structures both the vertical forces due to gravity and the lateral forces due to earthquakes are transmitted to the whole structure even when the force is concentrated on some part of it. If there is some weak point in the structure, some of the load at that point will be carried to stronger part of the structure, and finally, all the forces will be transmitted safely to the bearing soil layer.

In this type of structure, the antiseismic structural elements such as rectangular rigid frames and shear walls, are usually connected at every floor level by reinforced concrete slabs which have high rigidity in plane so that seismic forces are transmitted to those structural elements in proportion to their rigidity. For instance, a larger amount of force is transmitted to the shear wall which has high rigidity, than to the rectangular rigid frame which has relatively low rigidity. The stresses in each member of the frame due to these distributed horizontal forces are combined with those due to vertical forces, and the total stresses produced result in proper amount as compared with the size of the member. Thus, the reasonable proportioning of the structural members can be made. This is the most important characteristic of the monolithic structure, and furthermore, the reinforced concrete structure has good fire-proof and durability qualities. This is the reason why this type of structure has been developed as a representative earthquake resistant structure.

As the members of reinforced concrete structure have high rigidity and are rigidly connected to each other, the deflection of the structure as a whole due to external forces is in general fairly small. Further, the ductility of reinforced concrete structures is large enough to meet the requirements of earthquake resistant structures, provided that the reinforcing bars are properly arranged, although their ductility is not so large as those of steel skeleton structures or of steel framed reinforced concrete structures. Especially, attention should be paid to the arrangement of shear reinforcements, bond of reinforcements and anchor length of reinforcements, so as to get large ductility for shearing forces, because the reinforced concrete members have enough ductility for bending moments but they have comparatively poor ductility for shearing forces.

The heavy density of concrete is a disadvantage in reinforced concrete structures. In the ordinary reinforced concrete the density is 2.4 and the weight of the building per sq. m of floor area including the dead and live loads is 1.0 to 1.5 tons, of which the live load is only 0.2 to 0.3 tons. In this respect, the reinforced concrete structure may be said to be very inefficient.

Recently, the light weight concrete using light weight aggregates has become popular in Japan, and some type of this kind of concrete has the density of 1.5 and has the strength nearly equal to that of ordinary concrete. The use of this type of concrete will be beneficial for future development of reinforced concrete structures because it reduces earthquake forces as well as the cost of foundation construction as the result of reducing the weight of buildings.

6.1.2 Shape

As the reinforced concrete structure is constructed by casting concrete into a mould which can be formed in any shape and has the reinforcing bars arranged in advance, the section, length, and strength of members can be quite freely chosen. This is one of the merits of reinforced concrete structures. Therefore, it is difficult to point out the suitable or maximum scale of buildings for this type of structure. However, considering the restrictions in the present building code for design, building materials and workmanship, such as the limit of span length determined by excessive deflection of slabs, difficulties in casting concrete owing to crowed arrangement of reinforcements in columns and girders, especially at the joint of columns and girders, economical restriction in execution of works, the scale of structures may be limitted within the following ranges :

Number of stories of buildings
- above the ground : ordinarily less than six stories, but 11 stories has been constructed in the past.
- below the ground : no restriction, 5 stories has been constructed in the past.

Span length
- Span length of girders : 4～10 m
- Area of the floor carried by one column : 20～50 m^2

Height of stories : 3～4 m

Length of a block of buildings : 50～60 m

6.1.3 Examples of Earthquake Damage to Reinforced Concrete Structures and Their Causes

Some reinforced concrete structures have been damaged by past earthquakes. The damaged structures and damaged parts of structures were as follows :

i) The reinforced concrete structures which had masonry walls, especially partitions and curtain walls made of bricks.

ii) The reinforced concrete framed structures which had few shear walls and the columns and girders with small dimensions.

iii) Those which did not constitute rigid frames.

iv) Those which were insufficient in the foundation design.

v) Those which were inadequate in the arrangement of reinforcing bars such as inadequate position of tensile reinforcing bars, insufficient amount of stirrups, hoops and other supplementary bars, deficient arrangement of lapped splices and anchoring of reinforcing bars.

vi) Those of bad workmanship, especially inadequate mixing and placing of concrete.

vii) Those which had suffered from fire before the earthquake.

Reinforced concrete structures are often subjected to the following types of damage due to permanent loads and others, and these types of damage lead to serious damage on the occasion of earthquakes.

viii) Damage due to thermal stresses at a portion which is subjected to excessive change in temperature, such as concrete crack and rust of reinforcing bars in parapets and chimneys.

ix) Deflections of long-spaned slabs, cantilever beams and cantilever slabs due to creep of concrete.

Some of these types of damage are illustrated in Photo 6.1~Photo 6.7.

As all of these types of damage mentioned above resulted from deficiencies in design and workmanship, it is possible to avoid them with sufficient care and to make reinforced concrete structures earthquake resistant. Damage due to thermal stresses, creep and fires as well as cracks in walls and the portion with stress concentration caused by earthquakes of small or intermediate scale, should be repaired as soon as possible. Especially, the damage by fire leads to reduction of concrete strength and bond strength between concrete and steel and to

Photo 6.1 Failure of Faulty Framing (Mexico Earthquake)

Photo 6.2 Collapse of N-Building (Kanto Earthquake)

Photo 6.3 Collapse of D-Building (Fukui Earthquake)

Photo 6.4 Failure of a Column

Photo 6.5 Damage of T-Building

Photo 6.6 Damage due to Unequal Settlement (Niigata Earthquake)

Photo 6.7 Failure of a Structural Member due to Liquefaction of Ground

the development of neutralization of concrete, and as a result, the strength of whole structure is much reduced, and it has the danger of suffering serious damage on the occasion of severe earthquakes. Therefore, detailed examination of strength of those structures and their repair is required.

6.2 Structural Design

6.2.1 General

In order to maintain the safety of reinforced concrete structures against earthquake, it is necessary to design them to be safe not only for vertical loads but also for lateral forces. In the structural design this should be emphasized. The following principles are most important when designing reinforced concrete structures to resist lateral forces :

The first principle is to make rigid joints between main columns and girders and to form rectangular rigid frames. As the stresses in a frame due to lateral forces take maximum values at the joint of members, the whole frame can be made earthquake resistant by making the joints rigid and giving them enough strength. The rectangular frame is a statically indeterminate structure of high degree and the number of indeterminancy increases as the number of rigid connections increases. Such a structure does not completely collapse by seismic forces until rigid connections become plastic hinges one after another and finally the whole structure becomes unstable. In this respect, it can be said that to increase the degree of redundancy is a good way to make rectangular frames earthquake resistant.

The second principle is to arrange rigid and strong shear walls and connect them with open frames by reinforced concrete slabs which are cast integrally so that the structure as a whole can resist lateral forces. Shear walls are classified into reinforced concrete walls (with or without openings), walled frames with wall girders and wall columns, and reinforced concrete frame with diagonal bracings. They may be chosen in accordance with the general planning.

As reinforced concrete buildings have heavy weight, the design of foundation construction must be carefully made taking account of additional stresses due to lateral forces combined with those due to vertical loads. The design seismic forces are calculated by multipling the total weight of every story by the seismic coefficient. In the preliminary design, the weight of every story per unit floor area including live loads can be approximated as follows :

$$W=1.0\sim1.5\ \mathrm{t/m^2}$$

6.2.2 Design of Frames

a) Arrangement

It is desirable to arrange columns to be on the straight line in plan in both directions and to be along the vertical straight lines in elevation, and also desirable to arrange girders at every floor and beams if necessary, so that columns and girders form a rectangular space frame. On the other hand, the frames of irregular shape produce irregular and uncertain stresses due to seismic and other forces. In case that the location of columns in upper and lower stories does not coincide with each other producing the eccentricity, especially, the lack of columns at the lower stories or having a large hall which occupies the space extending over several stories

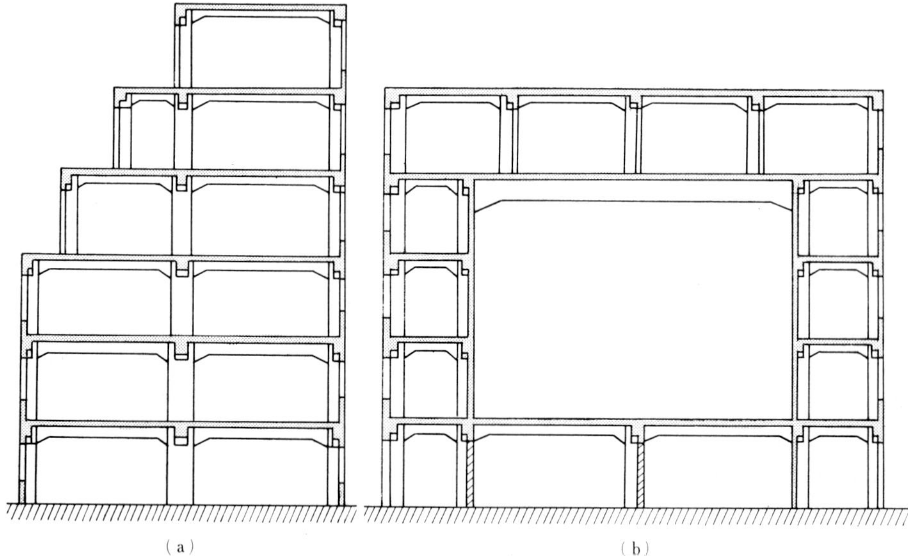

Fig. 6.1

(see Fig. 6.1), sufficient attention should be paid.

b) Rigidity

Reinforced concrete buildings of ordinary scale are expected to have high rigidity to lateral forces. In general, as the rigidity increases, the natural period of vibration of structures becomes shorter. It seems that the buildings having shorter natural period are subjected to larger seismic forces, but the antiseismic elements, such as properly arranged shear walls, can resist these forces, and the seismic energy is absorbed by local failure in the antiseismic elements (wide cracks in walls and others) and the resulting plastic deflection, and as the result, the buildings are kept safe. It is desirable that every part of building has a uniform rigidity. Otherwise, dynamic characteristics in various parts become different, and in case that the floors are flexible, the deflection in various parts of the building during an earthquake will be different from each other and the floors will be distorted. Further, if the floors are rigid and the frames are neither uniform nor symmetrical, the building will be twisted and unfavorable stresses will occur.

It is favorable that stiffnesses of columns and girders of a frame be uniform in the same story and vary gradually from story to story. And it is advantageous to have the stiffness ratio of girder to column as large as possible. In general if the stiffnesses of every part in the same story are not uniform, the stresses will concentrate on the high stiffness parts, and if there is large amount of change in the stiffnesses of successive two stories, the distribution of stresses will become irregular and stresses will concentrate on the story with the high stiffness.

As the stiffness ratio of girder to column is higher, the inflection point of column moves to the center of story. In case that this ratio is low, the inflection point of column is located at the lower part for upper stories and at the upper part for lower stories, and the differences in the top and bottom moments of columns in a story become large (see Fig. 6.3). This is in general disadvantageous.

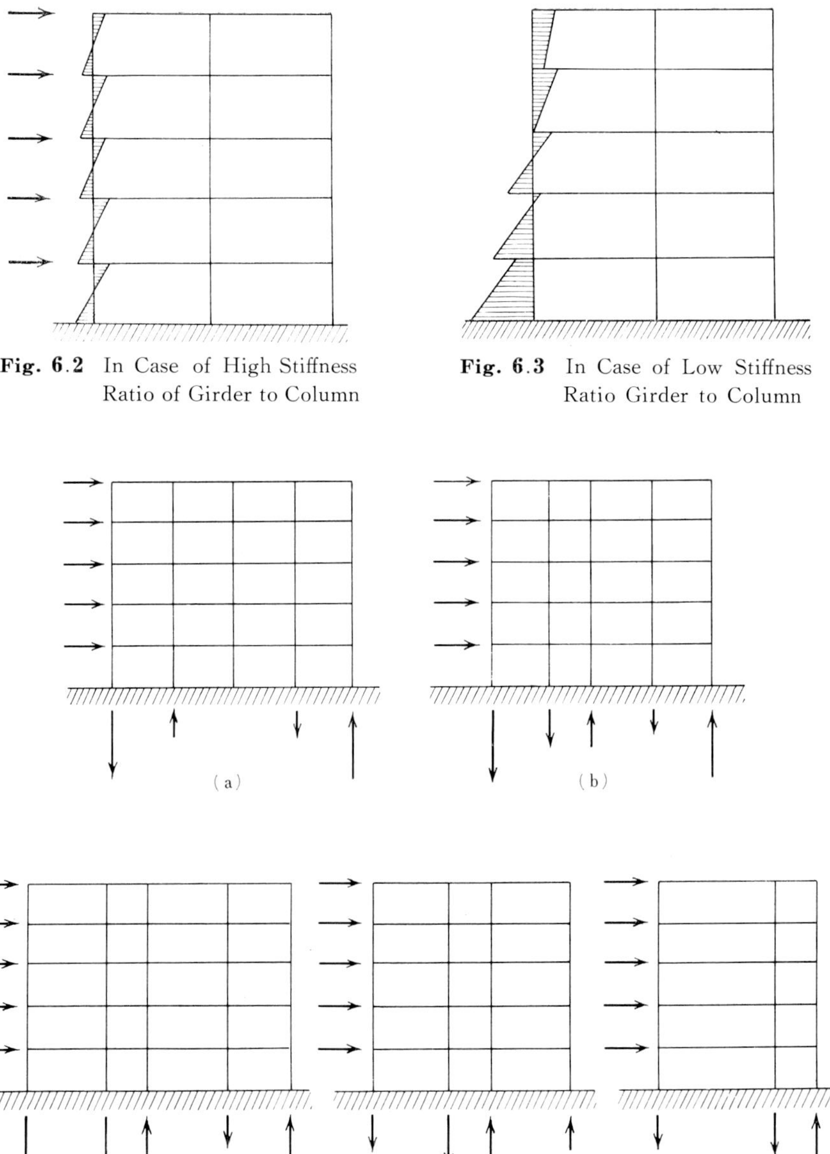

Fig. 6.2 In Case of High Stiffness Ratio of Girder to Column

Fig. 6.3 In Case of Low Stiffness Ratio Girder to Column

Fig. 6.4

In uniform frames, vertical reactions at the bottom of exterior columns are large while those of interior columns are very small (Fig. 6.4 (a)). When the frame has a short span, columns of both sides are subjected to large axial reactions (Fig. 6.4 (b)). Especially, if the stiffnesses of the short-span girders are high, the reactions will become very large (Fig. 6.4 (c), (d)). Taking the frame with one sided corridor which has the short-spanned and the long-spanned girders connected to each other, for example, if the same dimension of the section is used for both girders, the large amount of shear in the short-spanned girder and the axial force in the outer column will result due to lateral forces, because the stiffness of the short-spanned one is much larger than that of the long-spanned one (Fig. 6.4 (e)). And sometimes it is difficult to design such a frame within the elastic limit. In such a case it is advisable to redistribute the excess stresses or reactions by consideration of the plastic deformation.

c) Connections

When a rectangular frame is subjected to lateral forces, the maximum bending moment is produced at the ends of girders and columns, and after combining these moments with those due to vertical loads, moments at these ends usually take maximum values for the member. When these ends are difficult to design, it is recomended to make haunches at the end of girders (Fig. 6.5). Haunches have the merits of increasing the strength and the rigidity of frames considerably by strengthening the connection and its neighborhood and reducing the deflections due to bending moments. When the frame is subjected to lateral forces, considerable change in stresses, that is, from tensile stress to compressive stress, occurs in the reinforcing bars at the joint of columns and girders, and large amount of bond stresses between steel and concrete inside the joint are produced as well as large amount of shearing stresses in the concrete of joint. When the haunches are provided, the stresses are smoothly transfered and the bond stresses and the shearing stresses can be reduced. Furthermore, they may provide have some space for anchorage of the reinforcing bars (Fig. 6.6).

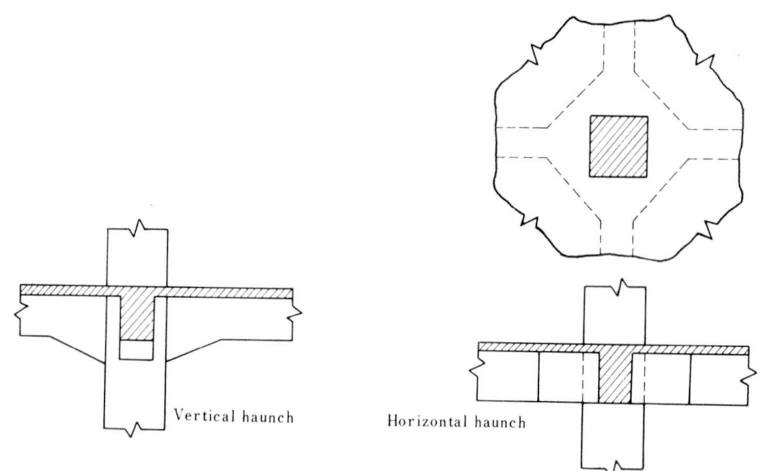

Fig. 6.5

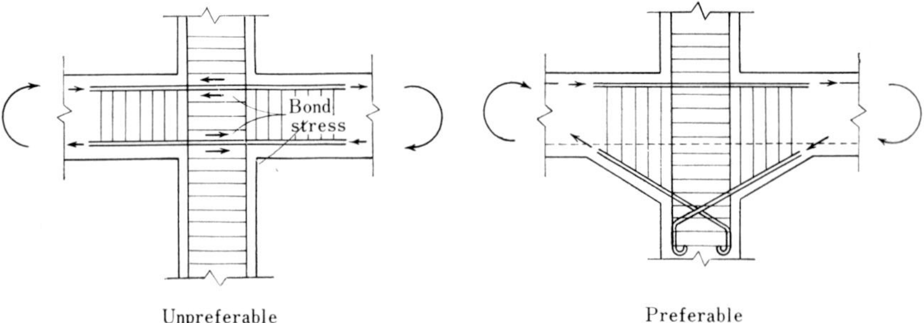

Fig. 6.6 Stress Conversion of Reinforcements at the Joint and Bond Stresses

6.2.3 Design of Shear Walls

a) Type of Shear Walls

(1) Reinforced Concrete Shear Walls

Reinforced concrete walls cast integrally with columns and girders are effective structural elements to resist lateral forces irrespective of exterior or interior walls. Among these reinforced concrete walls, those which are given enough thickness and reinforcements to resist seismic forces are called the "shear wall" or the "quake-resisting wall", and are considered to be one of the most effective antiseismic elements (Fig. 6.7). The stiffness and strength of reinforced concrete walls are much higher than those of open frames, provided that the surrounding columns and girders are properly reinforced. Therefore, they can carry a part of seismic forces acting on other portion of the building through the slabs connected to them. Thus they can resist seismic forces with the cooperation of other frames or other kinds of shear walls. Arrangement of shear walls is illustrated in Fig. 6.7. The shear wall should exist until the end of the service term of the building, having no possibility of being removed in the future. Openings in a shear wall are permissible, but it should be kept in mind that the stiffness and strength are considerably reduced due to openings.

(2) Walled Frames and Frames with Reinforced Concrete Bracings

Walled frames which consist of deep girders and wide columns, are also one of the most effective antiseismic elements. The exterior wall with openings usually takes this type of frames by giving certain amount of thickness to the wall. The rigidity of the walled frames to lateral

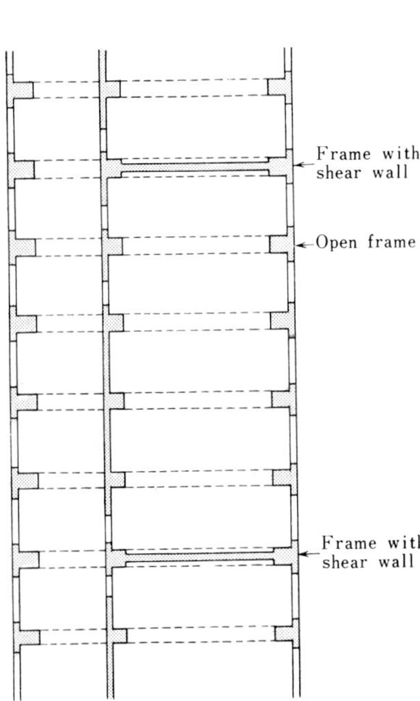

Fig. 6.7 Frame with Shear Wall

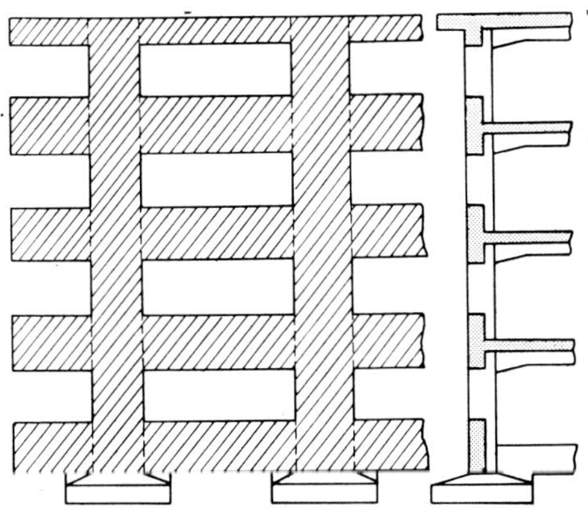

Fig. 6.8 Walled Frame with Openings

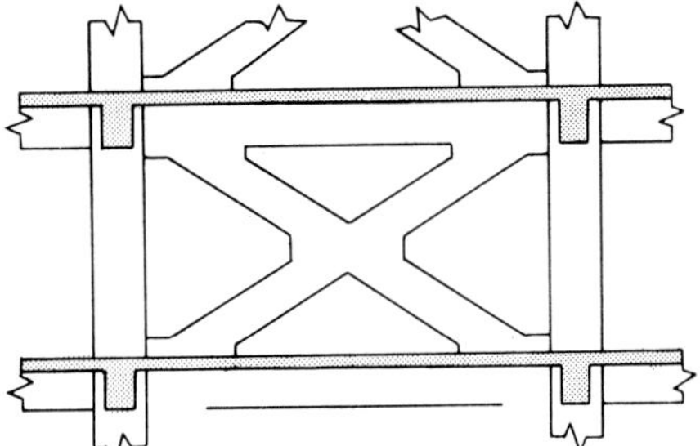

Fig. 6.9 Frame with Reinforced Concrete Diagonal Bracings

forces are ranked between those of shear walls and ordinary open frames. As the deflection pattern of the walled frames due to lateral forces is similar to that of ordinary frames, both the upper and lower parts are almost equally effective, while the upper part of the shear walls is not so effective because of large deflection due to cantilever action.

Reinforced concrete braced frames which have strong reinforced concrete diagonal bracings of about same size of sections as that of columns or girders in ordinary frames, have a similar effect to the reinforced concrete shear walls (Fig. 6.9). In these frames, attention should be paid to the concentration of stresses at the end of bracings and anchorage of reinforcing bars.

(3) Masonry Walls Surrounded with Reinforced Concrete Frames

In general ordinary masonry walls made of concrete blocks are not considered as antiseismic element because of their heavy weight and low shearing strength. They only add loads to the girders as a burden rather than contribute to resist lateral forces. But they have been used as a partition wall in many practical cases because of low cost. They should be properly reinforced with steel bars in order to prevent the collapse during earthquakes. Masonry walls which are surrounded with reinforced concrete frames are considered as an antiseismic element to some extent and they may be used in low buildings (Fig. 6.10).

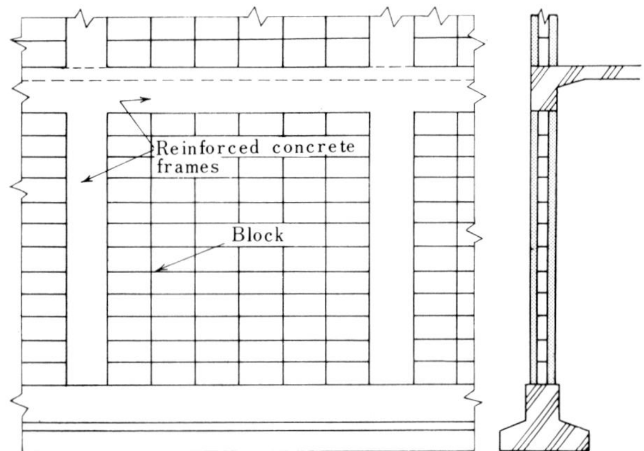

Fig. 6.10 Concrete Block Shear Wall Surrounded with Reinforced Concrete Frames

b) Arrangement of Shear Walls

(1) Planning in the Horizontal Plane

In planning, it is desirable that shear walls be arranged as uniformly as possible in every story and that the rigidity for lateral forces are balanced in order that the torsion will not occur (Fig.

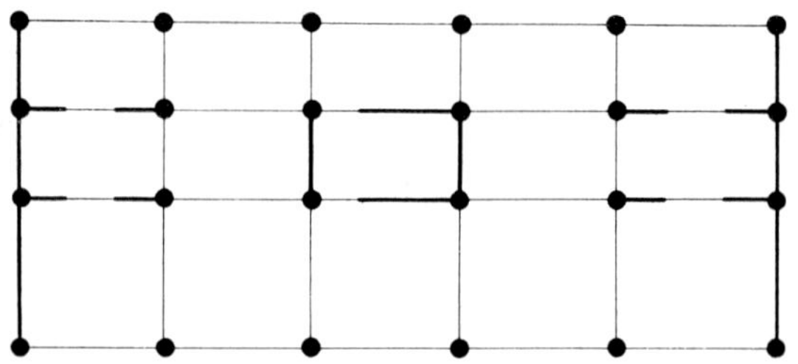

Fig. 6.11 Arrangement of Shear Walls

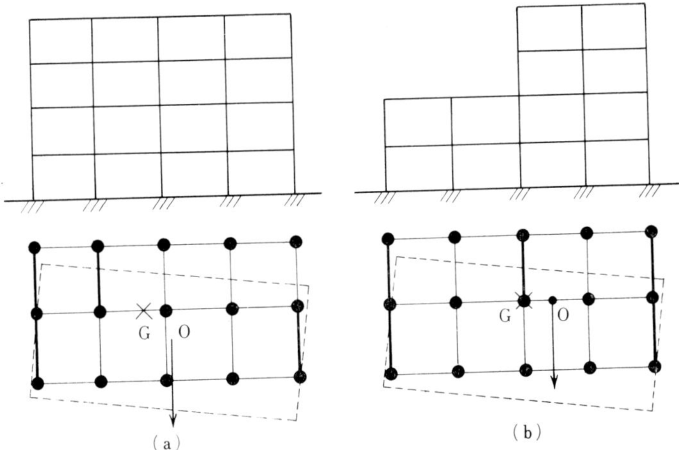

Fig. 6.12

6.11). If the center of gravity of seismic forces does not coincide with the center of lateral rigidity of structure, the whole building will be twisted (Fig. 6.12). In case the shear walls are eccentrically arranged (Fig. 6.12(a)) and that the shear walls are not eccentrically arranged but some parts of building are eccentrically tall (Fig. 6.12 (b)), the center of gravity does not coincide with the center of rigidity. In case that the distance of eccentricity is very great in proportion to the scale of the building, the influence of eccentricity must be considered. In such a case it is advisable to arrange the shear walls of high rigidity at both ends in the perpendicular direction in order to resist torsion.

(2) Planning in the Vertical Plane

The efficiency of the shear wall increases rapidly with increase of its width. So one consecutive two-spanned shear wall is more effective than two separate one-spanned shear wall

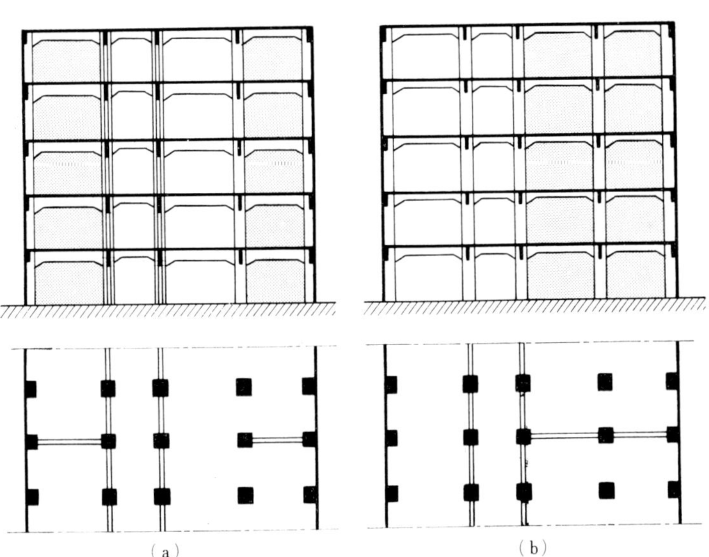

Fig. 6.13

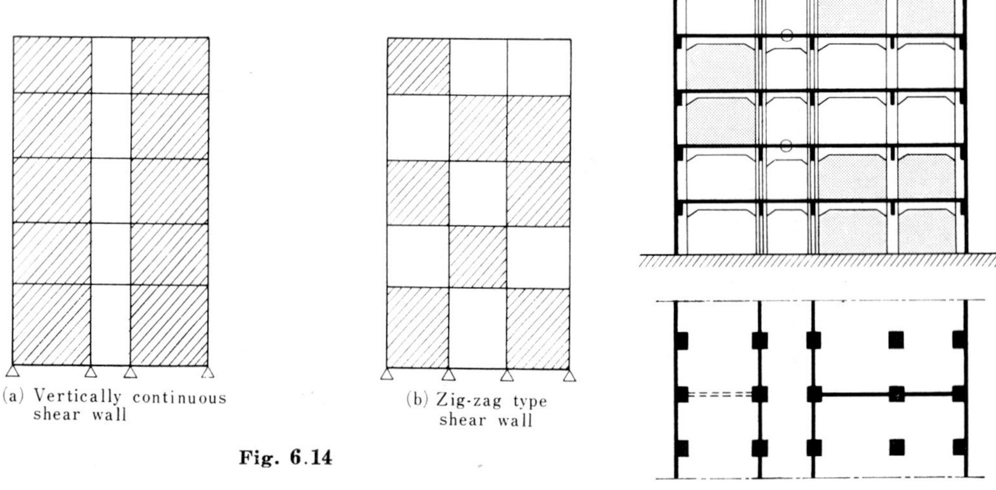

Fig. 6.14

Fig. 6.15

(Fig. 6.13). The ways of arrangement of shear walls are classified into two types, vertically continuous type and zig-zag type (Fig. 6.14).

The zig-zag type shear walls have advantages that they are uniformly effective through the height of the building and that the reaction below the walls at the lowest story are not dominant. However, in the case of a wall such as one shown in Fig. 6.15, girders marked by a circle are subjected to larger stresses and required to have much reinforcement because of the requirement for shear transmission between two side walls.

In case of vertically continuous shear walls, all the forces acting on them are transmitted to the foundation through them. And the deformation of superstructure is produced mostly by the cantilever bending rather than by the shearing force, while the deformation of open frame is produced mostly by the shearing force, and therefore, the lateral rigidity of shear walls is large for lower stories but becomes markedly small for upper stories.

Moreover, reaction forces under the basements which support these tall antiseismic walls become considerable, and the shear wall may deflect rotationally by lack of bearing capacity of the soil or by pulling up the column due to lack of vertical load (δ_R in Fig. 6.16). On the

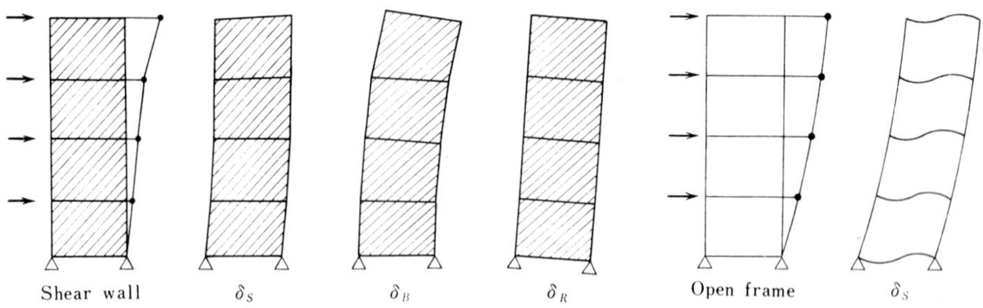

δ_S: Shearing deformation δ_B: Bending deformation as a whole
δ_R: Deformation due to foundation rotation

Fig. 6.16 Deformation of Open Frames and Shear Walls

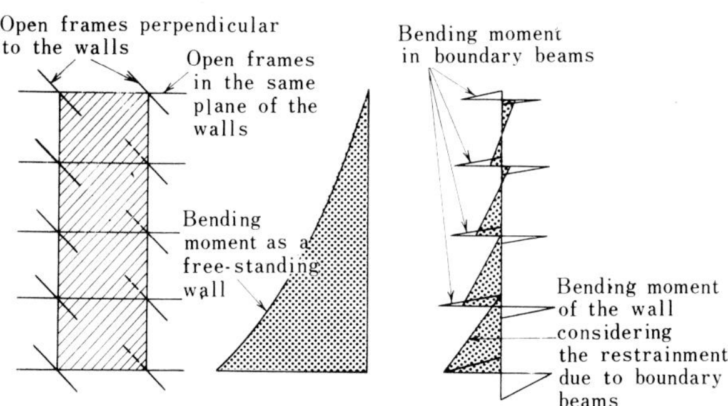

Fig. 6.17 Boundary Effect

other hand, there are boundary elements such as girders or footing beams which connect with the shear walls in the same plane or perpendicular to the plane of the walls to reduce the bending deformation or rotational deflection of the tall shear walls.

When the large effect of these boundary elements to reduce the deformation of the shear wall is expected, the rigidity of the shear wall increases and the foundation reaction is reduced. However, it should be noticed that the stresses concentrate on the boundary girders and other boundary elements.

The rigidity of shear wall increases in proportion to the increase of its thickness. It is general to increase the thickness of the walls as they go downwards. It is possible to control the variation of lateral rigidities of the wall along the height by means of the control of wall thickness.

The strength and the rigidity of the shear walls with openings are smaller than those of the solid shear walls. The decrease of rigidity may be considered to be in proportion to the periphery ratio of the opening*. The circumference of openings should be sufficiently reinforced.

6.2.4 Planning of Slabs

In order to make reinforced concrete buildings antiseismic, the reinforced concrete slabs, which are cast integrally with vertically arranged antiseismic elements such as shear walls and open frames, are required. The reinforced concrete slabs cast in situ are not easily deformed due to shearing forces in plane and are considered as a rigid plate. These slabs are indispensable to bind the various antiseismic elements and to let them resist the seismic forces as one body. In this respect the slabs can be called one of the antiseismic elements.

The floor system, which is made of prefabricated members such as precast concrete units resting on the reinforced concrete girders, does not have the ability to make various antiseismic elements monolithic. Therefore, sufficient consideration must be paid to determine the shear distribution to shear walls and open frames other than to make joints between the slab units tight.

The following are the comments on the design of reinforced concrete slabs (rigid slabs) :

The slabs which may possibly have large amount of shear forces in plan, are required to be highly reinforced. These cases are the slabs near the shear wall which is located at a

* Periphery ratio of the opening $=\sqrt{\dfrac{\text{opening area}}{\text{gross wall area}}}$

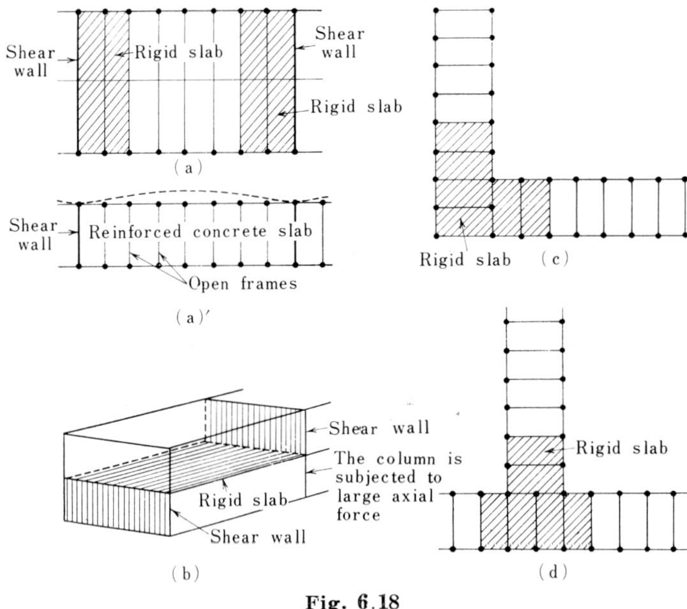

Fig. 6.18

distance from another shear wall (Fig. 6.18 (a)), or the slabs connecting the shear walls in upper and lower stories which are not in the same position in plan (Fig. 6.18(b)).

When the distance between shear walls is great, it should be kept in mind that the shearing forces of frames distant from shear walls are greater than those of frames near them due to the effect of deformation of the slab in the horizontal plane.

Slabs at the joint of two building blocks, whose rigidities are not same, e.g. in case of plan of L type or T type, should be highly reinforced (Fig. 6.18 (c)(d)). It is recommended to make an expansion joint between two blocks when the transmission of earthquake forces through the slab is considered to be inadequate, provided that there is no objection in the architectural design.

Though the slabs are fundamentally designed for vertical loads, the thickness of concrete or arrangement of reinforcing bars should be checked for lateral forces at the places mentioned above because the shear forces at those places due to lateral forces become markedly large. Especially, in case that there are openings in the slabs which are located near the walls, a check of the reinforcement is very important.

6.2.5 Planning of Foundations

The footings are classified into isolated footing, continuous footing, mat foundation, etc. (Fig. 6.19). Generally, isolated footings are used below columns and continuous footings are used below shear walls. In case that isolated footings are used alone, supporting condition at the bases of columns is nearly pin, and the column bending moment due to lateral forces becomes great. So it is desirable to arrange tie-beams with high rigidity between the columns, so that bending moments may be distributed to the columns (Fig. 6.19 (b)). Tie-beams are also effective in preventing settlement.

As reinforced concrete structures have heavy unit weight, their foundations should be

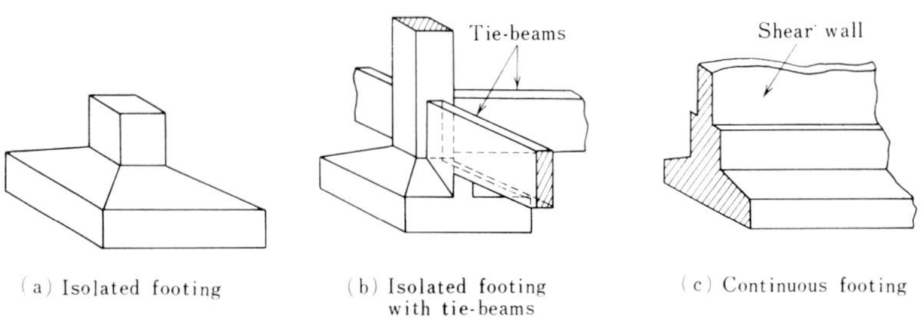

(a) Isolated footing (b) Isolated footing with tie-beams (c) Continuous footing

Fig. 6.19

especially designed to be strong and rigid. Any unequal settlement is, of course, unfavorable. It is desirable that the whole building be as uniform as possible. In case that there are few shear walls and they are arranged at a distance to each other, their footings should be designed with sufficient care considering the balance with other footings to avoid the unequal settlement. Generally, it is effective to arrange walls which are extending horizontally in the lowermost story, in order to prevent the unequal settlement, if there is no objection in architectural design.

6.2.6 Miscellaneous

Though flat slab structures are effective for vertical loads, slabs or connections of slabs and columns are weak for lateral forces. They should be used together with the above-mentioned antiseismic elements.

6.3 Rules on the Stress Analysis

6.3.1 General

In ordinary cases, stresses of a reinforced concrete structure under horizontal forces (seismic forces) are computed in accordance with the following assumptions and principles :

i) Horizontal forces are assumed to act separately in longitudinal and transversal directions of the building.

ii) Horizontal forces are assumed to act concentratedly at the floor levels of the structure. However, when horizontal forces acting on the middle part of the story produce a bad influence, the amount of the influence must be added.

iii) In ordinary cases, floors are assumed to be rigid in the horizontal plane. Consequently, columns and shear walls in the same story are assumed to have equal horizontal displacement. But when floors cannot be assumed to be rigid (e.g. in the case of precast concrete floors, or in the case that the distance between two adjacent shear walls is too long) or when the eccentric distance between the centroids of shear forces of the story and of rigidity of the structure are not negligibly small these effects must be considered.

iv) Shear forces of each story (story shear force, i.e. the total of horizontal forces acting on the floors above the story considered) are assumed to be distributed to the columns and other antiseismic elements of the story. The ratio of distributed shear forces and the stress distribution of each part are determined based on the rigidity of members by means of common methods of elastic calculation. But with reference to the member on which stress concentrates, rigidity is appropriately reduced considering plastic deformation, and the distributed stress of the member can be adequately reduced. In this case, sufficient ductility must be given to the member from a structural standpoint. For example, a short column or girder with large rigidity shall be sufficiently reinforced by stirrups and hoops in order to avoid brittle failure due to shear forces during the plastic deformation by bending moments.

v) Generally, the ratio of distributed shear forces is determined assuming that foundations are not displaced by rocking or rolling or sinking etc. due to horizontal forces or overturning moments, but in some cases it is desirable to revise the calculated value considering the influence of the displacements mentioned above.

vi) When the construction is composed of some large blocks, the calculation can be done for each block, dividing the whole structure properly. But in this case, the design must be made to balance the lateral rigidity of each block in both longitudinal and transversal directions. Comparison of the lateral rigidity of each block can be obtained as the ratio of the story shear force to the sum of shear force distribution coefficients of each block.

vii) Design of the section for each member is carried out combining the stresses due to horizontal forces with those due to vertical loads. In this case, the short term allowable stresses of materials can be used.

6.3.2 Distribution of Shear Forces

The essential point in the stress analysis is to calculate the distribution of shear forces acting on columns and shear walls as accurately as possible (cf. 6.3.10 (4)).

The stress analysis of rigid frames under horizontal forces is generally done by the use of structural mechanics. In Japan excellent approximate methods of calculation for rectangular frames such as "D-method" by Dr. K. Muto and "Modified portal method" by Dr. T. Naito were proposed and they have been widely used in practical design. Details of these methods will be found in the references of this chapter.

If the story shear force is distributed to open frames and shear walls strictly in proportion to each rigidity, it may occur, when many shear walls exist, that most of shear force is distributed to shear walls and the distributed shear force in open frames becomes very small. Even in such a case, certain amount of story shear which is larger than that of calculated shear must be distributed to open frames considering that the calculation methods include many imperfect assumptions.

a) D-method

This is the method to determine the distribution ratio of shear force to columns in open frames and to shear walls and to distribute the story shear force to them in proportion to these distribution ratios. The distribution ratio of shear forces is called the shear force distribution coefficient (or simply, distribution coefficient) and is denoted by D. D shows the rigidity against shear force, and is given as shear force Q (t/cm) with unit relative displacement (1 cm) at the story. In the practical calculation, D-value is represented by a practical unit $12\,EK/h^2$ (t/cm) and expressed as a non-dimensional number (h : story height (cm), E : Young's Modulus of concrete (kg/cm^2), K : standard stiffness (cm^3)). For example, the distribution coefficient of a column whose stiffness ratio is k_c is given by

$$D=k_c\left(\text{unit in}\frac{12\,EK}{h^2}\right)$$ when the top and bottom of the column are fixed.

$$D=ak_c\left(\text{unit in}\frac{12\,EK}{h^2}\right)$$ in general.

The value "a" depends on the degree of restriction against joint rotation imposed by girders which are framed into the column at the top and bottom, and "a" takes the value ranged as $0<a<1$, given by a simple formula. The location of inflection point, which is required to determine the moment distribution of the column, is given by the table based on the girder to column stiffness ratio, number of stories, location of the story, variation of the height, distribution pattern of horizontal forces, etc.

The shear distribution coefficient of shear walls can be determind by the following equation.

$$D=\frac{Q}{\delta}$$

Where Q is an assumed shear and δ is a relative displacement caused by Q, and it is desirable that the vertical distribution of assumed shear forces is similar to that of actual ones. In determining δ, displacements due to shear, bending, foundation rotation and foundation movement must be considered. And futhermore, the influence of openings, boundary effects of beams framing into the wall, and the effect of plastic deformation by the occurence of shear cracks in the wall must be considered if necessary.

b) Modified Portal Method

This method is applicable with small error to a nearly uniform open frame. The assumptions are as follows :

i) In each story, shear forces are equally distributed to interior columns, and those in exterior columns are 50% of those in interior columns, except that in the lowest story, shear forces in exterior columns are 80% of those in interior columns.

ii) The inflection point is assumed to be located at middle point of the column. But in the lowest story, $0.6\,h$ from the bottom (h is the column height).

iii) Axial forces of columns due to horizontal forces appear only in exterior columns and no axial force in the interior columns.

6.4 Notes on Structural Materials and Members

6.4.1 Concrete

Concrete used for reinforced concrete structures is usually classified into ordinary or normal weight concrete and light weight concrete. Concrete aggregates for normal weight concrete are river sands, river gravel or crushed stones, while those for light weight concrete are natural or artificial light weight aggregates which have been used recently. Any strength of concrete may be chosen for structural design. However, the standard design strength, F_c, should not be less than 135 kg/cm^2 for normal weight concrete and 120 kg/cm^2 for light weight concrete (AIJ Structural Standards).

In the case of long-spanned or tall reinforced concrete structures, these minimum strengths are inadequate and higher standard design strength is necessary. In such cases concrete strength, F_c, is required to be more than 225 kg/cm², and if proportioning of concrete design and execution of the work are properly made, this strength can easily be obtained not only for normal weight concrete but also for light weight concrete. When relatively many reinforcing bars are used in small sections, concrete with sufficiently good workability should be used. According to JASS 5, the range of required concrete slump is shown in the Table 6.1. It is desirable to use surface active agents such as air entraining admixtures in order to keep the required strength, durability, and workability, this being indispensable in the case of light weight concrete.

Table 6.1 Standard Range of Required Slump

Location	Slump (cm)	
	Place concrete by hand	Place concrete by vibrator
Foundation, slab, beam, and girder	15~18	5~10
Column and wall	18~21	10~15

To construct a rigid frame, simultaneous concrete placing must be made from the bottom to the top of a column including slab and beam in a body. In case of exterior wall girders as one of antiseismic members, concrete at the portion above the floor level is preferably cast at the same time. On the other hand, high slump concrete may produce bleeding and hollows under top reinforcements of beams or slabs, and also may produce cracks in the concrete which covers top reinforcements, and as the result, bond strength will be reduced. So, it is desirable to consolidate and re-tamp concrete before it hardens.

6.4.2 Reinforcing Bars

a) Quality

Reinforcing bars shall be steel bars prescribed in JIS G 3112 as "Steel Bars for Reinforced Concrete", or JIS G 3111 as "Rerolled Carbon Steel". These standards show 5 grade bars of the minimum yield stress, 24, 30, 35, 40 and 50 kg/mm². Reinforcing bars having suitable strength can be chosen according to design requirements. In the case that the yielding stress required is to be more than 30 kg/mm², it is desirable to use deformed bars having superiority in bond and anchorage. Table 6.2 shows the size and nomenclature of plain and deformed bars.

b) Hooks

Every plain bar shall terminate with a hook. This is a device to prevent plain bars from

Table 6.2 Size of Plain and Deformed Bars

Plain bar	Nomenclature	6 ϕ	9 ϕ	13 ϕ	16 ϕ	19 ϕ	22 ϕ	25 ϕ	28 ϕ	32 ϕ			
	Diameter of bars (mm)	6	9	13	16	19	22	25	28	32			
Deformed bar	Nomenclature	D6	D10	D13	D16	D19	D22	D25	D28	D32	D35	D38	D41
	Nominal diameter of bars (mm)	6.35	9.53	12.7	15.9	19.1	22.2	25.4	28.6	31.8	34.9	38.1	41.3

slipping, as plain bars have relatively poor bond strength to concrete. In ordinary cases, deformed bars are required to have no hooks. However, as they are inclined to break covering concrete by drawing forces, hooks are required in some places such as corner bars in beams and columns with thin concrete covering in two directions and bars in a reinforced concrete chimney shaft in which concrete easily suffers deterioration by fire. It is also desirable for deformed bars to have hooks at the ends of bars where sufficient anchorage can not be expected because of the size limitation of members, such as free end of cantilever loaded at that end. Stirrups and hoops shall also terminate with hooks. Fig. 6.20 shows the dimensions of hooks. At the ends of reinforcing bars of 16 ϕ and above shall be made a semi-circular hook, and that of 13 ϕ and below may be made 90° or 135° hook, as bars of small diameter have better bond strength than those of large diameter. The internal diameter of a hook should be more than three bar diameters (3 d) to avoid crushing of concrete inside it and it should be larger when bars with higher strength or light weight concrete are used.

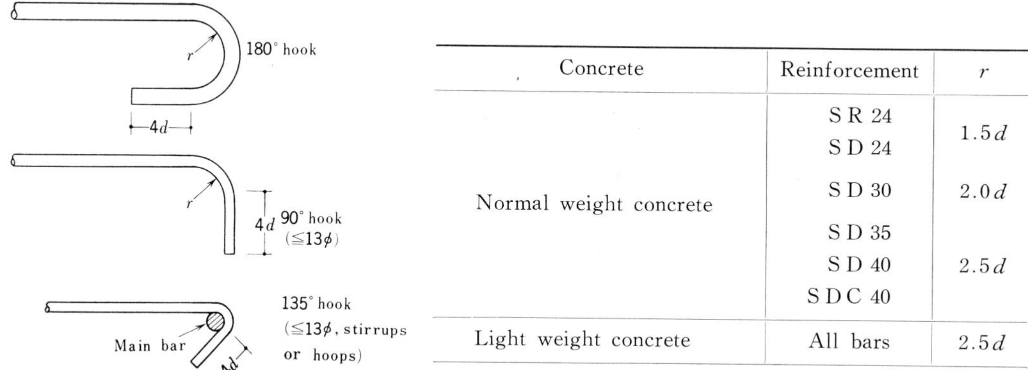

Concrete	Reinforcement	r
Normal weight concrete	SR 24 SD 24	1.5d
	SD 30	2.0d
	SD 35 SD 40 SDC 40	2.5d
Light weight concrete	All bars	2.5d

Fig. 6.20 Hooks

c) Bent-up Bars

As considerable stresses are produced in the bent-up part of a bar, this part should not be bent sharply.

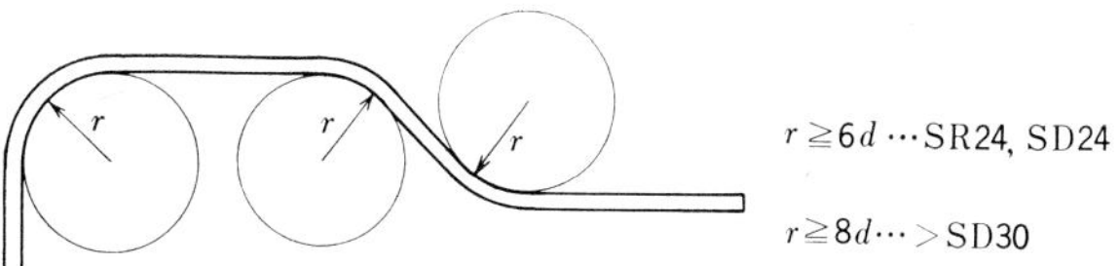

Fig. 6.21 Bend in the Middle of Reinforcing Bars

d) Splices

The method of splices of reinforcing bars are classified into lapped splices and welding splices. The details of these methods are illustrated in Fig. 6.22. The location of a splice should be, as a rule, selected where the stress of member is small, and the position of splices in a member

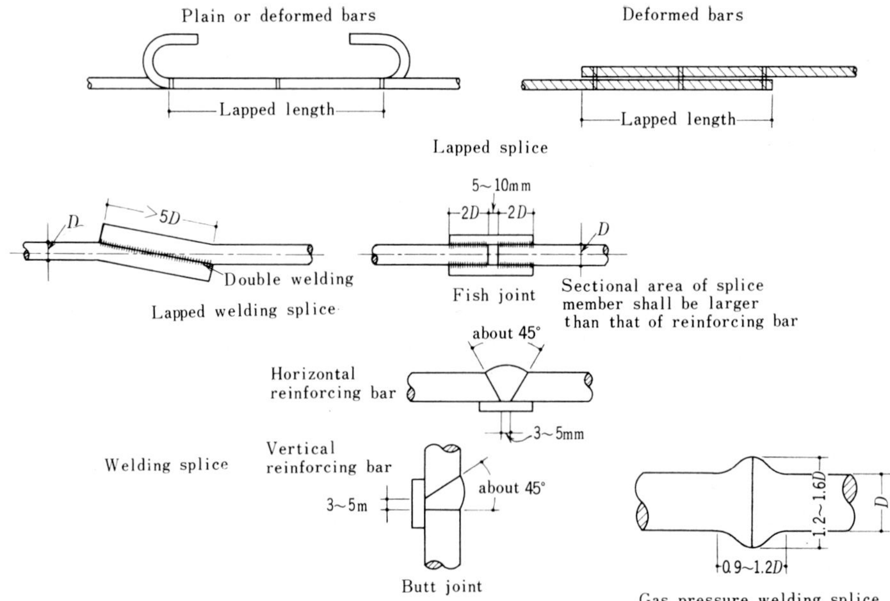

Fig. 6.22 Splices of Reinforcing Bars

should not meet in the same section but should be staggered as well. The lapped length is preferably determined by calculation so that the existing forces in the lapped bars can be transmitted to each other at the bond stresses less than their allowable values. Without this calculation, it is advisable to use those lengths of joint shown in the Fig. 6.23. However, small lapped length is indicated from the calculation, the lengths S_2 shown in the Table 6.3 should be chosen as the minimum length, considering the alteration of stresses and the imperfection of execution.

e) Anchorage

As the tensile stress of a bar which is anchored from a beam into a column shows almost as much as the allowable stress, enough anchorage length should be chosen. It is preferable to determine the length by calculation so that the maximum existing forces of the bar can be transmitted to surrounding concrete at the bond stress less than their allowable values. Without this calculation, it is advisable to use those length shown in the Fig. 6.23. However small anchorage length may be indicated for tensile reinforcement from the calculation, the length S_2 shown in the Table 6.3 should be chosen as a minimum length, considering the alteration of stresses and the imperfection of execution.

f) Clearance between Bars

Clearance between parallel bars shall be, in principle, more than 2.5 cm, 1.25 times the diameter of coarse aggregate and 1.5 times the bar diameter, so as to properly place concrete.

g) Concrete Cover for Reinforcements

Though it is desirable to have thick coverage to keep good durability and fire-resistance of a member and good bond property, thick coverage makes the member heavy so thickness of coverage should be chosen properly. The minimum value of the thickness is prescribed as shown in the Table 6.4 (a) and 6.4 (b) according to the kinds of concrete, quality, existence of finishing

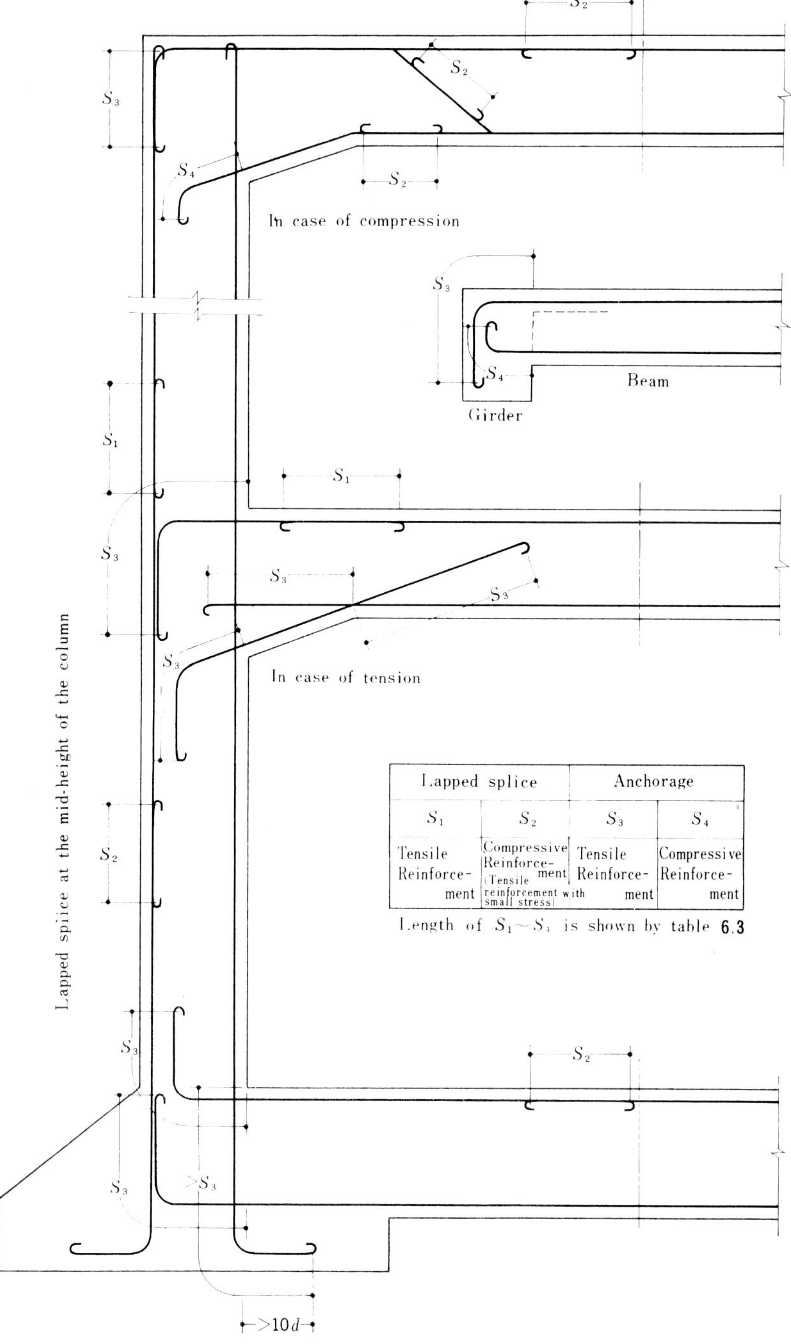

Lapped splice		Anchorage	
S_1	S_2	S_3	S_4
Tensile Reinforcement	Compressive Reinforcement (Tensile reinforcement with small stress)	Tensile Reinforcement	Compressive Reinforcement

Length of $S_1 \sim S_4$ is shown by table 6.3

Fig. 6.23 Location and Length of Splices or Anchorages of Reinforcing Bars

Table 6.3 Length of S_1, S_2, S_3, S_4

Concrete	Reinforcements		Hook	Lapped splice S_1 Tension	Lapped splice S_2 Compression	Anchorage S_3 Tension	Anchorage S_4 Compression
Normal weight concrete	Round	SR 24 SR 30	○	$40d$	$25d$	$40d$	$30d$
	Deformed	SD 24	○	$30d$	$15d$	$30d$	$20d$
			×	$45d$	$25d$	$45d$	$30d$
		SD 30	○	$35d$	$20d$	$35d$	$20d$
			×	$50d$	$30d$	$50d$	$30d$
		SD 35 SD 40 SDC 40	○	$40d$	$25d$	$40d$	$20d$
			×	$60d$	$35d$	$60d$	$30d$
Light weight concrete	Round	SR 24 SR 30	○	$50d$	$30d$	$50d$	$35d$
	Deformed	SD 24	○	$30d$	$20d$	$30d$	$25d$
			×	$45d$	$30d$	$45d$	$35d$
		SD 30	○	$35d$	$25d$	$35d$	$25d$
			×	$50d$	$35d$	$50d$	$35d$
		SD 35 SD 40 SDC 40	○	$40d$	$30d$	$40d$	$25d$
			×	$60d$	$40d$	$60d$	$35d$

Note
d : Nominal diameter of reinforcements

Table 6.4 (a) Minimum amount of Coverage for Reinforcement

Parts of Structures			Thickness of coverage (cm)
Parts not in contact with soil	Floors and walls other than bearing walls	Finished	2
		Unfinished	3
	Bearing walls, columns, beams	Interior, exterior and finished	3
		Exterior and unfinished	4
Parts contacting soil		Walls, columns, beams, floors	4
		Foundations and retaining walls (exclusive of subslab concrete)	6

Note :
1. This table shows the values in case of using Portland cement, Type A blast furnace slag cement, Type A silica cement, Type A fly ash cement.
2. When using Type B blast furnace slag cement, Type B silica cement or Type B fly ash cement, the minimum thicknesses of coverage indicated in Table 6.4 (a) shall be increased by 1 cm each. However, when using surface active agents with these cements, the above increases will not be necessary.
3. When using Type C blast furnace slag cement, Type C silica cement or Type C fly ash cement for parts contacting soil, surface active agents shall be used and the minimum thicknesses of coverage indicated in Table 6.4 (a) shall be increased 1 cm each.
4. For retaining walls not in contact with soil thickness of coverage shall not be less than 4 cm.

Table 6.4 (b) Minimum Values of Thickness of Coverage for Reinforcement in Lightweight Concrete (cm)

Type and part of structure \ Type of cement			Portland cement, Blast furnace slag cement Type A, Fly ash cement Type A, Silica cement Type A	Blast furnace slag cement Type B, Fly ash cement Type B, Silica cement Type B	Blast furnace slag cement Type C, Fly ash cement Type C, Silica cement Type C
Not in direct contact with soil	Floors, walls other than bearing walls	Finished	2	3	—
		Unfinished	3	4	—
	Column, beam, bearing wall	Interior and finished	3	4	—
		Exterior or interior and unfinished	4	6*	—
In contact with soil	Wall, column, beam, floor		5	6	7
	Foundation, retaining wall (excluding sub-slab concrete)		7	8	9

Note :
* 5 cm in case of class 1 and 2 lightweight concrete.

materials and importance of the member.

6.4.3 Columns

a) Design of Sections

Large amount of shear forces and end moments other than axial forces are produced in the columns of a rigid frame subjected to lateral forces. This must be considered in the design of section. The size of columns in a reinforced concrete antiseismic structures is determind mostly by stresses caused by seismic forces. In ordinary case, bars are arranged symmetrically in the concrete sections considering that alternate bending moment is produced due to seismic forces. Tensile stresses are often expected to occur in main bars of columns during an earthquake. It is quite difficult to show common sizes of columns, but the following method is commonly used in practical design. As for the column of the lowest story, the size of concrete section of a column is determined so that the value of stresses expressed as (axial force in the column)/(sections area of the column) should not exceed allowable compressive unit stress of concrete for permanent loading, and as for upper stories the depth of column is reduced by around 5 cm for every ascending two stories.

b) Main Bars

The amount of main bars is determind by the method mentioned above, but is also regulated by the following limitations :

i) Diameter of main bars shall be not less than 13 mm and their number shall not be less than 4.

ii) The ratio of total sectional area of main bars to that of concrete (total sectional area of main bars/total sectional area of concrete) shall not be less than 0.8% and 1.0% in case of using normal weight concrete and light weight concrete, respectively. When high strength reinforcements or high strength lightweight concrete is used, the above limitations may be reduced appropriately.

c) Hoops

As hoops have the role of resisting shearing forces, to prevent the main bars from buckling and to increase ductility of columns, spacing of hoops shall be close and be arranged to wrap all main bars. In the case of columns with large sectional area, hoops should be arranged not only outside but also inside of main bars so as to prevent them from buckling (Fig. 6.24).

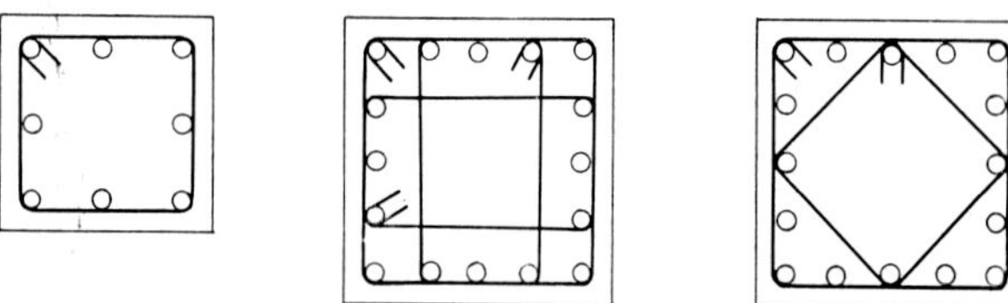

Fig. 6.24 Arrangement of Hoops

In ordinary cases, bars used for hoops should not be less than 9 mm in diameter and spacing of hoops should be chosen as shown in the Table 6.5. But at the top and bottom portion of a column in the range of its maximum width if possible, in the range of 1/4 times column length, spacing of hoops should be reduced 1/2 times the values shown in the Table.

Table 6.5 Spacing of Hoops

	Spacing of hoops
Normal weight reinforced concrete	Not more than 30 cm, minimum depth of the column, nor 15 times the diameter of main reinforcement
Light weight reinforced concrete	Not more than 25 cm, 3/4 of the minimum depth of the column, nor 15 times the diameter of main reinforcement

As the beam reinforcements are often anchored at the joint of beams and columns, hoops in a joint should be arranged with a close spacing to strengthen the joint. When the size of columns changes from below to above the joint, the main bars of the column should be bent

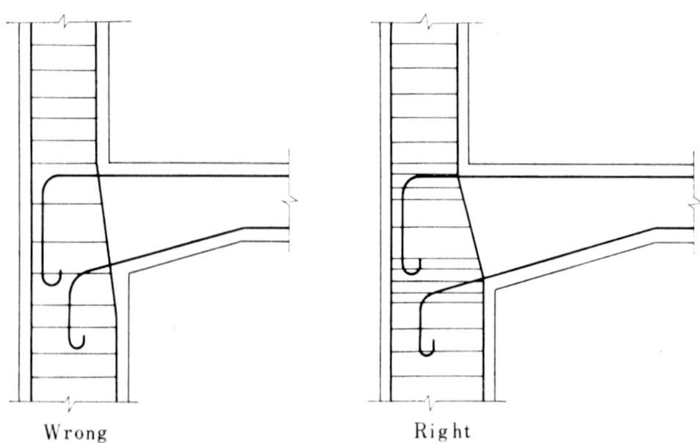

Fig. 6.25

within the joint, and enough hoops should be arranged in this portion (Fig. 6.25).

d) Long Columns

As for a slender column, the influence of buckling should be considered. The section is determined by increasing axial forces and bending moments according to the slenderness of the column, as shown in Table 6.6.

Table 6.6 Increasing Factor of Bending Moments and Axial Forces for Long Columns

Ratio of minimum depth or diameter to the distance between main supporting points	Increasing factor	
	Normal weight reinforced concrete	Light weight reinforced concrete
1/10	1.00	1.0
1/15	1.00	1.2
1/20	1.25	1.5
1/25	1.75	—

6.4.4 Beams and Girders

Horizontal forces produce both positive and negative bending moments at the ends of a beam, and even when combined with bending moments due to vertical loads, in most cases both positive and negative bending moments occur at the ends and even near the middle of the beam (Fig. 6. 26). For this reason, the principal beams should be doubly reinforced over the entire span. Tie-beam sections are mainly determined by positive and negative bending moments subjected to horizontal forces and in ordinary cases tie-beams have the same arrangement for the top and bottom reinforcements.

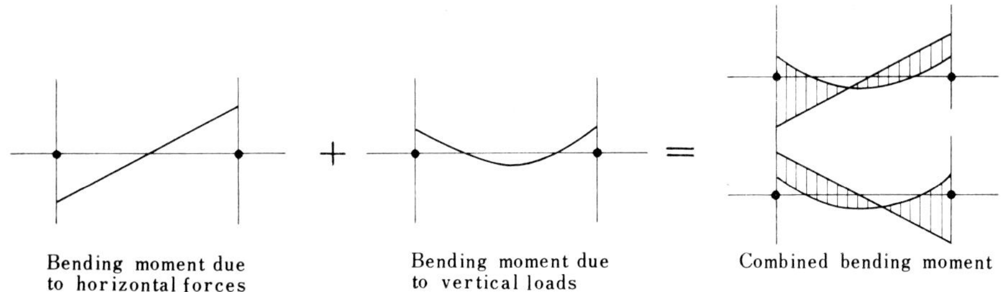

Fig. 6.26 Bending Moment in Girders

The approximate value of a beam size is as follows :

Depth of a beam or girder

around 1/12 times span length for beams

around 1/10～1/8 times span length for girders

Width of a beam or girder

around 25～40 cm

As for wall girders, the width is around 20～30 cm, and the bars are doubly reinforced and arranged in 2～3 layers at the top and bottom.

When the above mentioned depth is not enough, haunches are made at the ends of the beam. The amount of main bars in beams is determined in accordance with the ordinary design principles considering that the critical sections are at both ends and center of beam and at the beginning of haunch if it exists. It is desirable for beams to be regulated by the following limitations.

i) Diameter of main reinforcing bars should not be less than 13 mm.

ii) The principal beams should be of double reinforcement over the entire span, and double reinforcement ratio should not be less than 0.4 in reinforced light-weight concrete beams.

iii) The tensile reinforcement ratio should not be less than 0.4% in beams where maximum positive and negative bending moments occur.

iv) Arrangement of main bars should not exceed two layers, except in special cases.

Stirrups are arranged in beams to resist shearing forces. The amount of stirrups is determined by calculation for shearing force, and also should be regulated by the limitations shown in Table 6.7. The stirrups should not be less than 9 mm in diameter, and should be arranged to wrap both compressive and tensile main reinforcements, and should be bent not less than 135 deg. at the ends (Fig. 6.27).

Table 6.7 Maximum Spacing of Stirrups

	Shearing stress (τ)	Maximun spacing of stirrups	Remarks
Normal weight reinforced concrete	$\tau >$ allowable unit shearing stress of concrete	2/3 D or 30 cm	When diameter of stirrups is greater than 13 mm, the maximum spacing 30 cm and 25 cm may be changed to 45 cm.
	$\tau \leq$ allowable unit shearing stress of concrete	3/4 D or 30 cm	
Light weight reinforced concrete	Independent of τ	2/3 D or 25 cm	

Note :
D : Over-all depth of beams

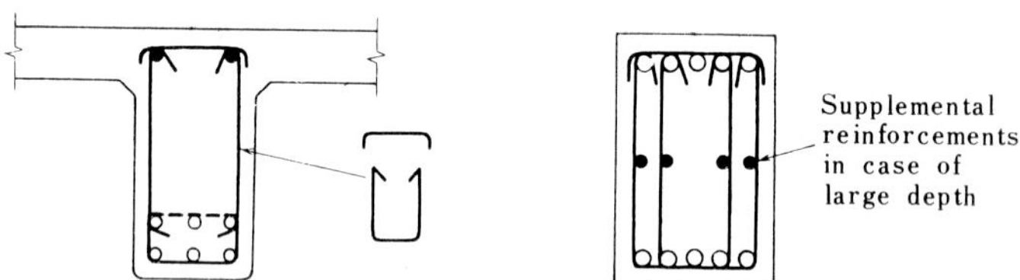

Fig. 6.27 Arrangment of Stirrups

There is a limiting value in the shear resistance ability (about 0.6% in shear reinforcement ratio), even though many stirrups are arranged. So, when unit shearing stress for permanent stress exceeds 1/12 F_c or the unit shearing stress for temporary stress exceeds 1/8 F_c, the size of beam sections should be increased.

It is efficient to use bent-up bars for shear reinforcement. But bent-up bars resist shear force in one direction and are of no use in the reverse direction, because cracks along the bars are easily developed. Therefore, bent-up bars are required to be used together with stirrups. Bent-up bars should be bent gently as shown in Fig. 6.21.

The bottom bars in haunch subjected to compressive stresses shall be bent along the concrete surface with closely spaced stirrups, but those subjected to tensile stresses (this happens often in the middle and lower stories during an earthquake) shall not be bent because the tensioned bent bars sometimes break the concrete and lead to serious damage. Therefore, it is important to use two straight bars instead of a bent bar and to anchor them separately. In the case of bent beams, the same considerations are required in the bent-up part (Fig. 6.28).

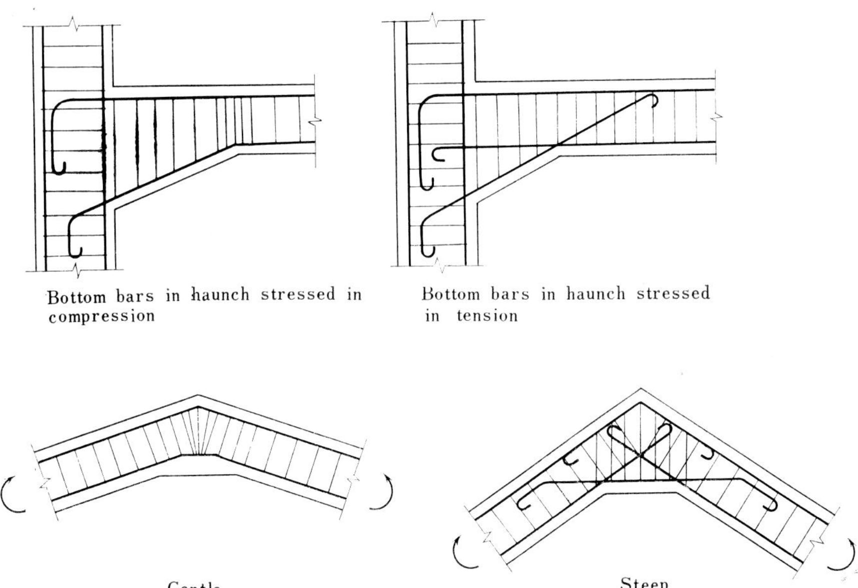

Fig. 6.28 Arrangement of Reinforcements for Haunched Beam and Folded Beam

In general, holes in the web of a beam should not be made for plumbing purposes. But when it is inevitable to make holes, following considerations are required;

i) The holes do not have great influence on bending moments. But for shearing forces, the strength and rigidity of a beam with holes is reduced considerably. So it is better to make holes in the middle part of the web when necessary.

ii) As for bending moments, it is better not to make holes in the compressive side of a beam, but if it is inevitable to make holes, compressive side should be properly reinforced.

iii) For shear force, reinforcements are required around the hole, especially as shear force concentrates on the remaining parts of the web in the upper and lower side of the hole and sufficient reinforcement is required in those parts. Sufficient anchorage length is also required for hole reinforcement bars.

iv) Diameter of the hole less than 1/3 times the depth of the beam is desirable and spacing of holes more than 3 diameters of the hole when several holes are perforated in a line.

6.4.5 Beam-column Connection

In the beam-column connection, a large shear force occurs during an earthquake and as a result large shear and bond stresses are produced in concrete and reinforcements (Fig. 6.29).

On the other hand, both diagonal cracking stress and shear strength in this part are relatively

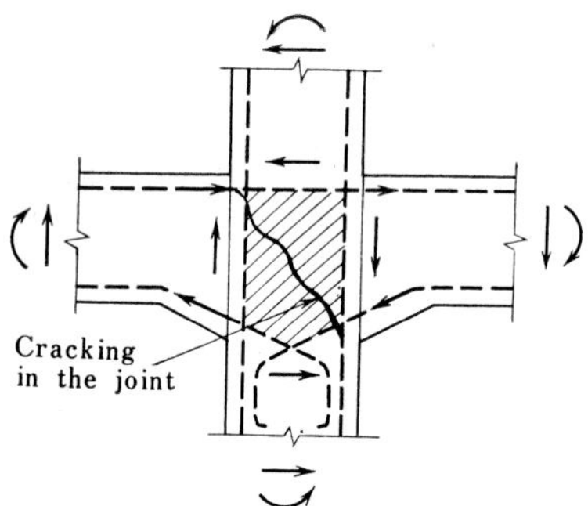

Fig. 6.29 Beam-column Connection

high, around two times as strong as in ordinary beams. This part should be designed considering the above mentioned facts. Vertical haunches or horizontal haunches may be arranged if necessary, and joint part should also be reinforced by stirrups as mentioned in 6.4.2. As considerably large bond stresses occur in bars of this part, main bars of beams should be anchored separately in the column, or anchored into the opposite side of the beam through the joint part. Or bond stresses are calculated as straightly arranged bars, and if bond strength is not enough, then the end of the beam should be designed as single reinforcement neglecting compressive reinforcements. In the case of deformed bars, anchorage strength of joint part is relatively large, so the depth of column is enough for anchorage length even if it is straightly arranged.

6.4.6 Quake-resisting Walls

Quake-resisting walls or shear walls are required to be thick enough so as to keep shear stresses as small as possible. Shear walls are desirable to be constructed monolithically with surrounding rigid beams and columns. If shear walls are surrounded with rigid frames, considerable strength (shear stress = about 1/8 F_c) can be expected even though the concrete wall is not reinforced. On the other hand shear walls are often subjected to larger amount of shear forces during an earthquake than expected in the design, and often cracks occur. So walls should be reinforced sufficiently so as to obtain ductility against horizontal forces and to prevent wide open cracks. Generally, shear stresses of a wall are calculated assuming that shear forces distribute uniformly in a horizontal section of the wall, and if opening exists, the values are increased according to the opening dimension. In ordinary cases, the reinforcements should be arranged according to the limitations listed in Table 6.8, depending on the shear stress that may be larger or smaller than the temporary allowable shear stress.

In case the size of surrounding columns and beams is large (sectional area of columns and

Table 6.8 Structural Limitations of Shear Wall

i) Thickness of wall shall not be less than 12 cm.

ii) In case of the thickness of wall is more than 20 cm, shear reinforcing bars is provided in double shear reinforcement.

iii) Diameter of shear reinforcing bars shall not be less than 9 mm, and their spacing shall be in accordance with the following table.

iv) Diameter of reinforcing bars in the perimeter of openings shall not be less than 13 mm.

	Normal reinforced concrete	Light weight reinforced concrete
Spacing of bars	Not exceed 30 cm. When double shear reinforcement is provided in zigzag arrangement, spacing of bars shall not exceed 45 cm in each surface of the wall.	Not exceed 25 cm. When double shear reinforcement is provided in zigzag arrangement, spacing of bars shall not exceed 40 cm in each surface of the wall.

beams is not less than $xt/2$ and minimum depth or width of columns and beam is not less than $\sqrt{xt/3}$ and not less than $2t$, where x=smaller one of clear span and clear height of the wall, t=thickness of the wall), the reinforcements are allowed to be arranged accordingto Table 6.8, even if shear stress of wall exceeds temporary allowable unit shear stress. When the unit shear stress of concrete wall exceeds $1/8\ F_c$ the concrete section must be changed. There are three ways of arranging reinforcements in the wall, vertical and horizontal arrangement (Fig. 6.30 (a)), 45° diagonal arrangement (Fig. 6.30 (b)), and combination of these two arrangements (Fig. 6.30 (c)). Arrangement pattern shown in Fig. 6.30 (b) is the most desirable for resisting shear force, because rigidity and strength are high after cracks occur, but construction is difficult. In general, the arrangement patterns shown in Fig. 6.30 (a) and 6.30 (c) are adopted. In both cases, the wall reinforcements should be anchored completely in columns and beams.

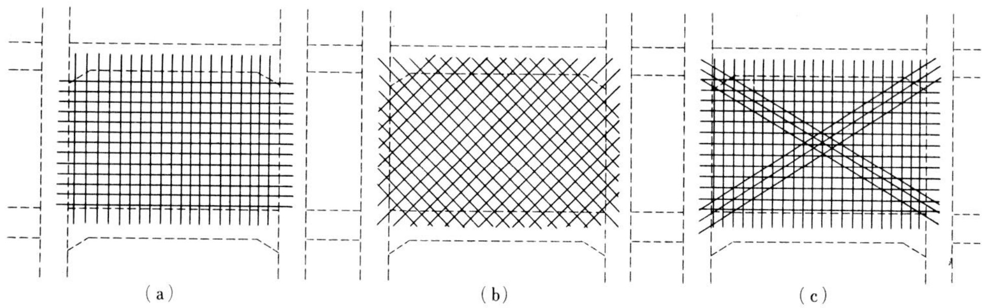

Fig. 6.30 Arrangement of Reinforcements in Shear Walls

When an opening exists in the wall, diagonal tensile force which would occur in the opening part of the wall, concentrates at the corners of opening. Therefore, sufficient diagonal reinforcements should be arranged at these corners. Tensile reinforcements should also be arranged along the perimeter of opening because this portion is subjected to tensile stresses due to bending moment caused by shear force. It is required for these reinforcements to have sufficient anchorage length from corners of opening. Columns and beams surrounding the wall should be reinforced to be completely safe from axial forces, bending moments, and shear forces. Especially columns are required to be reinforced sufficiently against axial forces which are produced due to bending moments in the cantilever action of the shear wall in addition to those due to vertical loads. As

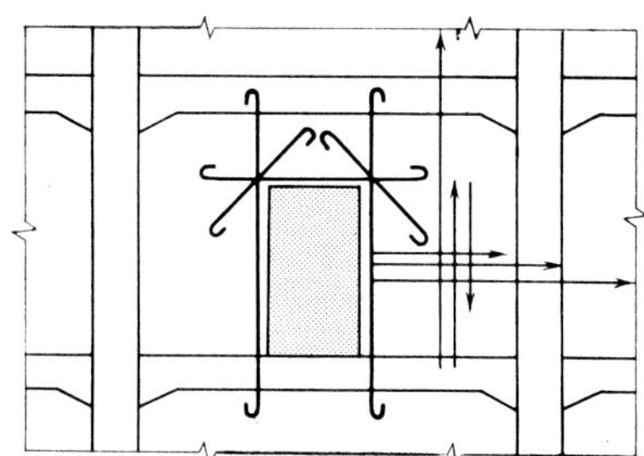

Fig. 6.31 Arrangement of Reinforcements Around an Opening in Shear Walls

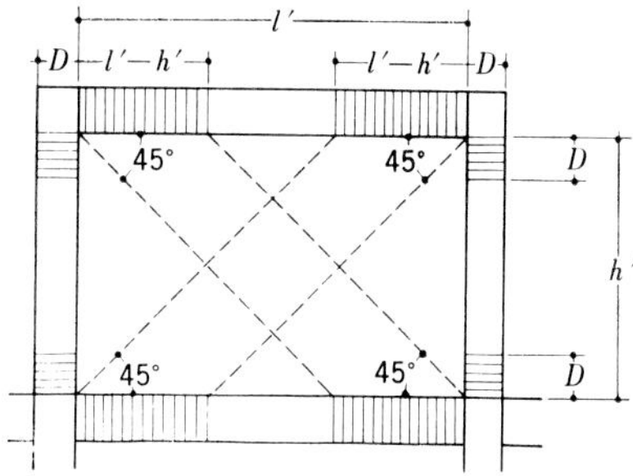

Fig. 6.32 Portion of Beams and Columns where Web Reinforcement is Efficient

beams and columns prevent development of cracks in the wall, it is quite efficient to provide web reinforcements at the ends of these members as shown in Fig. 6.32.

6.4.7 Floor Slabs

As the floor slabs are subjected to bending stresses by vertical loads and to horizontal shear stresses by earthquake forces, plane stresses as well as plate stresses should be considered. In case of ordinary floor slabs, slabs are often safe against shear force by earthquake motions when slabs are designed to be safe against vertical loads. But such parts of floor slabs as where shear forces concentrate on it, for example, floor slab through which large shear forces are transmitted and the slab near the openings such as a staircase, slab should be treated in the same way as a shear wall and should be made thick and reinforced sufficiently. Sizes of slab sections should be determined by computation for the design and by limitations shown in Table 6.9.

Table 6.9 Structural Limitation of Slab

i) The thickness of slab shall be as shown below in the table.

ii) The tensile slab reinforcements shall be not less than 9 mm in diameter, the spacing of these reinforcements subjected to maximum positive or negative bending moments shall be in accordance with the following table.

		Normal weight reinforced concrete	Light weight reinforced concrete
Thickness of slab		Not less than 8 cm, nor less than 1/4 of the effective short span length	Not less than 10 cm, nor less than 1/40 of the effective short span length
Spacing of reinforcing bars	Short span	Not more than 20 cm	Not more than 20 cm
	Long span	Not more than 30 cm, nor more than three times the thickness of the slab	Not more than 25 cm

Note : The effective span length is the inside dimension between supporting beams or other supporting members.

At the corners of the slab, the diagonally arranged reinforcing bars are effective for shear forces in plane and for vertical loads. Cracks in the concrete slab due to shrinkage occur easily and these cracks reduce the rigidity of slabs in the horizontal plane considerably. In order to prevent these shrinkage cracks, proper amount of reinforcements should be provided even at places where the bending moment is not large.

The ratio of reinforcements of this kind shall be not less than 0.25% for plain bars and 0.20% for deformed bars. As shrinkage cracks are apt to occur at the joint of concrete placing in the slab, sufficient dowel reinforcements are required to be provided. During the placing of slab concrete, reinforcements are required to be kept their exact position. Top reinforcements especially those at the fixed end of cantilever slabs should be set exactly at the same position as indicated in the drawings. For this purpose, it is better to use suitable chair and to use steel bars with large diameter such as 13 ϕ and 16 ϕ at important points. When a span is too large, deflections caused by vertical loads, especially by concrete creep, are conspicuous, so it is desirable to divide a large panel into suitable small panels by using beams.

6.4.8 Footings and Foundations

a) Footings and Foundations of Reinforced Concrete Structures

In general cases, the dimensions of footing slabs or foundation slabs are determined according to the stresses by vertical loads, but additional axial forces or bending moments transmitted through the column during an earthquake should also be considered. Generally tie beams of footings resist most of the bending moments from the bottom of columns, so in designing sections of footing slabs or foundation slabs, the stresses other than axial forces transmitted through columns need not be examined. But in designing a footing slab of chimney or of a cantilever column, bending moments from the bottom of the structure should also be considered. Footing slabs and foundation slabs are generally designed as slabs which are subjected to concentrated downward axial load through columns and to the soil pressure or concentrated force of piles upwards as the reactive force.

If the top of piles (reinforced concrete piles or steel pipe piles) are sufficiently anchored in to the footing slab or in the foundation slab, pulling forces of columns are transmitted to the ground through footing slabs, foundation slabs and piles. Main bars of footing slabs or foundation slabs are generally arranged crosswise at the bottom of the slab, but in case when bending moments are transmitted through the slab or in case when pulling forces act on the column, additional reinforcements are required to be provided as shown in Fig. 6.33 because tensile stress occurs at the top of the slab.

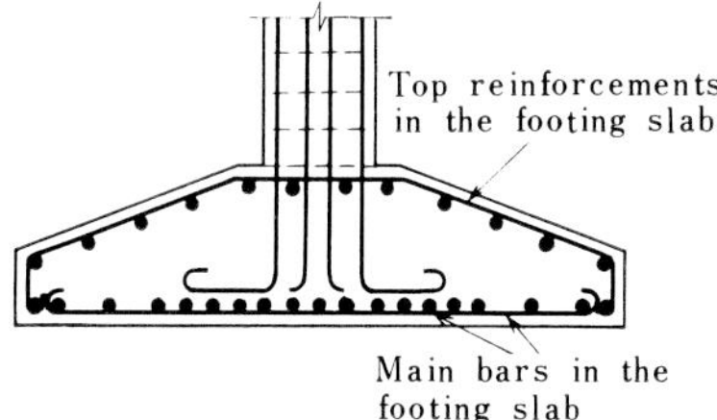

Fig. 6.33 Isolated Footing

b) Footings and Foundations of the Other Structures

Footing slabs and foundation slabs not only of reinforced concrete structures but also of other types of structures are usually made of reinforced concrete because of its long term durability. The design of footing slabs and foundation slabs in case of other type of superstructures should be treated in the same way as that for case of reinforced concrete superstructures and furthermore, the superstructure and the slabs should be connected tightly by anchor bolts so as not to slide or become disjointed each from the other due to shear forces and withdrawal forces transmitted through the superstructure. At the end of anchor bolts, hooks should be made and anchorage length of bolts should be determined as in the case of tensile reinforcing bars. In case when large pulling forces occur in the anchor bolts, anchoring beam and the like should be made at the end. Bolts should be held with double nuts so as not to be loosened

(Fig. 6.34).

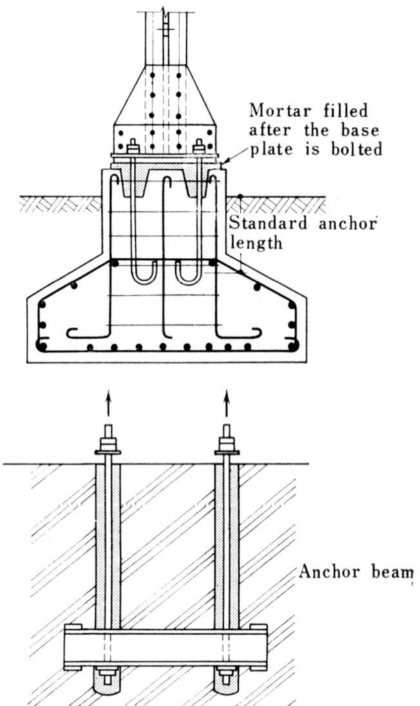

Fig. 6.34 Isolated Footing of Steel Structures

7. STEEL REINFORCED CONCRETE STRUCTURES

7.1 Outline

7.1.1 General

In steel reinforced concrete structures, longitudinal reinforcing bars are placed around the steel skeleton framing and concrete is poured so that they may function together.

The origin of this type of structure is rather new. In the Kanto Earthquake of 1923, a building which had been strengthened by reinforced concrete with American-style rolled H-sections was not severely damaged. This experience led to the origin of this type of structure. Since then the style of steel skeletons consisted of steel angle flanges and latticed webs considering web action of concrete. It is ten years since welded I-section with solid web began to be used, due to the development of better welding techniques. Presently, rolled H-sections are being used. It seems that the style is going back to the beginning. Now various types are used in practice, such as steel concrete with stout skeletons or reinforced concrete with slender skeletons. Thus, steel reinforced concrete structures are many. The common features are to provide longitudinal reinforcements and stirrups and hoops around the steel skeletons and arrange the steel so that concrete may be easily poured. So, in planning of steel reinforced concrete buildings, advantages and disadvantages must be studied well beforehand because it can be freely determined by the designer in order to make a consistent structural plan.

7.1.2 Special Features

In steel reinforced concrete structures, steel and concrete are used so this type of structure possesses both durability and fire-proofing qualities and mechanically, it possesses continuity of reinforced concrete and the toughness of steel. They improve the deficiencies of each material and this is the special feature of this type of structure. In other words, it possesses rigidity and strength as reinforced concrete does, but it furthermore possesses toughness against destruction by its ductility. Namely, it has a much greater capacity for absorbing energy before destruction than the usual reinforced concrete. This is an advantageous feature against earthquakes. This results from the fact that steel skeletons provide rigid steel frames.

Minoru Hamada (1929) and Shuzo Takada (1957) studied the ductility of steel reinforced concrete beams to destruction, and they were tested to shear failure. These studies showed that in the case of shear failure by load from one side, steel reinforced concrete structures can still sustain the load when the deformation of beams was about ten times the final deformation of reinforced concrete with the same quantity of steel as the steel reinforced concrete. Recently we have the experimental studies by Hajime Umemura and Seiji Kokusho. These studies include the tests of steel reinforced concrete columns subjected to compressive stresses of 100 kg/cm^2 or 40 kg/cm^2 (in regard to concrete section) and the alternate lateral forces. As the result, in the case of the column of 100 kg/cm^2 there was not any difference between steel reinforced concrete and reinforced concrete. But in the case of the column of 40 kg/cm^2, although reinforced concrete could not carry alternate forces at the deformation two times as much as at yielding, steel reinforced concrete could sustain the load carrying capacity at the deformation eight times as much at yielding. It is thought that the usual column of continuous frames will be subjected to compressive stress of 40~50 kg/cm^2 when an earthquake occurs, and therefore the latter will be a general feature of steel reinforced concrete. In these tests, as the steel skeletons were of tie plate type (see Fig. 7.4 (a)), much greater ductility would be expected in the case of truss type or solid web sections.

So far, the discussion is based on the feature of steel reinforced concrete that it is similar to reinforced concrete. In the case of heavy steel skeletons, concrete is available to prevent local buckling of steel skeletons and therefore it is considered that concrete is available to increase the ductility of steel structures. In this case as concrete bears axial compression of columns, the thickness of steel skeletons can be reduced.

The defect of steel reinforced concrete is that it is very heavy. Generally, in 10-storied buildings, it weighs from 1.4 to 1.5 t/m^2 of the floor area. Therefore it may be supposed that buildings more than 10-storied can hardly be built with steel reinforced concrete. However, by using light concrete of natural aggregate, its weight may be decreased from 15% to 20%. And by using light-weight concrete of artificial aggregates, the weight may be reduced to 1.0 tons per sq.m.

Considering that concrete in beams has little contribution to the strength and rigidity of frames, steel beams with solid webs may be used without concrete and only steel columns may be cast with concrete. As the result, weight can be considerably decreased, for example to about 0.8 t/m^2. Thus if the weight of the steel reinforced concrete is decreased, it can be used for skyscrapers without difficulties. This is a desirable type of construction with economical merit for construction of such works.

7.2 General Remarks on Structural Planning

7.2.1 General

Steel reinforced concrete is usually used as continuous frames. Slabs are built with reinforced concrete. Shear walls are usually built with reinforced concrete but according to circumstances, they may be of steel braces covered with reinforced concrete. Plans for the arrangement of frames and shear walls, plan for slabs, prevention of movement and differential settlements of foundations are almost the same as that for reinforced concrete structures.

7.2.2 Quantity of Steel Materials

Though the consumed quantity of steel material varies according to the span, height and scale of the building, it is important to know the quantity of steel consumed by the traditional buildings.

The quantity of steel per 3.3 m^2 floor area is shown in Fig. 7.1, divided into four periods; pre-Kanto Earthquake, after the Earthquake, after World War II, and after 1955. It shows steel materials were used increasingly after the earthquake and decreased after the war. Fig. 7.2 shows the ratio of the quantity of steel skeleton to that of the total steel material. It shows that the quantity of steel skeletons decreased and that reinforcing bars increased.

Recently the variation of the quantity of steel for each building is decreased as compared with the quantity in 1950~51 and the average of the quantity is 0.4 t per 3.3 m^2. The quantity of reinforcing bars is a little more than that of steel skeletons. But these reinforcements include a part of pure reinforced concrete such as slabs and walls. As to the main frames, steel skeletons consume more than half of the total.

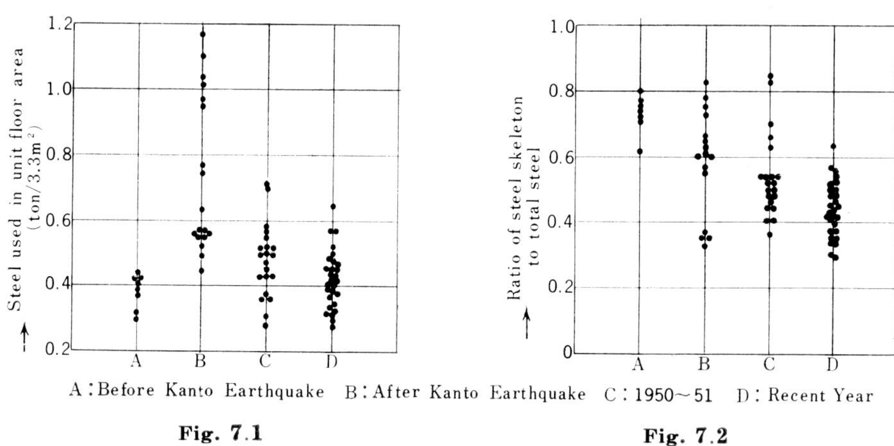

A: Before Kanto Earthquake B: After Kanto Earthquake C: 1950~51 D: Recent Year

Fig. 7.1

Fig. 7.2

7.2.3 Quantity of Steel and Web Type Steel Skeletons

The quantity of steel is an important element which determines the toughness of frames. Steel reinforced concrete of today was created on the basis of Kanto Earthquake experiences. In other words, the concept that steel skeletons are strengthened by reinforced concrete was developed to another concept that the steel skeletons act together with reinforced concrete. So strictly speaking, this is not a structural type that has experienced earthquake disaster. Therefore, in planning the frame, steel skeletons must be strengthened just as the original concept intended.

The quantity of steel has much influence on the state of shear failure of columns and beams which form the framing. The shear failure of main columns causes the complete destruction of frames, as seen in Fig. 7.3, showing that bending or bending failure destroys only a part but shear failure extends over a wide range.

Even when main columns have large shear, if steel skeleton is strong enough, it would make up for the decrease of web action caused by the destruction of concrete. So even in such a serious case, stable load carrying capacity could be expected. So that the frame would be safe.

Shear failure is influenced by the type of web in steel keletons. Tie plate type (Fig. 7.4

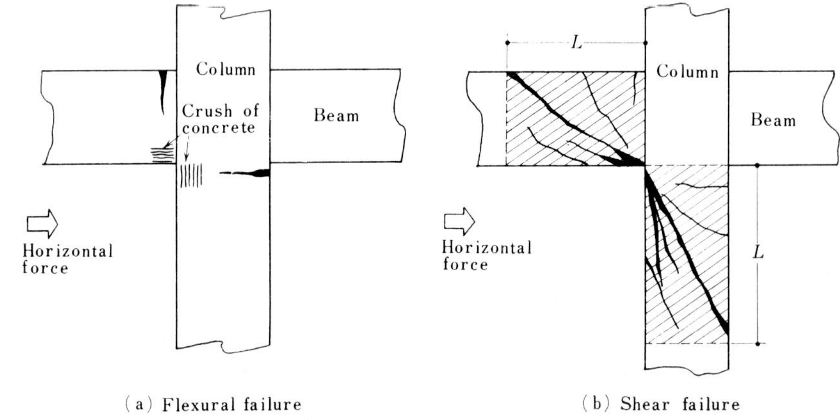

(a) Flexural failure (b) Shear failure

Fig. 7.3

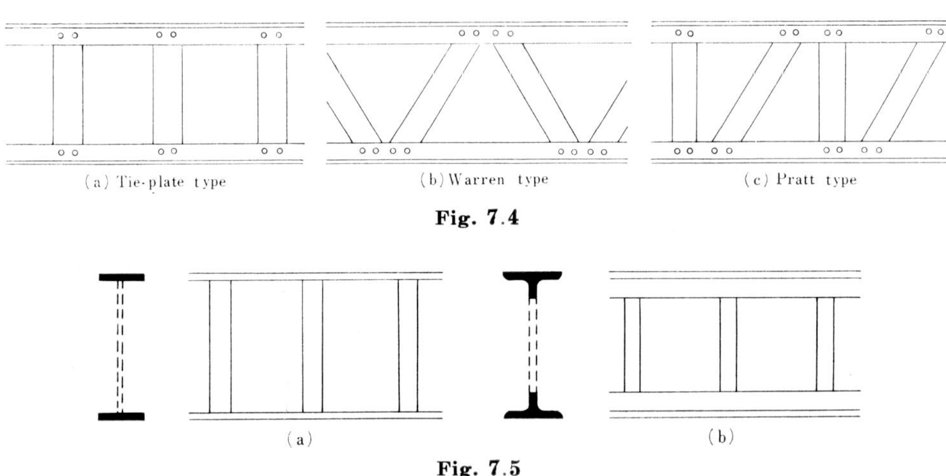

(a) Tie-plate type (b) Warren type (c) Pratt type

Fig. 7.4

(a) (b)

Fig. 7.5

(a)) depends upon the bending rigidity of flanges. Steel plates are not enough, and steel angle flanges or rolled T-section must be used (Fig. 7.5). As the strength of tie plate type against shear is weak, it can only be used when shear is rather small. It is common to use tie plate type only when the shear stress of concrete is less than the allowable shear stress ($1/15\ F_c$).

Besides this type, truss type of web and solid web are used. As the quantity of steel skeleton increases, compared with concrete, the web type must be strengthened. If the web is strengthened, steel skeleton can withstand the shear without depending upon the adhesive stress of concrete so the shear stress of concrete can be decreased. Therefore, the redistribution of steel stress pattern by shear crack or decrease of adhesive stress would have considerably less influence on the structural frames. So the flexural capacity of beams and columns would be increased and at the same time shear failure of concrete would hardly occur.

7.2.4 Notes on Parts of Structures

From the viewpoint of earthquake resistance, beam to column connections, column or beam joints and column bases require special examination. In regard to beam-column connections, the steel joints shall be rigid and stress must be carried smoothly and longitudinal reinforcements must be placed spaceously. Panel zones of connections are subjected to large shear in earthquakes and so this zone must be treated carefully (see paragraph 7.4.2(c)).

Also in column or beam joints, steel stresses shall be carried smoothly. In this location, the section of steel skeleton is often reduced. But on such occasion, gravity center of a section should not be shifted. If it is shifted, enough reinforcement shall be added (see paragraph 7.4.3)

Column bases join the reinforced concrete of lower part and the steel reinforced concrete of the upper part so that the transmission of stress from reinforcing bars of the lower part to the steel skeleton of the upper part must be carefully examined (see paragraph 7.4.4).

7.2.5 Structual Calculations

Steel reinforced concrete structures are calculated by the summation method. The allowable strength of steel reinforced concrete is calculated from the sum of both allowable strength of steel skeleton and reinforced concrete. For instance, for a beam section, M_a moment of allowable bending moment is the sum of ${}_sM$, moment of allowable bending of steel skeleton and ${}_rM$, the allowable bending moment of reinforced concrete (Fig. 7.6). This idea is adopted for convenience sake to estimate the flexural resistance of a section. This method is not based on the usual straight line theory but facilities evaluating the strength of a member. As for a beam of solid web ${}_sM={}_sf_b \cdot {}_sZ$ (${}_sZ$ is section modulus, ${}_sf_b$ is allowable bending stress of steel skeleton). If not a solid web, as in Fig. 7.6 ${}_sM={}_sa_t \cdot {}_sf_t \cdot {}_sj$ (${}_sa_t$ is sectional area of steel skeleton flanges, ${}_sf_t$ is allowable tensile stress of steel skeletons, ${}_sj$ is distance between centroids of both flanges). Usually longitudinal reinforcement of reinforced concrete part is less than the balanced reinforcement ratio. So ${}_rM={}_ra_t \cdot {}_rf_t \frac{7}{8} {}_rd$ (${}_ra_t$ is sectional area of tensile reinforcements, ${}_rf_t$ is allowable tensile stress of reinforcements, ${}_rd$ is effective depth of reinforced concrete part).

Column section is subjected to compression N, as well as bending moment M. This is

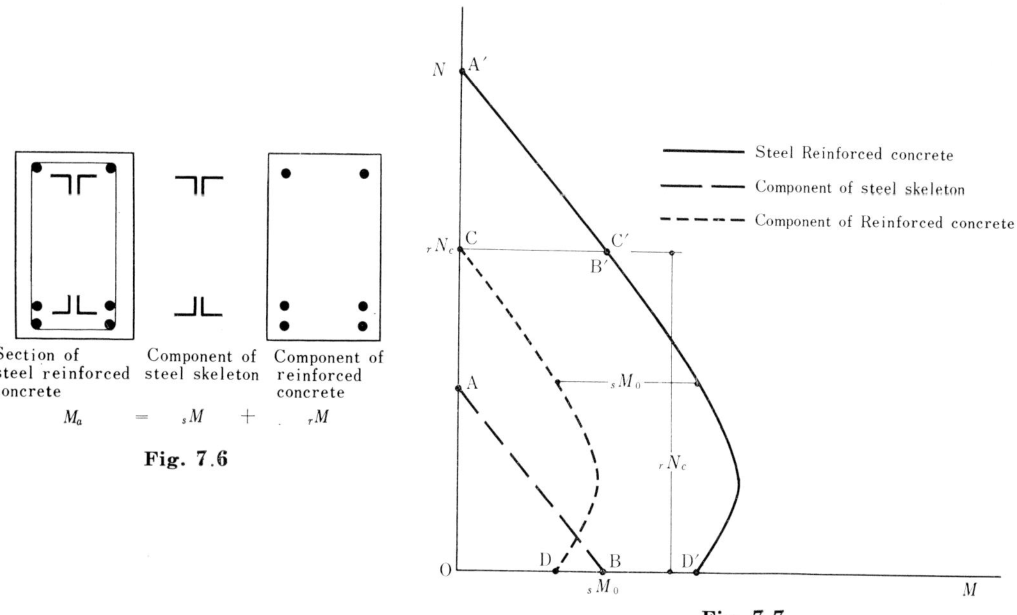

Fig. 7.6

Fig. 7.7

shown in Fig. 7.7.

$$\frac{{}_sN}{{}_sa}+\frac{{}_sM}{{}_sZ}={}_sf_c$$

(${}_sa$ is whole section area of steel skeleton, ${}_sf_c$ is allowable compressive stress of steel skeleton.) According to the formula above, the allowable axial compression ${}_sN$ and allowable bending moment ${}_sM$ (line ——) for the steel skeleton are determined. When the allowable axial compression ${}_rN$ and allowable bending moment ${}_rM$ are shown by the dotted line, the part of reinforced concrete, in the case of comparatively small compression ($N \leqq {}_rN_c$) or large bending moment ($M \geqq {}_sM_0$)

$$\left.\begin{array}{l} N={}_rN \\ M \leqq {}_sM_0+{}_rM \end{array}\right\}$$

For the case of comparatively large compression ($N > {}_rN_c$) or small bending moment ($M < {}_sN_0$)

$$\left.\begin{array}{l} N \leqq {}_rN_c+{}_sN \\ M={}_sM \end{array}\right\}$$

As shown in the figure, in the case of small compression, the broken line is shifted to the right by ${}_sM_0$, and for the case of large compression, the broken line is shifted upward by ${}_rN_c$. As a result, the interaction curve of steel reinforced concrete (full line) is obtained. In other words, the summation method is composed of the steel skeleton part which mainly carry bending moment because of the large flexural capacity and the reinforced concrete part which mainly carry axial compression because of its large compressive capacity.

The above method is for the common symmetrical sections. When the concrete covering of the steel skeleton is large, or when the section is not symmetrical, the formula is more complex than the above. These cases are discussed in detail in AIJ Standard for Structural Calculation of Steel Reinforced Concrete Structures and its Commentary (1963).

As for the calculation for shear, a modification relating to the working stress of steel skeletons and reinforced concrete is necessary. For the formula, T, working stress of total tensile steel is divided into two components, ${}_sT$, of steel skeleton and ${}_rT$, of reinforced concrete.

$$\left.\begin{array}{l} T=\dfrac{M}{j} \quad j \fallingdotseq \dfrac{7}{8}d \\ {}_sT=T\dfrac{{}_sa_t}{a_t} \\ {}_rT=T\dfrac{{}_ra_t}{a_t} \end{array}\right\}$$

Therefore

$$\frac{dT}{dx}=\frac{Q}{j} \qquad \left.\begin{array}{l} \dfrac{d}{dx}{}_sT=\dfrac{Q}{j}\cdot\dfrac{{}_sa_t}{a_t}=\dfrac{{}_sQ_0}{j} \\ \dfrac{d}{dx}{}_rT=\dfrac{Q}{j}\cdot\dfrac{{}_ra_t}{a_t}=\dfrac{{}_rQ_0}{j} \end{array}\right\}$$

(a_t is the total sectional area of tensile steel, d is the distance from the compressive edge of concrete to the centroid of tensile steel, and ${}_sa_t$ and ${}_ra_t$ are explained above.)

The formula above shows that shear, Q is divided into the steel skeleton part ${}_sQ_0$, and the reinforced concrete ${}_rQ_0$, part. As to steel skeletons, when the web strength of steel skeleton is ${}_sQ$, $({}_sQ_0 - {}_sQ)$ depends upon web action of concrete, and as for the reinforced concrete part, the whole ${}_rQ_0$ depends upon the web action of concrete. Therefore, web reinforcement is calculated by the following

$$({}_sQ_0 - {}_sQ) + {}_rQ_0 = Q - {}_sQ$$

in which

$${}_sQ_0 \geqq {}_sQ$$

(when ${}_sQ_0 < {}_sQ$, it is assumed that ${}_sQ_0 - {}_sQ = 0$)

Calculating in the same way, bond stress of steel skeleton may be known by the shear $({}_sQ_0 - {}_sQ)$. For steel skeletons of the tie plate type, the web strength ${}_sQ$ is equal to zero, and tie plate is dealt with as a web reinforcement (But the tensile strength of the tie plate is decreased to one-half or less). In this case, shear is completely subjected by web action of concrete, and bond stress of steel skeletons is calculated from ${}_sQ_0$. The calculation of steel skeletons is done about the adhesive area except the lower part on the occasion of filling concrete. As steel skeleton flanges of beams have smaller adhesive area than actual area and allowable bond stress itself is weak, so it is available in order to obtain good results that the web strength of steel skeletons is increased with trussed or solid web.

But if ${}_sQ > {}_sQ_0$, steel skeletons have more strength than expected by working stress, bond stress of steel skeletons is assumed to be zero.

Furthermore, beam-column connections, joints beams and columns and column bases are also calculated with the summation method. Besides the total strengths of these parts working stresses of steel skeletons and longitudinal reinforcements shall be considered, as a rule. And when the connection strength of steel skeletons parts is not enough for working stress, longitudinal reinforcements are available to complement the lack of strength, in skeletons parts. In this case bond stress shall be inspected considering the changing of stress from steel skeletons to longitudinal reinforcements.

As to beam to column connections, the section of beam or column end will be independently determined according to the design stress. Therefore, as the case may be, though steel skeletons may be of principal in the strength of beam sections, longitudinal reinforcements may be of principal in column sections. Even in these case, if the filling up of concrete of panel zones of connection is complete, the stress will be transfered from steel skeletons of beam to longitudinal reinforcements of column, and so any special examination is not necessary. But, practically, as the filling up of concrete is seldom perfect, it is necessary to plan intending to avoid the difference in the strength of steel skeletons between beams and colums.

When horizontal diaphragms or stiffeners are welded in order to strengthen the joint of steel skeletons, it is difficult to fill up concrete. So the stress of steel skeletons shall be planned on the basis that it is transmitted directly to the columns. In other words, steel skeletons of columns and beams shall be planned with consideration of the mutual transmission of stress. Therefore, the web plates of panel zones in steel frames shall also be, carefully examined. Recently welded structures using solid web steel skeletons consisting of rolled H-section or I-section are often designed based on this concept.

7.3 Beams and Columns

7.3.1 Beams

The sections of steel framing consist of H or I-sections. Longitudinal reinforcements are arranged at the four corners. This is because steel skeletons cannot easily pass through the joint parts if the longitudinal reinforcements are put right above or under them.

Webs of steel skeletons are of three kinds ; tie plate type, truss type and solid type. It is difficult to have concrete poured around the steel members with full web compared with those with open web. So it is better to have perforations in the web plate where the working stress is small. In this sense, "the castellated beams" which are made by cutting the web in a zigzag way are preferable for steel reinforced concrete (Fig. 7.8)[2].

Web reinforcements consist of stirrups. The distance between each stirrup is determined by the working shear with the following limitations.

When $\tau > f_s$

$\frac{2}{3}D$ and less than 30 centimeters,

when $\tau \leqq f_s$

$\frac{3}{4}D$ and less than 30 centimeters.

(D is total depth, τ is working shearing stress of concrete, f_s is allowable shearing stress of concrete.)

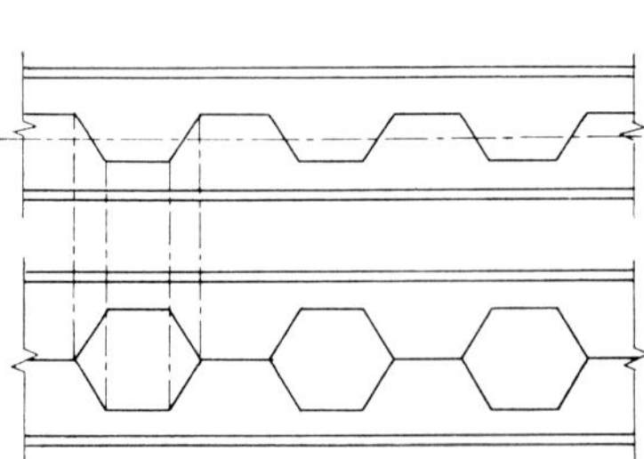

Fig. 7.8 Castellated Beam

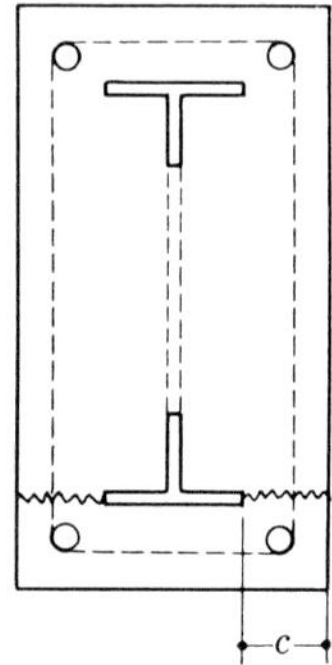

Fig. 7.9

The limit of covering thickness of concrete is more than 5 centimeters for steel skeletons, and more than 3 centimeters for reinforced concrete. Usually considerable clearance is provided between the steel skeleton and longitudinal reinforcements and then stirrups are arranged. So the steel skeletons usually have concrete covering from 9 to 12 centimeters. As the flange of steel members has considerable width, it often happen that the side covering of the flange edges becomes too small and causes cracks of theses parts subjected to shear (c in Fig. 7.9). Therefore it is necessary to provide sufficient clearance.

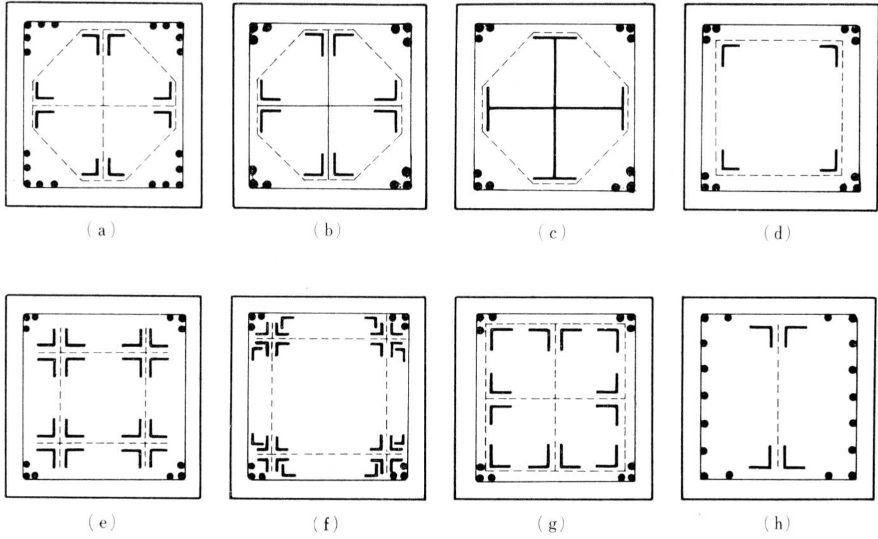

Fig. 7.10

7.3.2 Columns

It is a common rule to use the shape of column section which carries bending about the two principal axes, as shown in Fig. 7.10. Steel skeletons have two kinds of section, one is the cross and the other is the box-type. Considering the joints of steel skeletons and arrangement of reinforcing bars, a cross type can easily be dealt with. But for the cross section, if the web is not tie plate type or solid web, it is difficult to construct skeletons and also to use truss type which has an intermediate strength between tie plate and solid web types.

As mentioned before, the ultimate deformation of this type will be large compared with the other type of web. Therefore, it is preferable to plan with tie plate type when $\tau \leqq f_s$ and with a solid type of cross column or truss type of box column when $\tau > f_s$. The limitation of concrete covering is the same as those for beams, but practically it will often be thicker than those for beams. For the box-shaped steel skeletons, if the covering is not enough, it is difficult to arrange reinforcing bars and they will probably be steel reinforced concrete with few longitudinal reinforcements. As to the distance between tie-plate or lattice, the following condition shall be met in order to prevent the local buckling of steel flanges.

$$\frac{l_1}{i_1} \leqq 70$$

(l_1 is the fastening pitch of flanges (Fig. 7.11), i_1 is the smallest radius of gyration of each flange).

Longitudinal reinforcements are tied by hoops. The distance between hoops shall be closer than the smallest dimension of the concrete section, and fifteen times the diameter of longitudinal reinforcements and closer than 30 centimeters. For earthquake resistance, it is effective to increase the hoops in top and bottom parts of columns where large moments exist. In "Standard Calculation of Reinforced Concrete Structures", it is specified that spacing of hoops shall be one-half the usual spacing in the top or bottom of columns. This rule may also be applied here.

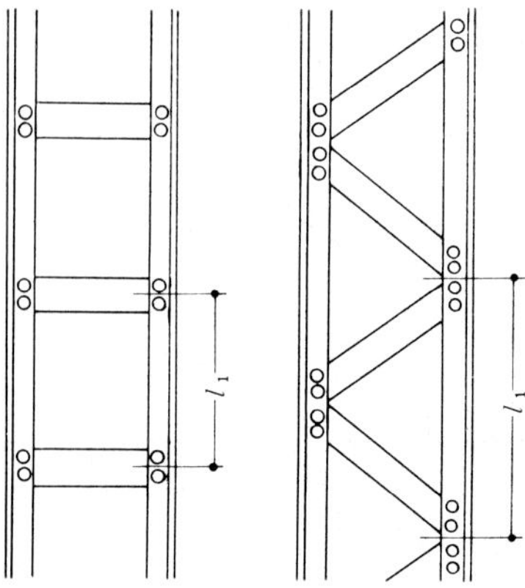

Fig. 7.11

Buckling of columns is examined on the basis of the concrete section. In the "Standard Calculation of Steel Reinforced Concrete Structures", buckling is not directly examined but the section is designed by increasing the bending moment and compression, when the columns are comparatively slender. The coefficient is shown by the following formula.

When

$$\lambda \leqq 50, \ \alpha=1$$

$$\lambda > 50, \ \alpha=\frac{100}{150-\lambda}$$

$\lambda=l_k/i$ (slenderness ratio).

i : minimum radius of gyration of concrete section.

l_k : distance between lateral supports ; the length of the column may be used.

The minimum amount of steel required is 0.8% of concrete section. There is no maximum limit. But if the steel used is too much, filling of concrete becomes difficult. So the normal is about 4%. If the quantity is more than this, steel skeletons must be strengthened and web must not be of tie plate type. And according to the quantity of steel used for the compressive part, the allowable compressive stress of concrete is decreased by the following formula.

$$f_c'=f_c\,(1-15{}_sp_c)$$

${}_sp_c$: compressive reinforcing ratio of steel skeletons.

f_c : allowable compressive stress of concrete.

According to the above formula, when ${}_sp_c=6.7\%$, $f_c'=0$ so that concrete is substantially disregarded.

7.4 Joints

7.4.1 General

In regard to joints, there are beam to column connections, beam and column joints and column bases. All are parts where joining of steel is required and careful consideration is necessary. Especially, beam to column connections have close relations with the shape of steel sections. So usually the standard type of connections is first determined and then the steel skeletons are determined accordingly.

The aim in planning connections should be to maintain strength and rigidity, easy passing of longitudinal reinforcements and easy filling up of concrete. These are necessary conditions so that joint parts of steel skeletons can act together with reinforced concrete. Therefore, as a rule, joint parts of steel skeletons must have continuity, and joint parts must be strong enough so that destruction would not occur at these parts.

7.4.2 Beam to Column Connections

a) In the Case of Rivet Joints

The method to use gusset plate is universal. Joining of gusset plates crossing in two ways may use steel angle flanges but it is better to weld to increase the rigidity. Through each gusset plate, holes must be bored for hoops which are tied together with longitudinal reinforcements passing through joint parts. The strength of joint parts of steel skeletons depends upon the section modulus of the gusset plates. For steel structures, the formula is:

$$Z=\alpha\cdot\frac{th^2}{6}$$

α : reduction coefficient for rivet holes

h : depth of gusset plates

t : thickness of gusset plates

However, for steel reinforced concrete, the following formula may be used.

$$Z=\alpha\cdot\frac{th^2}{4}$$

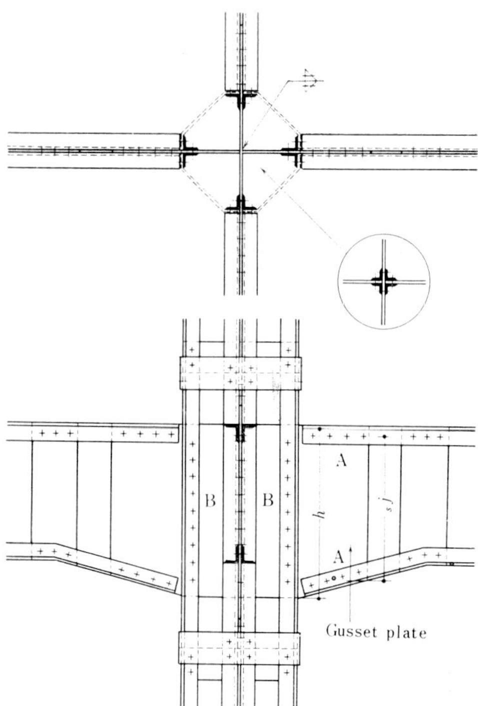

Fig. 7.12

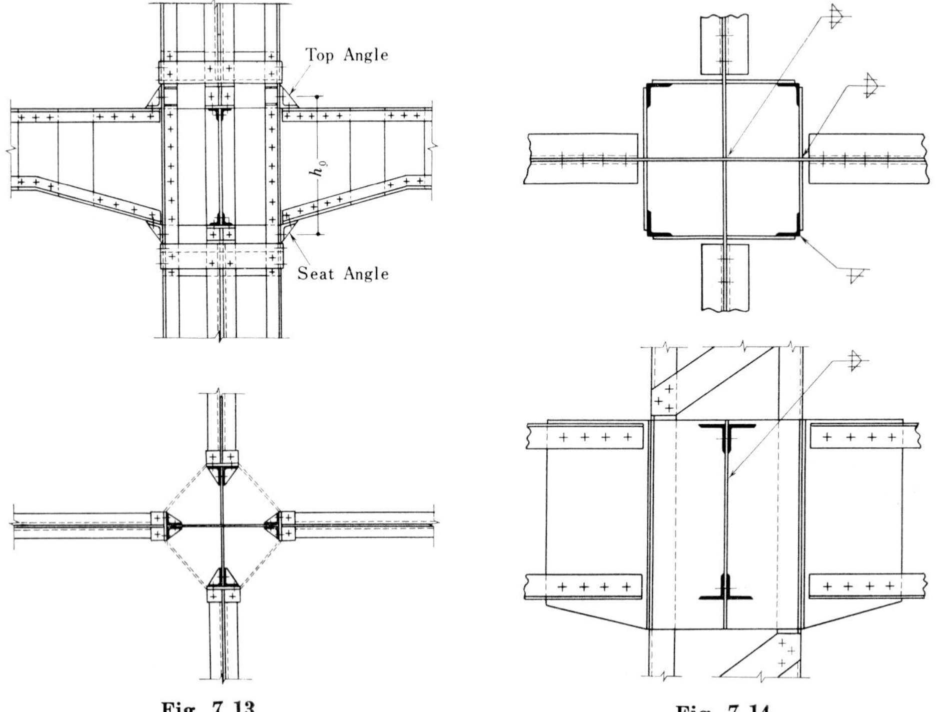

Fig. 7.13

Fig. 7.14

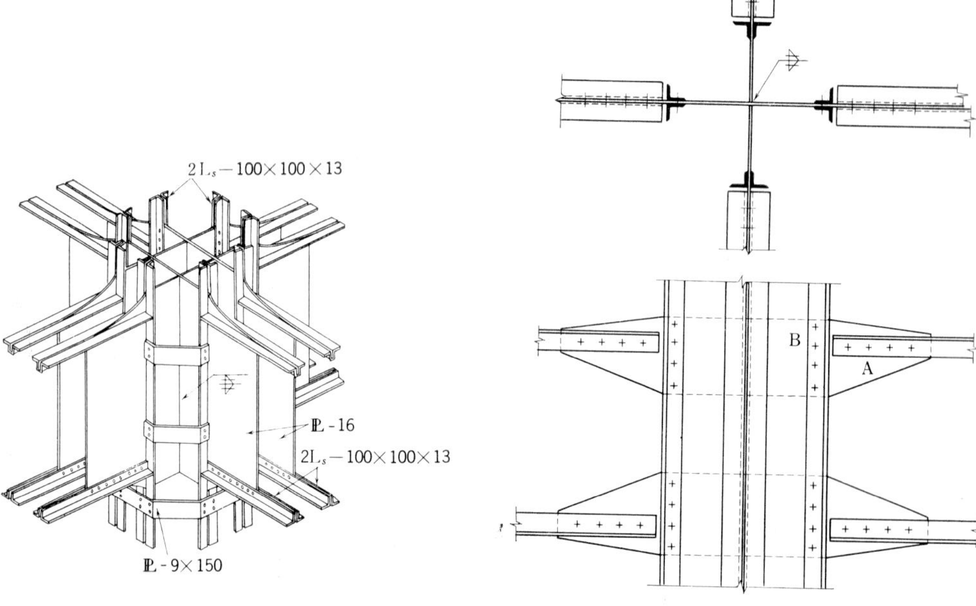

Fig. 7.15

Fig. 7.16

This equation is used only when tensile rivets in top angles are not used at the same time.

Gusset plates have poor section efficiency. So for connections of this kind, reinforcing bars of beams must be increased in order to level up the allowable bending moments.

In order to increased the allowable strength of gusset plates, tensile rivets in top angles may be used. But these tensile rivets are not appropriate because of lack of the continuity in these parts (Fig. 7.13).

If welding is used at the same time, it is possible to join box-shaped columns with I-section beams using gusset type connections. This is shown in Fig. 7.14. Fig. 7.15 shows an example of double arrangement of flanges.

In order to join high depth beams such as wall beams, gusset plates may be divided into two parts (Fig. 7.16). In this case, the shear of joint panels (explained in Paragraph 4.2.3) depends upon the concrete so careful attention must be paid to filling up of concrete.

For connections of solid columns and beams, high tensile strength bolts may be used. Several examples are shown in Fig. 7.17. These methods are generally used in the United States but these began to be used only recently and their use is not so widespread in Japan.

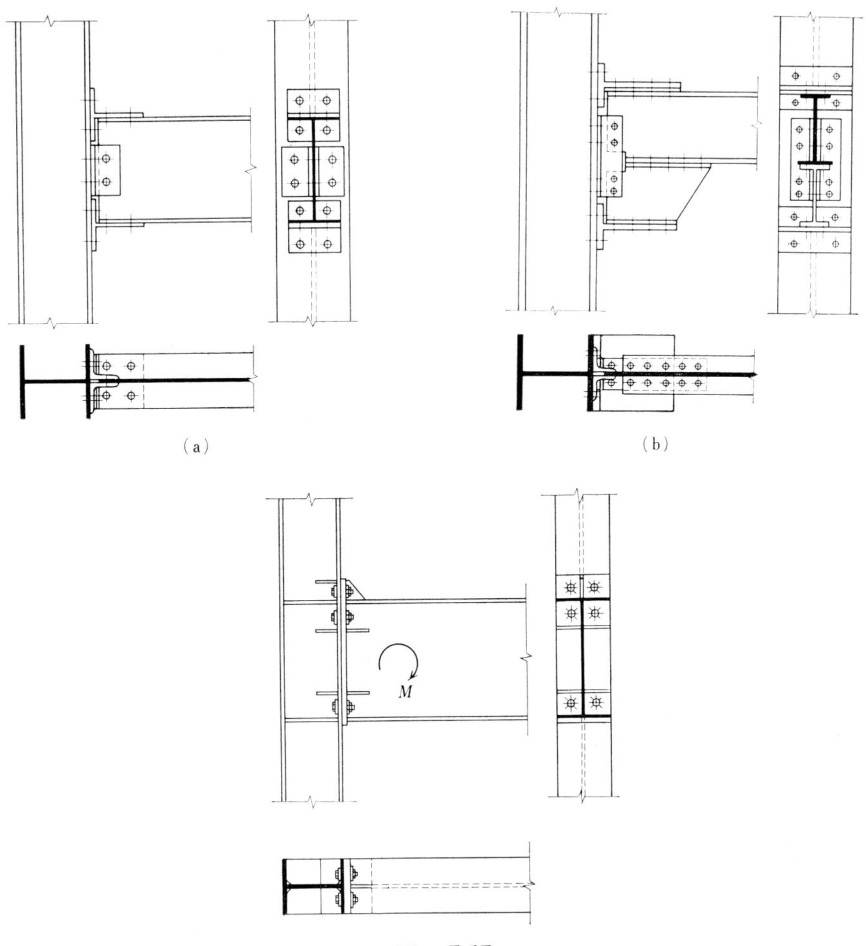

Fig. 7.17

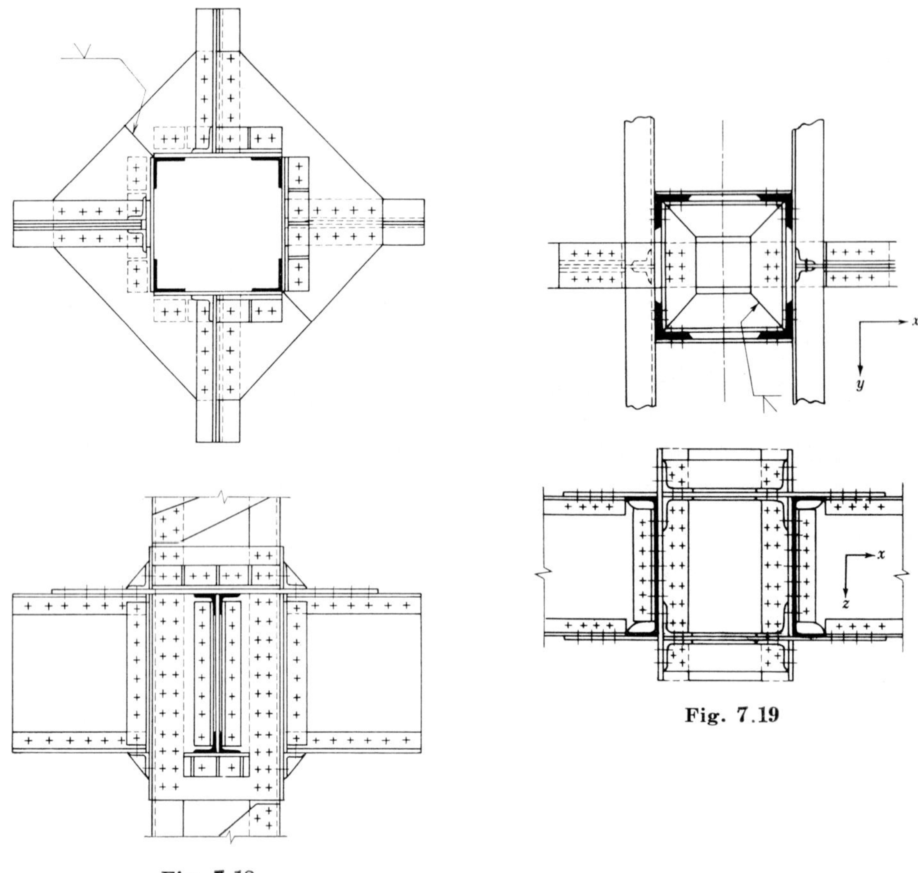

Fig. 7.19

Fig. 7.18

Therefore, when using these methods, it is necessary to make sure of the strength by tests. Other joining methods without the use of gusset plates are shown as examples in Figs. 7.18 and 7.19.

b) In the Case of Welding

Connections are classified considering which type of member is going through the connected part, columns, beams, or both columns and beams.

The type with columns going through is the most common. As shown in Fig. 7.20, horizontal diaphragm, rib or vertical stiffener are placed in the beam flange level in order to strengthen the column joint. Welded joint of beam and column flanges must be made thick in order to decentralize the stress. As this type of welding results in thick plates at the lower part of the building, a careful examination is necessary to select the basic materials. Also, as column flanges are subjected to tensile stresses in the direction of plate thickness, a close examination is necessary.

Examples of the type with beams going through are shown in Figs. 7.21 and 7.22. This type has two merits. One is that the amount of welding may be decreased because the column flanges are thinner than those of beams for buildings of approximately ten stories and the other is that the tensile stress in the direction of the plate thickness may be smaller than

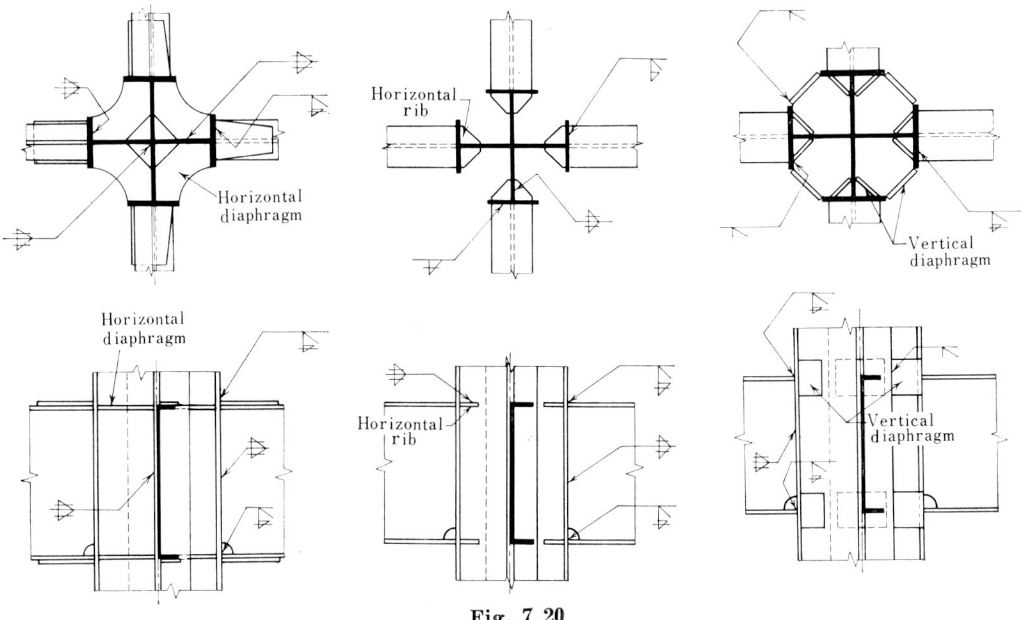

Fig. 7.20

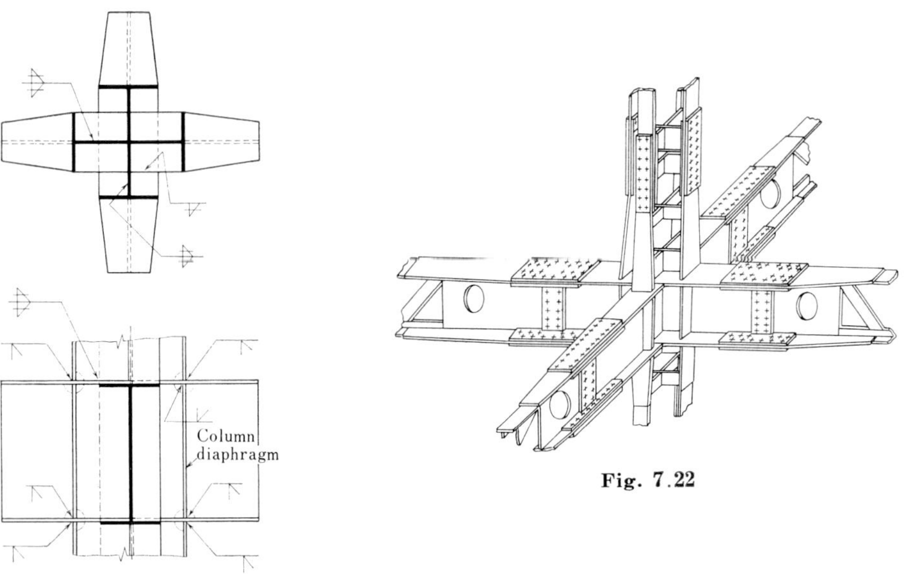

Fig. 7.21

Fig. 7.22

the type described above because the tensile stress of column flanges is small.

The type with both columns and beams going through is good for box-shaped columns and an example is shown in Fig. 7.23. Also, there is an example where horizontal and large gusset plates are used as shown in Fig. 7.24, but in this case, concrete cannot flow smoothly and therefore, this type is not appropriate in steel reinforced concrete construction.

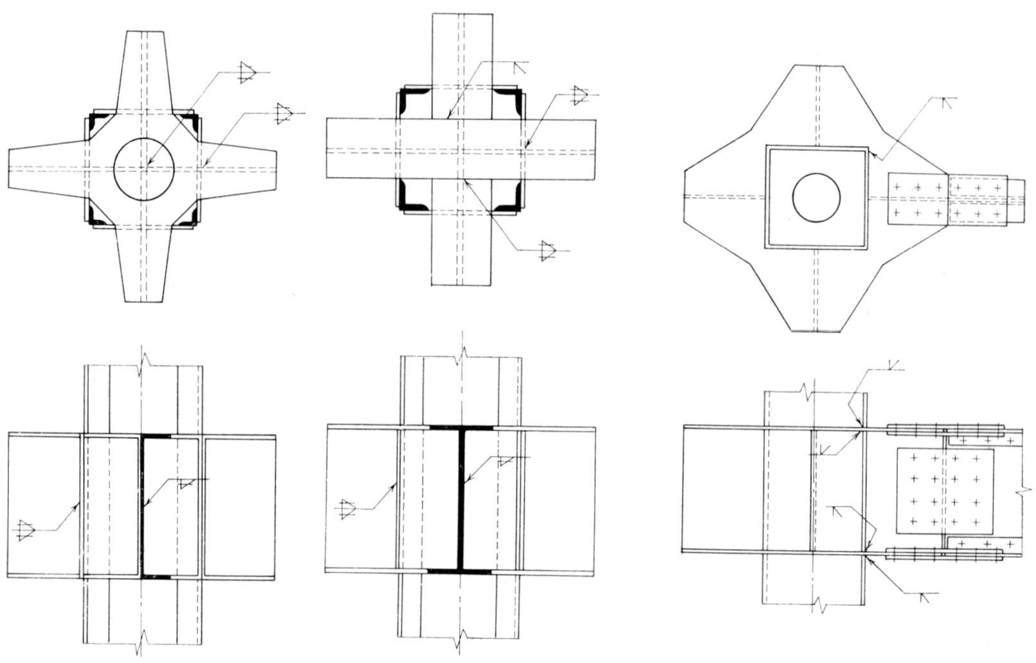

Fig. 7.23 **Fig. 7.24**

c) Connection Panels

Panel zones in the connection parts are subjected to a considerably large shear due to the horizontal load of an earthquake. If the magnitude of the shear is Q_p,

$$Q_p \fallingdotseq \frac{M_1+M_2}{0.8\,D_c}-\frac{Q_1+Q_2}{2}$$

This shear is carried by the concrete and web plates in the panel zone. So, if the filling up of concrete is good, concrete carries almost all the shear which is very advantageous. According to experiments, shear stress of concrete is

$$\tau=\frac{Q_p}{b\cdot\frac{7}{8}d}$$

b : width of the beam section

d : effective depth of beams

Calculated by this formula, shear cracks occur at $\tau \fallingdotseq 0.2\,F_c$, and complete destruction occurs at $0.5\,F_c$. These values are quite large when compared with the shear failures of usual beams. Therefore, concrete in this part must be filled up carefully in order to utilize this nature.

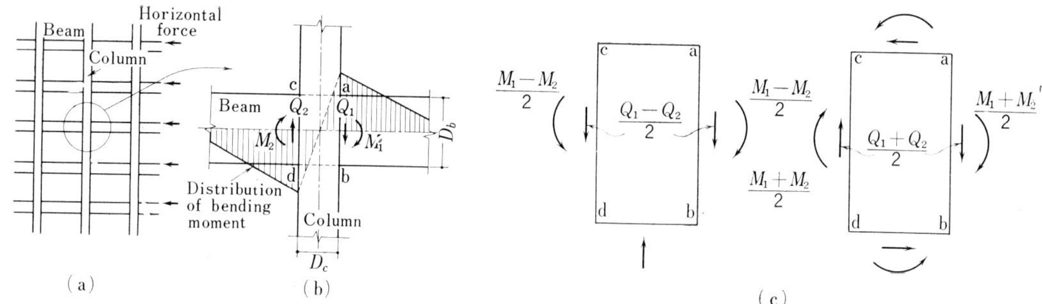

Fig. 7.25

7.4.3 Joints of Beams and Columns

a) General

It is a general rule to make joints where the stress are small so there is generally no earthquake-resistance problem. It is necessary to plan to obtain rigidity by maintaining continuity of members.

As mentioned in Paragraph 7.2.4, when the depth of the section at the joints of steel frame is changed, shifting of the section center is dangerous because it destroys the continuity. Also, if the center is shifted, eccentricity occurs which is usually disregarded in the frame calculations. Therefore, dangers are doubled so shifting of the center should be avoided.

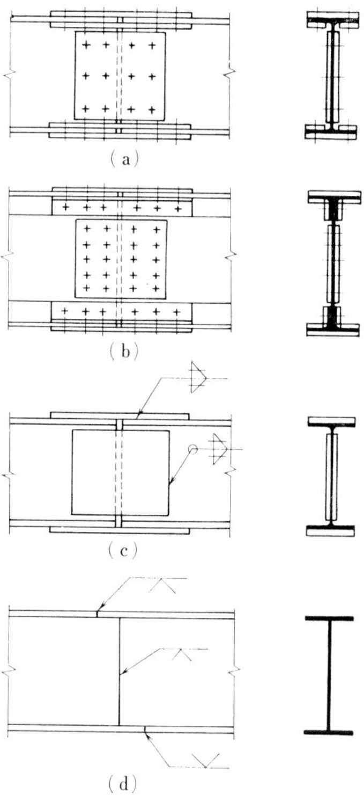

Fig. 7.26

b) Joints of Steel Skeletons

There are four fundamental joint types, as shown in Fig. 7.26. In the case of rivetted joints, flanges and webs are joined using splice plates, and in the case of welding, splice plates for fillet weld or butt weld are used. However, butt welds are not suitable for field work at the building site.

Typical beam and column joints are shown in Fig. 7.27 and 7.28. As for columns, the joint edges are usually finished at the factory and are butt welded at the building site.

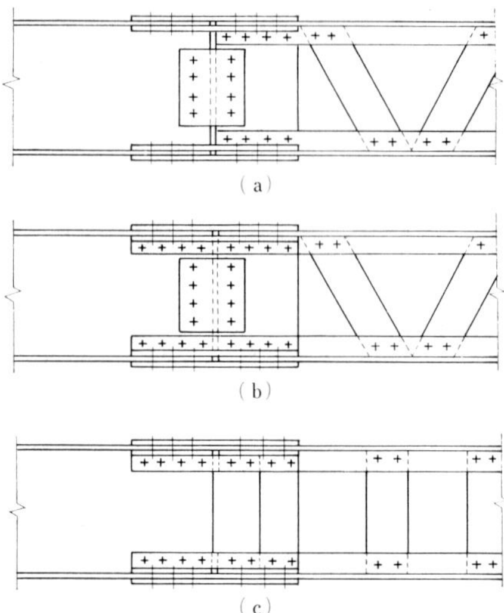

Fig. 7.27

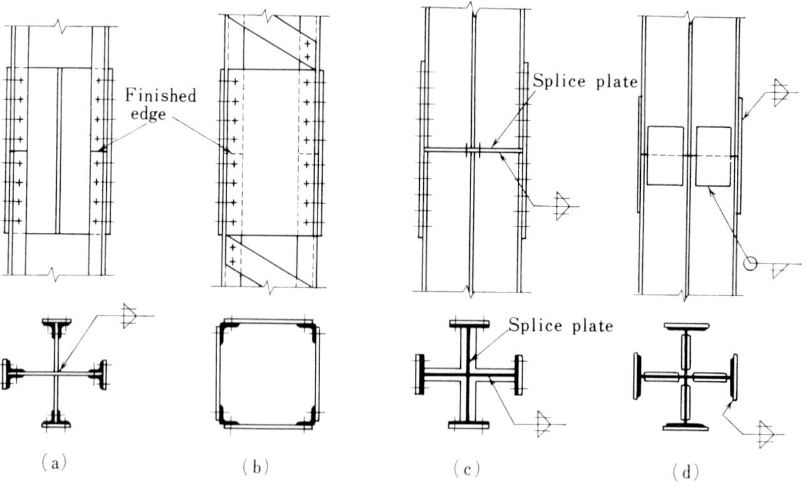

Fig. 7.28

7.4.4 Column Bases

a) General

The foot of columns of steel skeletons is put on the upper edge of the footing beam. This is a convenient way of construction but a large bending moment usually occurs in this

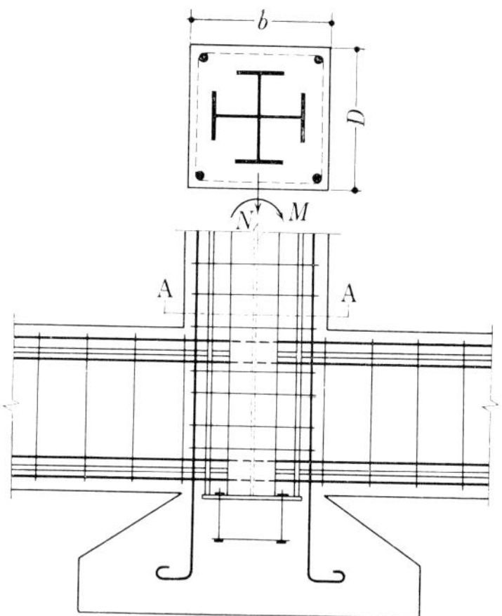

Fig. 7.29

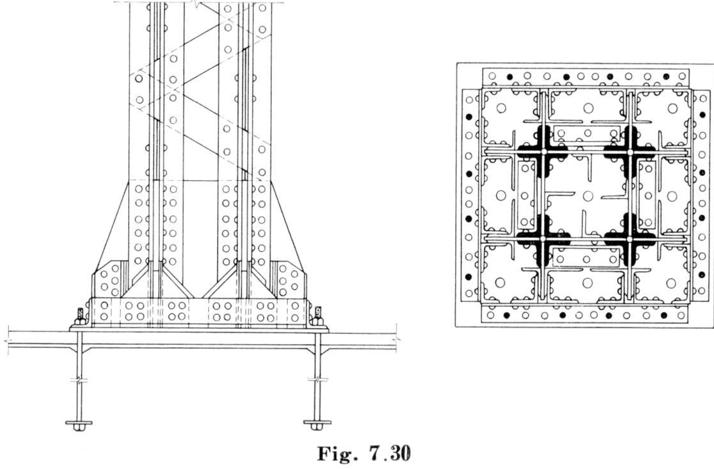

Fig. 7.30

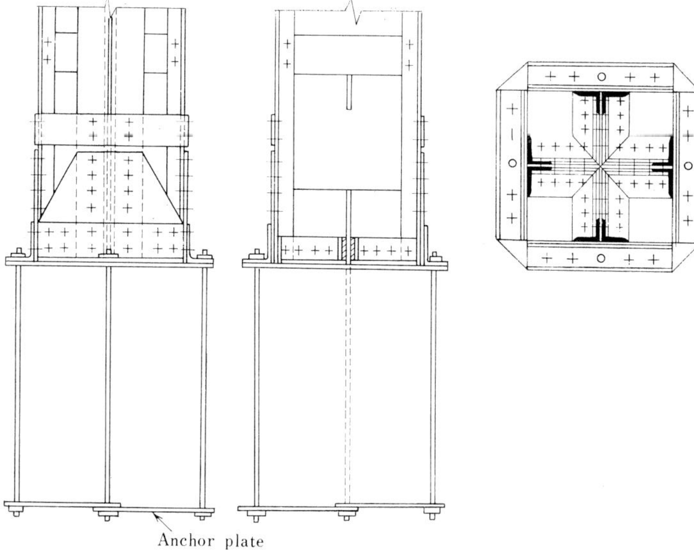

Fig. 7.31

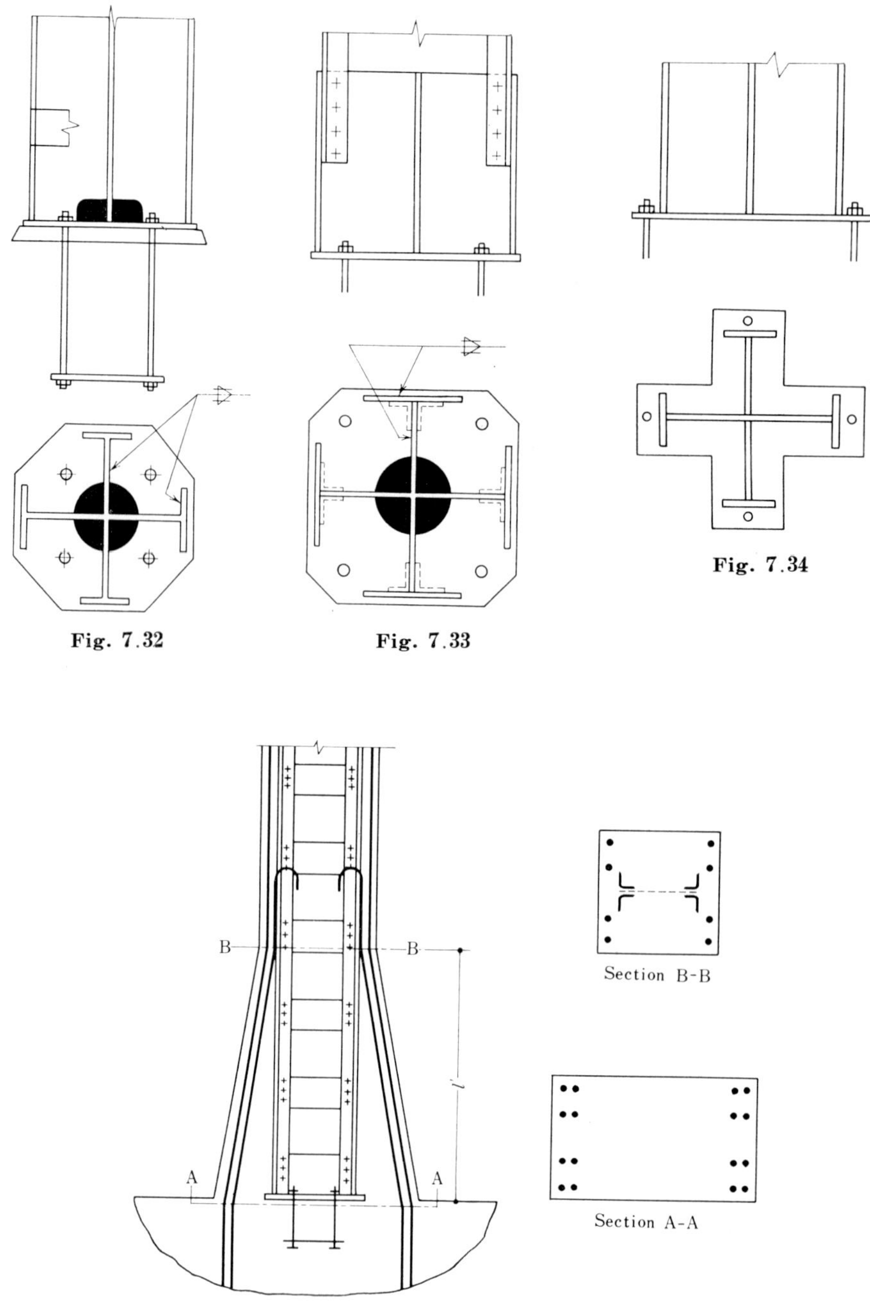

Fig. 7.32

Fig. 7.33

Fig. 7.34

Fig. 7.35

place so this is not the best procedure. A better way is to place the steel in the footing beam to which column base of steel skeleton is joined, as shown in Fig. 7.29.

Early details of column bases were as stout as details of structural steel, as shown in Fig. 7.30. At present, it is simple by considering the action of the surrounding concrete. Fig. 7.31 is a typical example using rivets. Usually base plates and wing plates are not calculated but these may be determined considering the convenience of transport and erection. Anchor bolts are used to determine the placement of steel columns and these are tied by anchor plates and buried in the concrete so that they would not be displaced.

When welding is used, base plates are welded directly to the columns. As this method is very simple, its use has spread rapidly. Even when the upper structure is rivetted, joints are placed right above the column base and are welded (Fig. 7.33). Base plates and anchor bolts are designed with the considerations mentioned before.

The action of base plates is, similar to hooks in reinforced concrete, to keep the steel skeleton from being pulled out from the concrete. Therefore, if there are no inconveniences in construction, it is possible to cut it down to the necessary parts near the steel skeleton section.

b) Stress Transmission at Column Bases

It is ideal to put steel sections in footing beams as there exists no problem concerning stress transmission. When column bases are put directly on footing beams, as shown in Fig. 7.35, reinforced concrete below section A-A must carry all the stresses and therefore the section of concrete becomes large and so much reinforcement is needed. As it becomes remote from the foot of columns, steel skeleton begins to work and stress which the reinforcements sustain at the column base is transmitted to the steel. This stress transmission is usually dependent on the bond stress between the concrete and steel. However, if reinforcements are welded to the steel skeleton, a more reliable stress transmission can be expected.

References

1) Shuzo Takada; Plan of Steel Reinforced Concrete. OHM Company.
2) Yawata Iron and Steel Company, Rolled H-Section Steel Design Manual.
3) Yawata Iron and Steel Company, Figures of Joint Parts of Rolled H-Section.

8. SHELL STRUCTURES

8.1 General

Shell structures mean the space structures consisting of thin curved plates or of a number of plates joined together on folded planes. However, in practice, there are cases where the curved or folded surfaces consist of combination of linear members instead of plates but in such cases, the same mechanical approach may be made as described below.

On these structures, since they consist of plates either on curved or folded planes, even against the normal load component to the plate surface, primary equilibrium can be maintained by membrane stresses (in the case of folded plates, the normal loading on each plate requires bending stresses and shearing force as a slab to maintain the equilibrium) and the whole structure acts as if it were a beam or a wall. Consequently, in comparison with the ordinary slab, which is bent in order to support normal load on plane, the whole of the structure has high rigidity and therefore can cover space with a comparatively large span.

The shells have their own mechanical characteristics, respectively, depending on type of curved surfaces and "in addition" a state of stress differs depending on boundary structures or methods of support. Generally speaking, however, in the case where the rise is small in comparison to the span, i.e. a flat shell structure, if rigidity of the whole structure is taken into consideration, the structure resists rather favorably the horizontal load rather than the vertical load and therefore as far as a static load is substituted for a seismic force, the structure may be, generally speaking, considered safe against seismic forces. Furthermore, when the vertical force component by the up-and-down motion in earthquake acts nearly uniformly on the whole shell, the vertical load from the dead and the live loads is either increased or decreased only, and therefore it can usually be ignored in the design. However, as no equilibrium can usually be kept by membrane stresses alone at the boundary of the shell, bending stresses occur in the curved slabs depending on boundary and supporting conditions, and consequently, these factors often govern the design of sections. There may be a case where local bending stress in the boundary zone, occuring due to an asymmetric load like a horizontal force, can not be considered negligible in design in relation with the value of the vertical load, and therefore although a sufficient allowance is given until when the shell collapses but, in practice, there have been examples of seismic fractures resulting from local bending (see Photos in 8.1 and 8.2).

In respect to the occurence of local stresses in the neighbourhood of concentratedly supporting

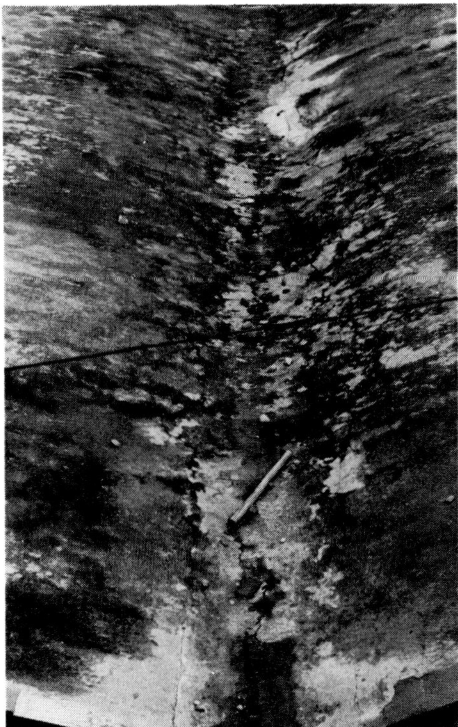

Photo 8.1 Cracks in Joints along the Edge Beam of Cylindrical Shell

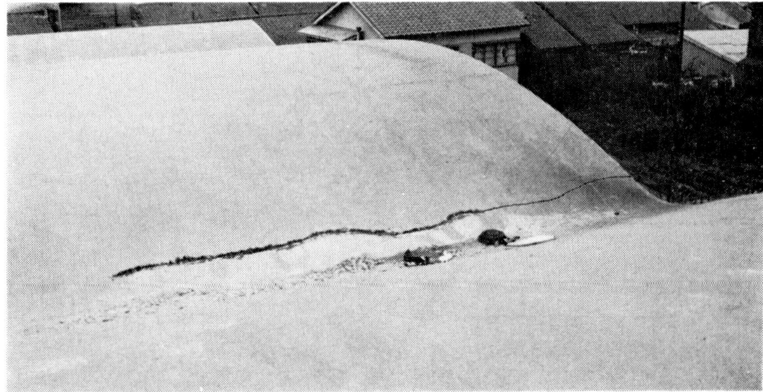

Photo 8.2 Cracks in Joints along the Edge Beam of Cylindrical Shell

points and openings, it is necessary that the design be made with care, taking into consideration what has been stated above. Furthermore, if deformation becomes larger, more becomes the risk of incurring buckling collapse, and in particular, in the case of a large span and flat, having small flexual rigidity, or where boundary supporting condition is unfavorable, special care is required. In designing of shell structures, the following items should be taken into consideration in addition to the type of curved surface.

i) The case where a perfect member is constituted into which an equilibrium with the given

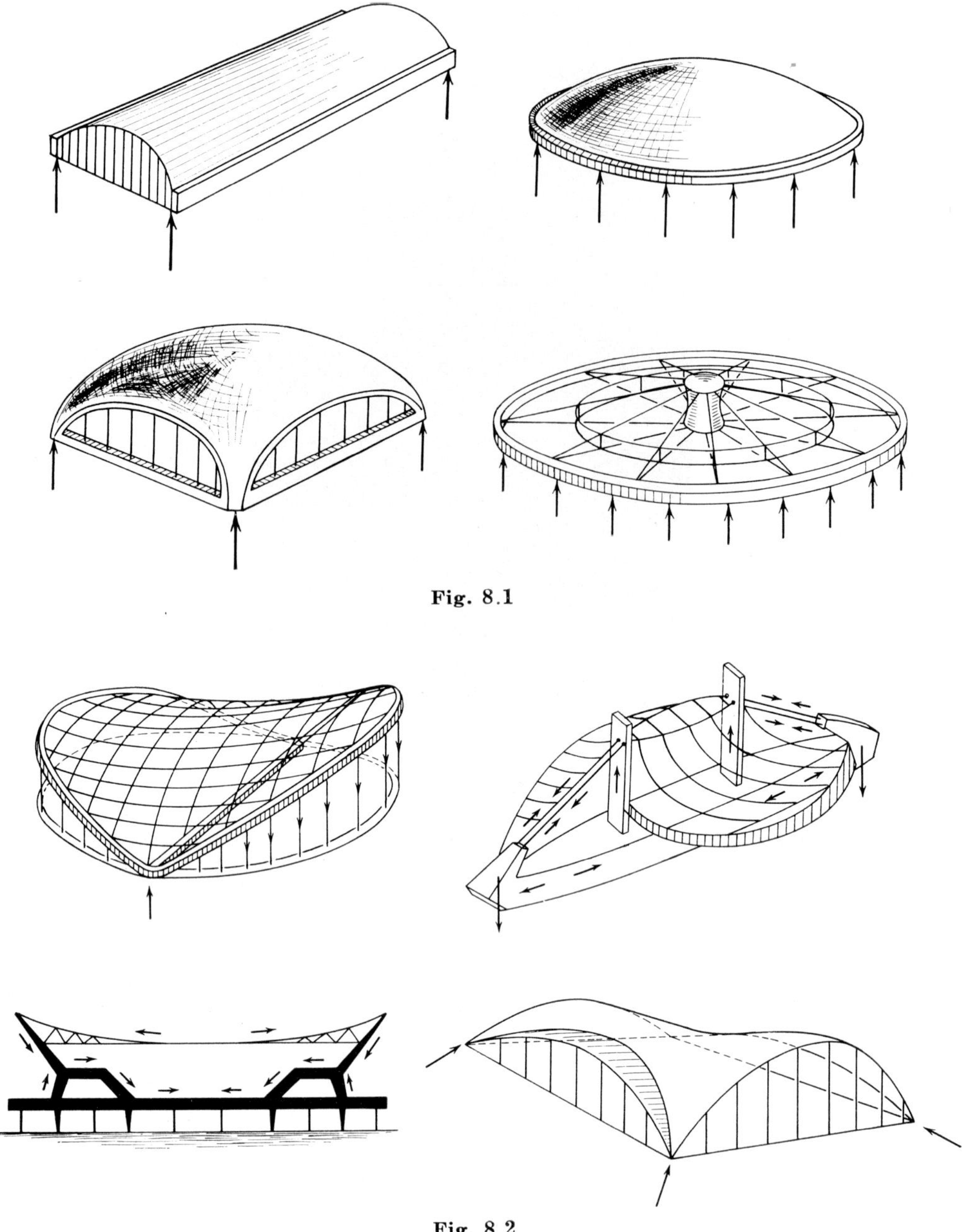

Fig. 8.1

Fig. 8.2

load can be kept by the shell including the edge beam alone, and supported by the substructure (Fig. 8.1).

ii) In the case where internal forces of the shell structure alone are not able to be in equilibrium but a stable structure is obtained only when the internal forces occurring in the structural parts supporting the shell are combined with the internal forces (Fig. 8.2).

In either case, in designing the connection between the shell supporting structure and shell, and the shell supporting structure, careful considration should be given in regard to their interaction to the shell.

Since the shells are in many cases used for large span structures, there are many instances where it is difficult to have the whole of the structure supported on the similar ground condition, and local or all-over differential settlement is liable to occur (refer to Photos in 8.3 to 8.5).

In these cases, since shell plates are thinner in comparison with the span, flexural deformation of plates allows a considerably large deformation as a whole.

Furthermore, since plates are statically indeterminate to a high order, if the stiffening of the edges is adequate, the whole structure functions efficiently and therefore even in the case where differential settlement is taking place, there is no fear of the whole structure collapsing

Photo 8.3 State of Differential Settlement of the Buckling Externally of the Gable Windows

Photo 8.4 Rupture due to Differential Settlement of the Shell Supporting Part

Photo 8.5 Cracks in the Vicinity of the Intersection of Edge Beams due to Differential Settlement of the Shell Supporting Part

unless the supporting condition becomes unstable. However, damage such as cracks in concrete, fracture or buckling of steel members do occur and furthermore, if deformation becomes too large, state of equilibrium is changed due to the change of curvature, and finally stability may be lost. Hence in the structural designing of the whole structures including the substructures, care should be taken to ensure that no uneven settlement occurs as far as possible, and at the same time it is desirable that in design, care should be taken to ensure that bases of columns are connected by tie beams (refer to Photos 8.6~8.9).

Photo 8.6 Cracks in Edge Beams due to Differential Settlement of the Shell Supporting Part

Photo 8.7 Rupture due to Differential Settlement of the Shell Supporting Part

Photo 8.8 Fracture of Shell (Truss) Members due to Differential Settlement of the Supports

Photo 8.9 Buckling of Shell (Truss) Members due to Differential Settlement of the Supports

8.2 Structural Design

8.2.1 General

Shell structures are used mainly as roof, wall, footing and also as combined wall and roof. There are cases where shell of a type is used individually and shells of the same or different types and/or same or different sizes are used continuously.

a) The Case where Shell is Used as the Roof

Mass which is the cause of seismic force is almost attributable to the dead weight and distributed over the roof surface but a ratio at which the seismic force is transmitted to every place on structure supporting the shell and the pattern of stresses transmitted (shear forces in and normal to plane, bending and torsional moments) are considerably different according to not only types of shells but also proportion or supporting methods, and likewise stresses in shells themselves vary being affected by these factors (Fig. 8.3).

Generally speaking, there are many cases where a large bending moment occurs locally in the boundary zone of the shell; nevertheless, in many cases an external force given to the shell surface is mainly transmitted, as forces in the plane of the shell, to the structure supporting the shell. If the substructure is designed with those points as mentioned above taken into consideration, a favorable state of stresses could be obtained.

b) The Case where Shell is Used as the Wall

The cases may roughly be classified as follows: where a shell has a closed section like cylindrical shells of a tank or a column part of mushroom type shell, where curved or folded plates in corrugated shape are used as an open section, and where the extended part i.e. edge zone of the shell roof becomes the wall.

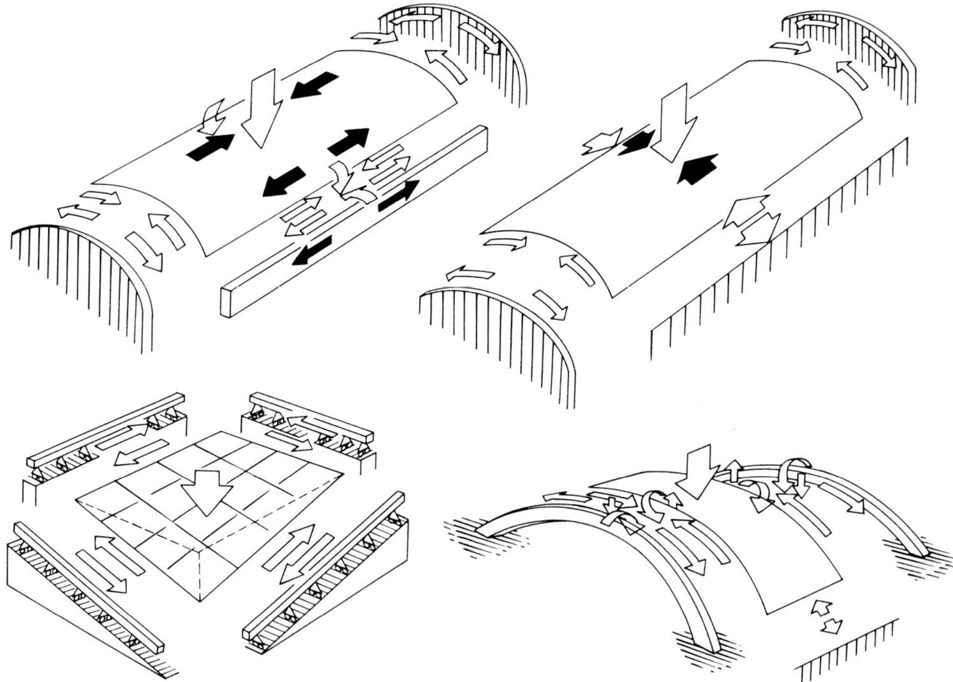

Fig. 8.3 Examples of Transmission of Loads and Stresses (Show Principal Forces only)

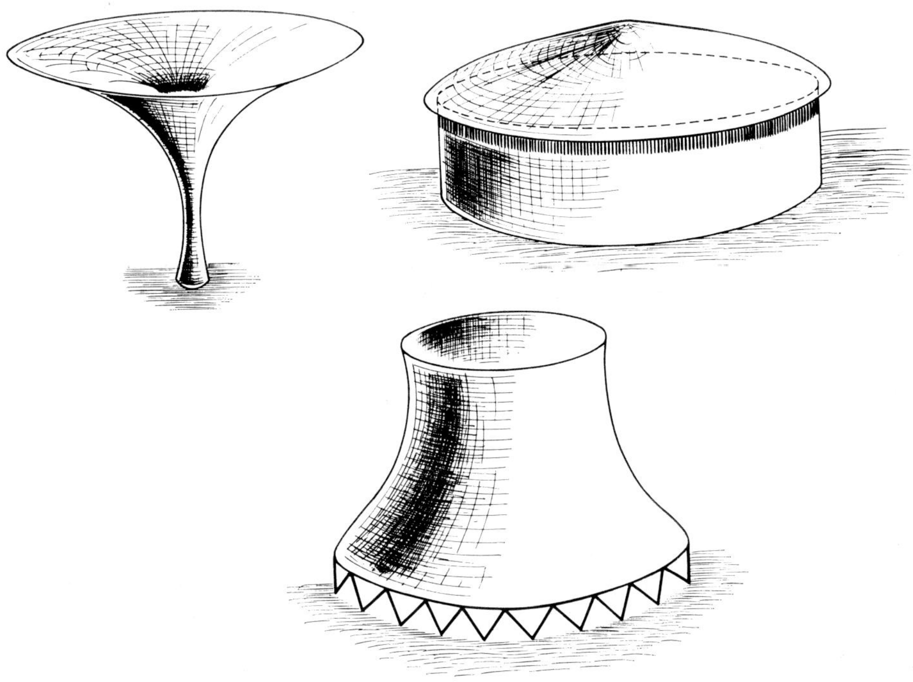

Fig. 8.4

(1) The Case of Closed Section (Fig. 8.4)

If the diameter of the section becomes larger, the whole structure does not vibrate monolithically, then arises the possibility of a plate being bent largely by deformation of the section and therefore in designing them, care should be taken as to fitting of stiffeners or providing of ring-shaped parts, whose thickness is to be increased, and spaced out appropriately. In particular, in case the top surface is open with free end, care should be excercised.

(2) The Case of Open Corrugated Section (Fig. 8.5)

Since the shell functions as a column, an axial force of course arises in the direction of generatrix but if the column shell is continued to the roof plate or beam, bending occurs and then it is possible to incur buckling because of the open type section. Hence, it is desirable to follow the designing procedure whereby the thickness of the plate is increased, as the necessity

Fig. 8.5

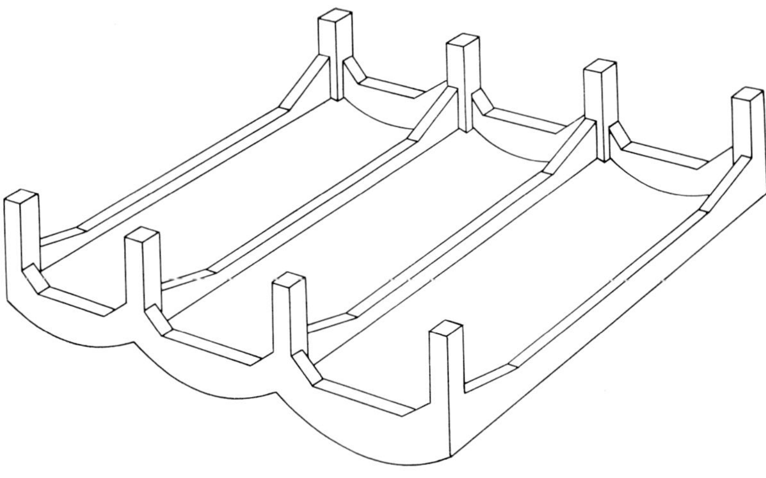

Fig. 8.6

arises, or the stiffening plate fitted so that the sectional shape is held to ensure that the shell functions monolithically, and furthermore, in case an opening is to be provided, it is necessary to make such reinforcement as to resist the local stresses in the boundary and at the same time to compensate for the reduction in rigidity of the whole structure.

(3) The Case of Edge Zone Concurrently Serves as Wall (Fig. 8.5)

The zone corresponds to the boundary zone and in many cases is liable to incur the highest degree of both membrane and bending stresses, and moreover, if it is positioned in the opening, a design should be made with the utmost care. In relation to the foundation structure, the utmost care should be taken in design to ensure that the projected boundary condition may be obtained.

c) The Case the Shell is Used as Footing (Fig. 8.6)

In the case where the shell is used as the footing, design shall call for the whole of the shell to bear stresses occurring in the columns, walls or footing, beams of overstructure and then transmitted to the underground. For these purposes, a design should be worked out to ensure that an axial force and bending moment exerted from the columns are transmitted not directly to the shell plate, but through foundation beams or diaphragm walls which are arranged appropriately so that large local stresses do not occur in the plate. And, since there is the possibility of subgrade reaction (contact pressure) being concentrated in particular part, it is desirable to work out a design having safety in reserve, taking into consideration the above factors.

8.2.2 Shell Plate

Since the shell plate is on a curved plane, load from any direction (including the case of horizontal loads also) can be supported mainly by membrane stresses, provided only when the boundary condition of the shell is suitable, but depending on supporting methods, bending stresses become considerable values throughout the shell.

On the other hand, in the case of the flat shell, membrane and bending stresses are co-existing in all parts of the shell, and therefore depending on the boundary condition, a ratio of the load components supported by the membrane and bending stresses changes. In general, in the boundary zone of the shell there are many cases where bending moment and corresponding shear force become large. These conditions have their own characteristics respectively, depending on the nature of the surface and, at the some time, are effected by rigidity of the edge beam and details of the actual construction method, and therefore when the values obtained by calculation are used in the design, it is necessary to carry out the design with flexibility in such a way as to ensure that deviation from the calculation values on the stresses and the deformations to the actual may be acceptable.

In case the boundary of the shell is supported by a number of columns, local stress arising from both the vertical and horizontal loads are liable to cause cracks around the columns, and therefore the procedure should be such as to increase the rigidity of the edge beam connected to the columns, thereby to disperse and minimize the local stresses, and to increase the thickness of shell about the columns and then to provide sufficient reinforcement, etc.

As stress also concentrates around openings for lighting etc., there is the possibility of occurrance and development of fracture at these places, procedure should be taken such as fitting stiffners and providing special reinforcements.

8.2.3 Edge Beam, End and Intermediate Supporting Beams (Wall) and Stiffening Beam, etc.

a) Edge Beam (Figs. 8.7, 8.8)

The edge beam to be provided on the circumference of the shell is to bear stresses occurring in the edge of the shell. Forces exerted from the plate causes deformation of the beam, but the beam is constituted monolithically with the shell plate and consequently acts as a part of the shell from continuity of the deformations. In general, since stresses in this part tend to become larger, if it is made of reinforced concrete, the edge beam serves to secure the section necessary for arrangement of the reinforcement. Furthermore the edge beam stiffenes the shell plate in order to withstand large local stresses in the boundary zone by an asymmetric load such as seismic forces and serves as a frame to ensure the shell acts as a monolithic body and thus prevents failure progressing from the boundary zone and from total collapse occurring. Also it acts to prevent buckling of the shell by confinement of deformations in the boundary zone.

In the case of such shell structure in which the edge beam is directly supported on some points, it means adding of the stresses to the beam incurred by acting as a continuous beam supported by the points.

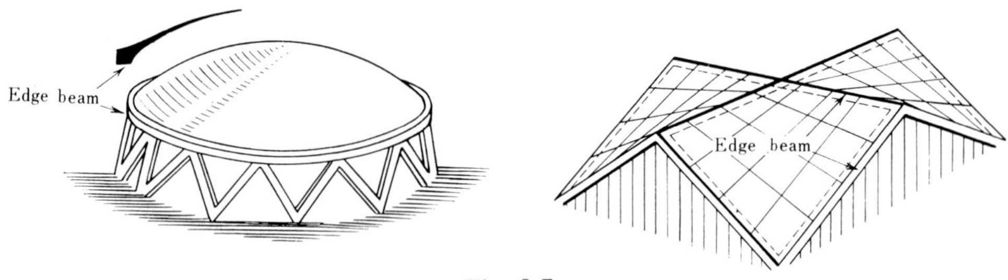

Fig. 8.7

b) End Wall (Beam) and Intermediate Supporting Beam (Wall) (Fig. 8.8)

These are important for keeping the fundamental shape, like a rise of the shell, deformations

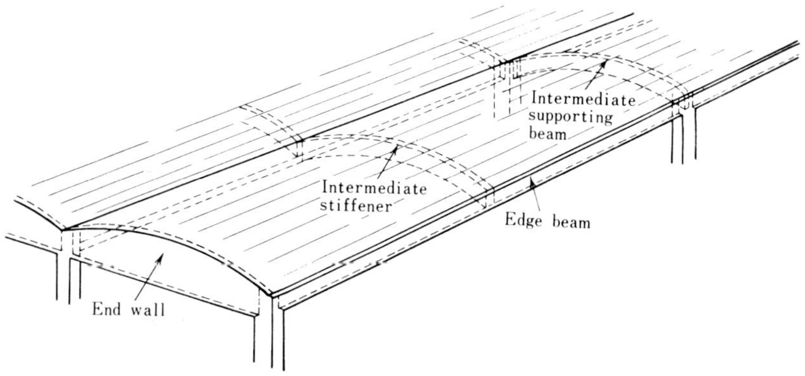

Fig. 8.8

of these parts will directly affect the mechanical function of the whole of the shell. Moreover, a seismic force on the shell is transmitted through these parts to substructures and simultaneously stresses from the substructure are also transmitted through these parts to the shell. Consequently, it is necessary likewise to design carefully in the case of the edge beam, to ensure that assumption of calculation is satisfied and that the shell acts as monolithic structure so that the structure may be aseismic.

c) Intermediate Stiffening Beam (Fig. 8.8) (Diaphragm Wall)

These are to be fitted in case where a span of the shell is considerably large, and serve to hold the sectional shape of the shell and to prevent buckling, etc. For these purposes tie-beams may be used if necessary.

8.2.4 Structure for Supporting Shell

Methods of supporting the shell may roughly be classified into two as stated under the heading "General". In case the shell is constructed as a unit member, if it is supported simply, it can be assumed that no additional stresses as bending moment etc. from continuity with the substructure, would occur.

In this case, it is required that designing of supporting points (lines) should be worked out with proper care to ensure that at the time of an earthquake, the shell would not slip off or come off the supporting structure. If methods of practical construction work are unable to make a perfect simply support, the case may arise where stress changes considerably from the calculated as described in 8.2.1∼8.2.3, and therefore, it is necessary to design, taking into consideration such effects.

In simply supporting cases, effects given on the shell from the substructure can be minimized, and the design is clarified. However, as far as the substructure is concerned, since its top comes near to the hinge or free end, the structure cannot be considered aseismic. Consequently, it is required to design the rigidity of each part to be as equal as possible ; to provide the aseismic wall, or connect by the underground beam between the bases of columns, thereby letting the substructure alone make the whole structure aseismic.

A design can also be worked out that the shell constructed as a unit member acts as a member of a rigid space frame by positively connecting to the substructure.

In this case, it is desirable to avoid a sudden change in the rigidity of the shell and the substructure at their connecting points, by substructure at their connecting points by connecting the heads of columns to the end or the edge beams. When earthquake occurs, there arise greatest possibilities that local crushing or tensile cracking may occur at the head of columns or shell plate about the connecting points.

In case the substructure carries thrust from the shell, tensile or compressive force of the shell, only when the substructure carries these forces, the shell is able to exhibit its mechanical function; otherwise there is the possibility of the total collapse of the shell. Hence, the design should be work out with the utmost care to ensure that the structure is adequately safe. In cases where the bottom of the footing is inclined and the thrust is borne by the carrying capacity of the subground, or the frictional force of the bottom of the footing or the carrying capacity of piles, it is necessary to investigate carefully the behavior of the ground at the time of earthquake. Hence, unless they are found particularly safe, it is desirable to depend on the use of underground beams.

8.2.5 Structures and Equipments Provided within the Shell Space

In respect to a structure to be built within the shell, it is better either to construct it as a monolithic body incorporating the structural part supporting the shell, or else construct it isolated from the structure constituting the shell. Intermediary construction between the methods mentioned above had better be avoided because the shell and internal structure are liable to suffer damage resulting from differences in vibration between them or from their differential settlement. In the former case, it is necessary to plan with particular care to ensure that no differential settlement between the outer structure supporting the shell and the inner structure would occur at times of ordinary and earthquake loadings. In case of poor ground condition, it is likely that the foundations of the outer structure supporting the shell are designed more carefully than that of the inner structure, so therefore this precaution should be taken. In the latter case, approach to analysis may be clear because stresses and deformations of the inner structure

Photo 8.10 Rupture of Joints of Outer Structure Supporting the Shell and Inner Structure

Photo 8.11 Rupture (Slipe out) of Joints of Outer Structure Supporting the Shell and Inner Structure

Photo 8.12 Rupture of Joints of Outer Structure Supporting the Shell and Inner Structure

give no effect to the shell. In this case, since a difference in dynamic behavior between the outer and the inner structures causes collision between them with resultant rupture, it is necessary to provide a considerably large space or expansion joints to ensure that no mutual collision between them would take place when earthquake occurs. In such a case, since the foundations would be mutually effected, and a severe differential settlement may occur due to a difference in design conditions, a design should be made, anticipating the occurrence of what has been stated above (refer to Photos 8.10∼8.12).

In case necessity compels connecting directly the inner structure to the shell structure, a design should be made with care because concentrated or line loads act locally on the shell and supporting condition of the shell is compelled to change the state of stresses.

The inner equipment also should be dealt with in the manner described in the case of the inner structure but care should be taken to ensure that differential settlement or a difference in dynamic behavior would result in an accident like breakage in pipelines or electric wiring.

8.3 Precaution on Structure and Execution of Work

Since stress or deformation states of the shell vary depending on the boundary condition, it is desirable to take constructional procedure on the supporting edge line to ensure that assumptions of the calculations are satisfied as much as possible. For instance, when the boundary is roller supported against the normal load, a case may arise where support for horizontal load becomes unstable, and therefore after deformation caused by a principal part of vertical load, as a deadweight, has been ended, supporting edge line should be set in hinge state or fixed.

In the arch-shaped opening which carries the shell, relative displacement (extension) of the arch bottom in the cord direction causes a change (reduction) of the rise as much as several times or more of the relative displacement. Although it is necessary to investigate the effects of the change of the rise on the stability of the arch or the entire shell, on the other hand, if no prior arrangment is made, part fitted with a sash shall have expansibility, damage sash as buckling of the sash in the vertical direction or breakage of glass would occur.

Rise of the shell is important in order to allow the shell display its function. In particular, in case of a flat shell, variations in rise give considerably large effects to distribution of stresses and in an extreme instance, buckling or collapse phenomena is liable to occur and therefore in construction of works measures should be taken to adjust the rise beforehand as much as the expected value of the deformation.

However, in respect to shells having negative Gaussian curvature like the H.P. shell, since the adjustment of a rise which is favarable on a principal curvature, is unfavorable on other of positive and negative principal curvatures of the shell, the planning of construction work should be done with care also in case of the adjustment of rises.

For other particulars, refer to general information given in the structural planning and prepare a design in accordance with the behavior of materials for construction of shell structures and observing instructions on each structure.

8.4 Instructions on Calculation

As mentioned above, many cases arise where the assumptions in calculations and actual structures do not agree with each other. Furthermore, there are many cases where the effects to stresses or deformations are so great that they can not be neglected. Hence, when an assumption in the calculation is made, it is necessary to give proper engineering judgement on variations in actual states of stresses and deformations and their extensions depending on the deviation

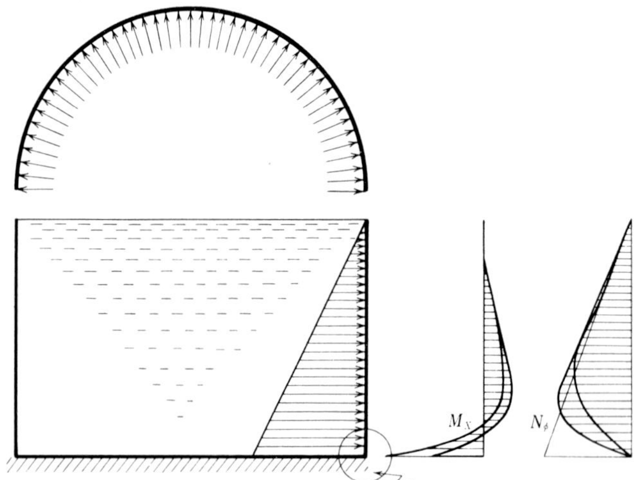

Fig. 8.9

from the assumptions, then design with allowance for modifying to the variations in stresses and deformations.

Presently, as no explanation can be given concerning buckling collapse except in some special cases, it should be avoided in any design, which is liable to incur a considerable large deformation even though the stresses are comparatively small.

In the event of extremely large local stress occuring (i.e. Fig. 8.9) by the theory of elasticity is concerned, if metalic material is used, it may be possible to take into consideration in the design that relaxation of stresses may occur by allowing yielding of a certain section occur.

9. PRESTRESSED CONCRETE CONSTRUCTION

9.1 General

9.1.1 Structural Characteristics

Prestressed concrete construction is a type of reinforced concrete structure with prestress introduced. As its structural characteristics, the following items should be noted.

i) This is an elastic structure under sustained loading and has a tendency to behave like a steel structure rather than as a concrete structure. It is to be noticed however, that the creep of concrete in this system is larger than in ordinary reinforced concrete.

ii) For temporary loading, the design method allows some cracks in the member section (refer to Code of Practice on Prestressed Concrete). Hence, after cracks occur, the system has almost the same structural behavior as reinforced concrete structure. However, so long as the stress of the steel used is within the allowable capacity, the residual deformation is remarkably small compared with reinforced concrete structure.

iii) For the design under sustained loading, through using high strength concrete and introducing large prestressing, the section of small size can have more load bearing capacity, and the ductility for the temporary loading may be lowered consequently. It is, therefore, advisable to have a balanced design for both sustained and temporary loadings. For applying the system, it can be used for poured in place structure or for a composite structure by joining precast members.

9.1.2 Building Scope

As a principle, prestressed concrete construction has no limitation on section shape and dimension in the same way as in reinforced concrete structure. Furthermore, depending on the layout of prestress tendon, a part or whole of the dead load can be balanced off so that a large span can be freely supported. In actuality, however, constructional availability and economy limit the use to some extent and also the building by-laws place stipulated restrictions. Hence, the following scope at present is generally observed.

Number of stories	up to 5 above ground
Span	10~30 m

9.1.3 Damage and Its Causes in Prestress Concrete

The number of actual examples of prestress concrete buildings is still small and therefore damage due to earthquake is little known. Nevertheless, here we describe some possible damage to be expected if any should occur and at what portions in the system. Also, the causes and their remedies will be discussed.

a) Fracture of Steel

This can be considered as due to corrosion of steel or insufficient elongation at fracture. To prevent this, a complete grouting should be performed and use of material which may cause "stress corrosion" is to be avoided. Also steel should not be anchored at the position where extreme flexure may occur under horizontal loadings.

b) Damage of Steel Anchorage

Photo 9.1 Damage of a Prestressed Concrete Lift Slab Construction in the Alaska Earthquake (1964)

Photo 9.2 Same as Photo 9.1

According to an American information this damage is presumably caused in the following sequence.

A great inclination of core part due to insufficient bond length of steel in the reinforced concrete core base → A large forced rotation occured at the juncture between core and slab → Fracture of tendon anchorage → Part of tendon jumped out and some pierced into the building at 10 m apart → Breakage of floor slab

If the anchorage becomes incapable of holding the stressed steel, a part or whole of prestressing will be lost, resulting in lowering the bearing capacity of the structural member. Thus the structural monolithic character of the whole system will be lost. In case of no grouting around the steel, the tendon may jump out of the concrete, causing secondary damage or a local fracture may induce damage of the whole structure[1] (e.g. damage of a prestressed concrete lift slab construction in the Alaska Earthquake (1964) Photo 9.1).

As the causes of damage, slip of tendon or fracture of tendon in the anchorage system and brittle failure of concrete around the anchorage due to inadequate design of the system are considered. It is, therefore, necessary to adopt an appropriate anchorage system and reinforce around the anchorage with considerations for possible rotation in concrete. In addition, complete performance of grouting seems desirable because it is believed to prevent propagation of fracture effectively when local breakage occurs.

c) Brittle Fracture of Concrete

When a member is subjected to bending fracture in a brittle fashion, it corresponds to the case of fracture due to shear in advance of bending. Since there are no actual appropriate examples, here we present a conceptual explanation based on test results. Fig. 9.1 shows an example of the computation result of load-deflection curves for prestress concrete members subject

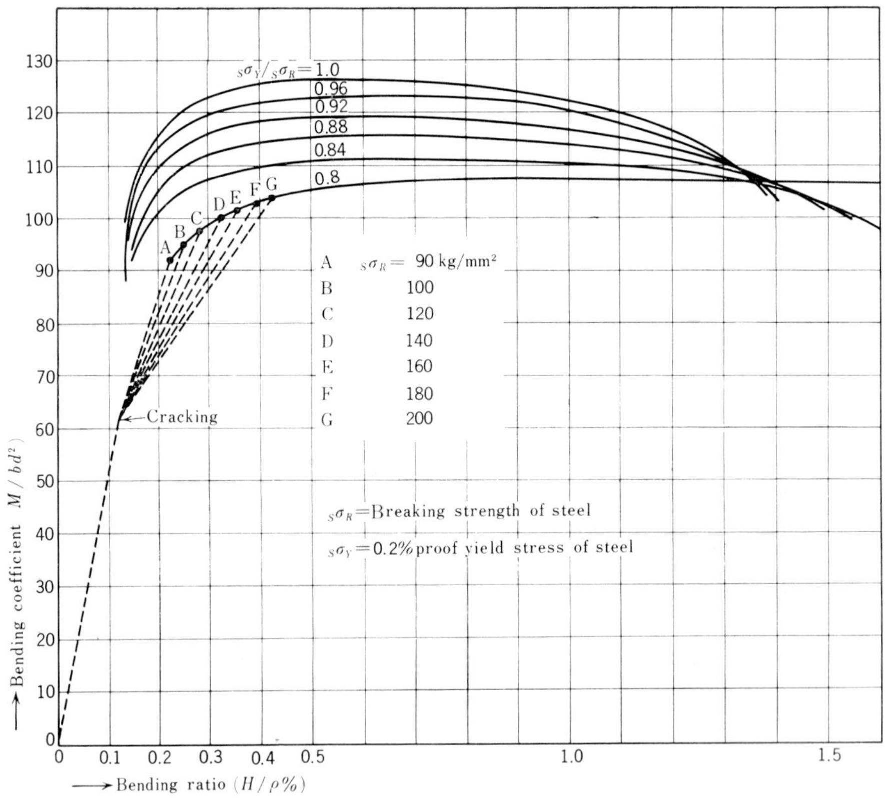

Fig. 9.1 Example of the Computation Result of Load Deflection of Prestressed Concrete Subject to Pure Bending

to pure bending[2]. These are rectangular sections with single tendon and the tensile strength of the steel used is varied between 90~200 kg/mm^2. The case, when other conditions are the same, is indicated in the figure and it is found that ductility of the member is affected by the design significantly. The factors influencing ductility to be provided are considered to be as follows: steel amount, strength, stress and strain relation of steel, amount of prestress, stress and strain relation of concrete and strain magnitude of concrete at the time when steel yields. Generally speaking, when the bearing capacity of compressive concrete is not sufficient, the

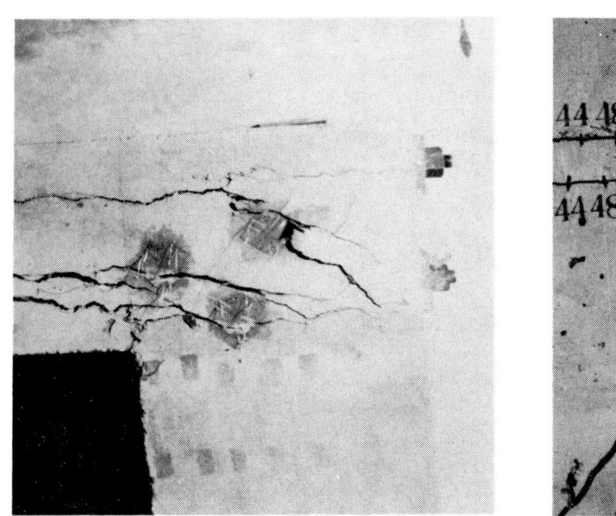

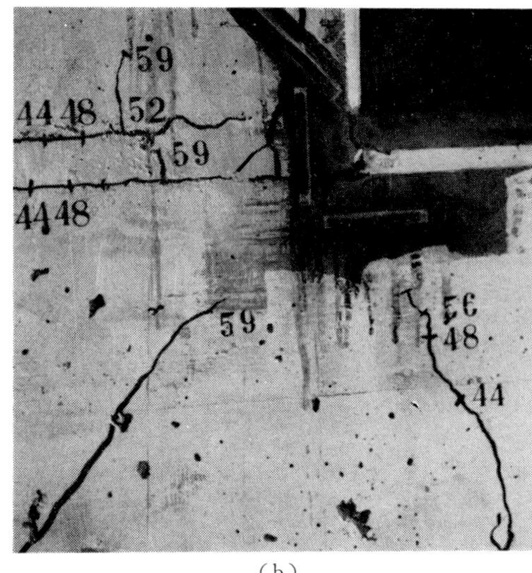

(a) (b)

Photo 9.3 Test Examples Showing Cracks at the Corner Part

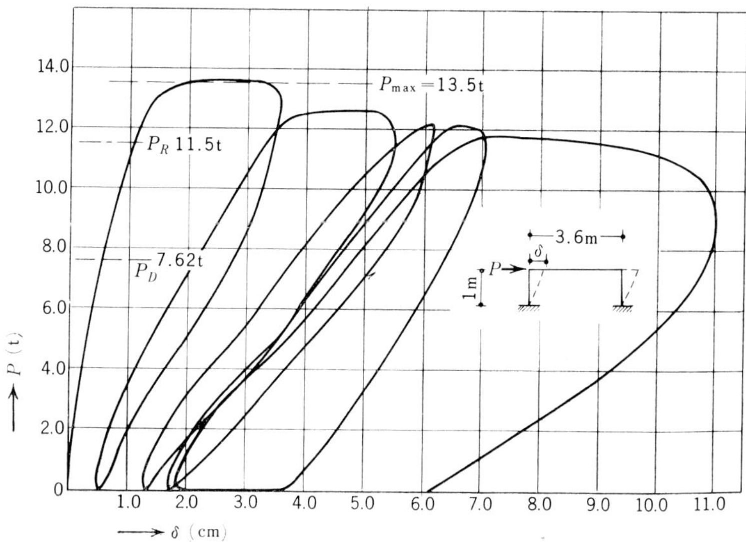

Fig. 9.2 Horizontal Loading-Deformation Curve for One Storied Prestressed Concrete Rigid Frame[4]

[This curve is plotted for the successive loading after the first loading (P_{max}=13.5 ton and the residual deformation is 5.4 mm)

provided ductility becomes poor and therefore excessive amount of steel and extremely high strength should be avoided.

Eliminating shear failure before bending failure takes place is a principle for designing as specified in AIJ Code of Practice on Prestressed Concrete. Since sometimes the corner juncture between column and beam may have cracks as shown in Photo 9.3, such portions should be carefully designed. Fig. 9.2 shows the experimental results of the horizontal loading-deflection curve of one storied rigid frame, indicating that sufficient ductility can surely be provided by appropriate design[4].

d) Loss of Monolithic Character of Structure and Collapse of Members

The experiments indicate that the ultimate strength of framing will be lowered, losing characteristics of rigid frames due to defects in splices and joints of the members[5]. Photo. 9.4 shows prestressed concrete members that fell down by insufficient bearing at the support position. This is an actual example in the Alaska Earthquake (1964). There are some other reported examples of damage in cases of prestressed concrete beams supported on fragile block walls.

Photo 9.4 Example of Collapse of Prestressed Concrete Beams by Insufficient Support Bearing

e) Damage due to Earth Movement

When earth movement and especially sway of columns occur, buildings with large spans and small number of columns are apt to collapse. It is desirable to connect column bases and foundations by foundation tie beams[6].

f) Miscellaneous

Since prestressed concrete framing has generally a longer natural period and lower damping characteristics, the story drift in severe earthquakes becomes larger compared with reinforced concrete and consequently damage to external and internal finish may be extensive. The juncture between shear wall and core structure may develop some fractures. Especially when different systems are used together in the same building, e.g. in case of prestressed concrete for the core and reinforced concrete for the framing, because of different vibration characteristics, the building tends to be subjected to this type of damage.

9.2 Structural Planning

9.2.1 General

As for rigid framing system, its characteristics of rigidness should be maintained with proper design of the joints until the whole structure collapses. For this purpose the joints should be sufficiently rigid and also ductile. Members of simple or hinged support should be arranged not to move or fall down and substructures supporting the members should also be adequately designed.

Though the layout of the resistive elements against earthquake in prestressed concrete buildings follows generally ordinary reinforced concrete structural systems, in large span halls, it may often be difficult to locate the regular aseismic shear walls and in application of precast floor slabs there may be some cases where joint action between the seismic elements and the open frames is not sufficient. Therefore, in planning of the aseismic elements, instead of following the usual concepts on reinforced concrete structure, it is necessary to proceed to design with considerations for the behavior of actual buildings in earthquakes.

As for foundations, since the weight per column to be carried tends to become larger compared with ordinary reinforced concrete, they should be designed carefully.

9.2.2 Framing Plan

a) Layout

There are two ways in laying out the framing plan.

In the one way, as shown in Fig. 9.3 (a), prestressed concrete beams are used in both x and y directions and in the other way, as shown in Fig. 9.3 (b), prestressed concrete beams are used in the y direction only and reinforced concrete beams are used in the x direction. In both layouts, columns are of either prestressed concrete or ordinary reinforced concrete. In any case, a symmetrical layout is desirable in both plan and elevation. In prestressed concrete construction, symmetry should be considered as an essential factor.

b) Rigidity

Prestressed concrete buildings designed by proper utilization of material characteristics have

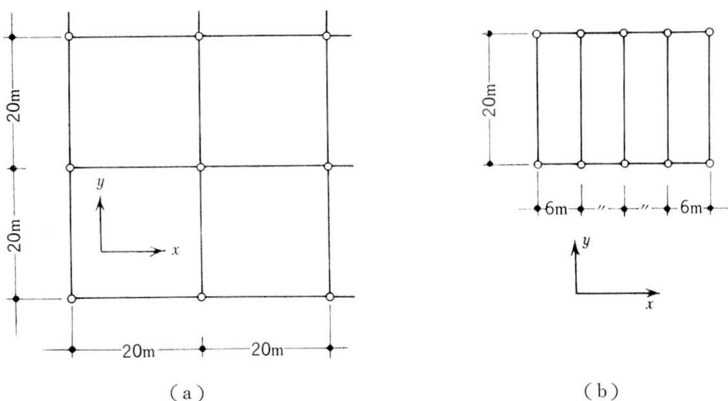

Fig. 9.3 Framing Layout

generally larger spans with relatively smaller sections and consequently the rigidity provided is lower than in reinforced concrete structures.

Fig. 9.4 shows the measurement of natural periods of actual buildings. Observing this figure, prestressed concrete building has less rigidity in the region of low stress than in ordinary reinforced concrete building. If the rigidity is forced to be close to the one of reinforced concrete structure, it is feared that the characteristics of prestressed concrete structure will be lost. In case where the column size is proportioned to be too slender, sway at the time of earthquakes becomes excessive and some secondary hindrance may occur even though the design is safe in the normal strength calculation[3].

The stiffness ratios between columns and beams are different from the ones in reinforced concrete and there are many cases where the beam stiffness is only a fraction of the column stiffness. When span is large, this corresponds to the above case and it is more economical to make the stiffness larger, combining the slab and the beam to form a monolithic unit.

The characteristics of prestressed concrete will be most effectively displayed for vertical loading and consequently in many cases horizontal loading will be largely taken in the resistive core against earthquakes. In these cases, the interaction between the core and the frames on deformation and the distribution of forces and also the safety of the joint should be investigated for all stages of loading.

The calculation of strength for the connection between column and beam is usually not

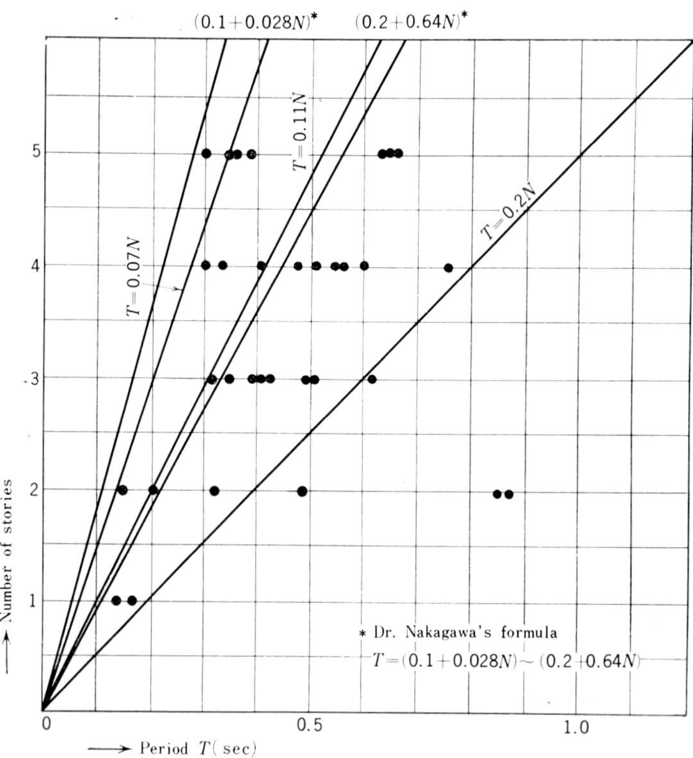

Fig. 9.4 Example of Measurement of Natural Period of Prestressed Concrete Buildings

performed. When the beam depth can be small as in prestressed concrete structure, dimension of the juncture panel is small for the amount of moment existing at the junction. The shear capacity of the juncture panel sometimes may be lower than the one of the column or the beam. Especially in case of joining column and beam by reinforced concrete method instead of introducing prestress, the design should be carefully carried out[3)5)].

9.2.3 Planning Arrangement of Resistive Elements against Earthquake

Though general provisions for reinforced concrete structures should be referred to, particular items for prestressed concrete are the following.

i) Since there are many cases where span is large and framing or floor system is prefabricated, the resistive elements against earthquakes should be so arranged that torsion does not develop.

ii) When the horizontal force distribution ratio between the frames and the aseismic elements is determined, the deformation characteristics, especially those beyond the elastic range for both systems should be considered. If one is reinforced concrete and the other is prestressed concrete, this item becomes very important.

iii) It is effective to introduce prestressing into aseimic shear walls for reducing the residual deformation of the building. However, since the maximum deformation in earthquakes sometimes becomes rather large, it is necessary to consider the influence on the framing connected to the shear walls or on the floor slabs and the exterior or interior finishes.

9.2.4 Slab Arrangement

It is a principle which requires to make a monolithic system combining floor slab with framing and to provide sufficient shear rigidity of the system as seen in reinforced concrete structure. However, since prefabricated floor slabs are often used in prestressed concrete, this principle can not always be followed. In this connection it is essential to regulate the shear forces to be transmitted to frames and walls based on necessary calculations and experiments. Fig. 9.5 shows an example of prefabricated floor system.

Although it is devised to transmit the horizontal shear forces by welding the inserted steel

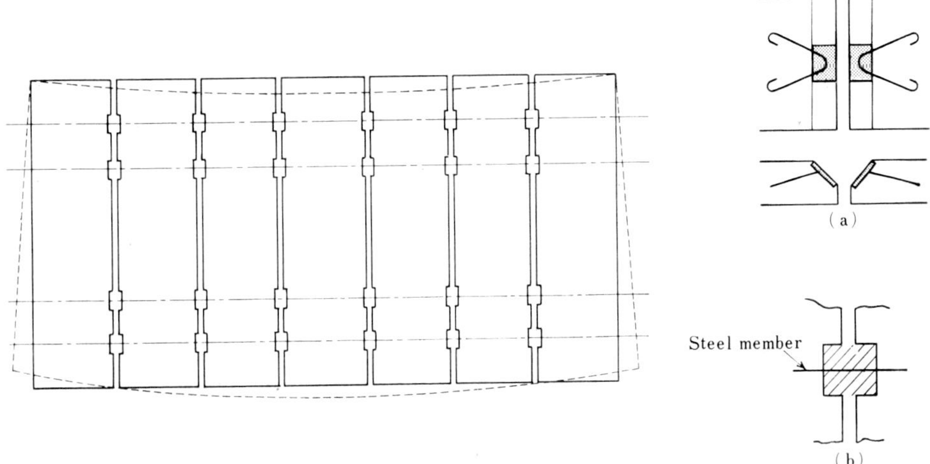

Fig. 9.5 Example of Prefabricated Floor System

plates in the adjacent units (Fig. 9.5 (a)), when the floor slab area is large and out-of-plane rigidity of the surrounding framing is not sufficient, the system tends to deform as shown by the dotted line. In this case, tension develops at some joints and if the dowel bar is not covered sufficiently by concrete, it will be tensioned out in the perpendicular direction to the slab. When the slab thickness is 5 cm or so, the difference in level between the adjacent slab units should be considered and given adequate treatment. Joining the steel bars located in the longitudinal lines crossing the units as shown in Fig. 9.5 and providing the shear keys to transmit the shear forces between the slab units, reliability is improved (Fig. 9.5 (b)).

9.2.5 Foundation Planning

In prestressed concrete structures there is a tendency to have more concentration of forces in the foundation than in ordinary concrete structures. Therefore, connecting the foundations by strong and rigid tie beams to distribute the reactions from one foundation to another and it is desirable to take into consideration the stresses as in rigid frames due to the rotation of the foundations. When tie beams can not be used, the whole system may have to be treated as a statically determinate structure, changing the basic concept of approach.

9.3 Seismic Calculations

9.3.1 General

The seismic calculation for prestressed concrete structure should be performed in accordance with the AIJ Standards and Code of Practice for Prestressed Concrete based on Building Standard Laws and also the official notice No. 223 of 23 Feb. 1960 issued by the Ministry of Construction.

In the seismic calculation, the following points differ from the ones in reinforced concrete structures:

i) The allowable stresses for temporary loading are not given.

ii) For the emergency loading, the breaking strength of the section should be checked whether it is greater than a specified stress to be obtained from multiplying the stress based on elastic calculation (for G, P, and K) by a load factor, e.g. $1.2(G+P)+1.5K$ where G, P and K are stresses due to dead load, live load and earthquake force respectively. Unlike in reinforced concrete, the design in which the stresses of steel and concrete are to be within the allowable amounts is not to be applied.

Apparent safety factor ($S.F.$) for the prestressed concrete structure, designed as described above, for example :

$$S.F.=\frac{1.2(G+P)+1.5K}{G+P+K}$$

is varied depending upon the relative magnitudes of G, P, and K.

This relation is indicated in Fig. 9.6.

From the figure $S.F.$ is 1.3 as the minimum value and the prestressed concrete structure is designed apparently safer than the ordinary reinforced concrete structure.

When the actual safety is considered, the following items need proper judgement:

i) Whether the same design seismic coefficient as applied in reinforced concrete structure

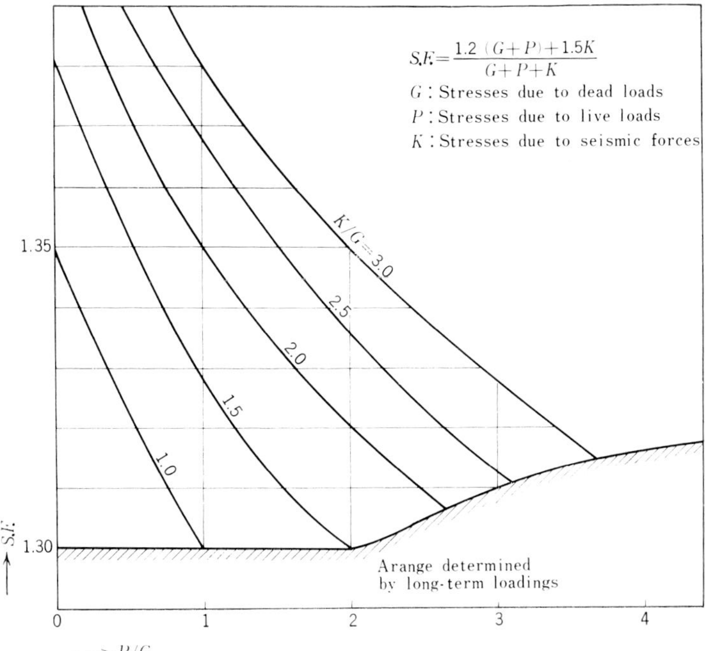

Fig. 9.6 A Simulate Safety Factor Against Earthquake of Prestressed Concrete Structure

is appropriate or not for prestressed concrete structure.

ii) In case where the concrete has not much reserve strength in the design and if the actual seismic coefficient exceeds the one used for the design, could the safety be considered in the same way as in reinforced concrete structure?

To adopt a more definite approach requires further research and investigations. At present, it is suggested to follow the above mentioned code stipulations and to design the structure with sufficient ductility. Besides, if the actual construction is properly perfomed, then the resulting structure is considered to be sufficiently seismic resistant.

9.3.2 Combination of Forces and Investigation of Member Section

Combination of forces to be considered for the earthquake loading should be as follows:

In general cases $1.2(G+P)+nK$

For cases where there is much snowfall $1.2(G+P+S)+nK$

where

$n=1$, when the sign of $G+P$ or $G+P+S$ is opposite to the sign of K

$n=1.5$, when the sign of $G+P$ or $G+P+S$ is equal to the sign of K

G, P, S, and K : stresses due to dead load, live load, snow load and earthquake force, respectively, based on the Building Standard Laws.

For the earthquake loading, it should be investigated that the ultimate strength of the section be larger than the total amount of the combined stresses described above. Besides, the AIJ Code stipulates that the ratio of ultimate strength to cracking strength in flexure should be more than 1.4 and shearing failure should not take place before bending failure. These are important for prevention of abrupt collapse.

9.3.3 Remarks on Stress Calculation

a) Elastic Modulus of Concrete and Stiffness of Member

Elastic modulus of concrete is to be taken as listed in Table 2.10 described in Chapter 2 of the AIJ Code on Prestressed Concrete.

In prestressed concrete, since the elastic modulus of concrete is used for the calculation of deformation and creep and also for the fabrication, it is desirable to obtain the elastic modulus corresponding to the working stress from constructing the stress strain curve by testing. Fig. 9.7 shows one example of the stress strain curve for a dense concrete using river sand and gravel.

When precast floor system is adopted, the addition of rigidity for the beam in the framing should be figured with considerations for the joint details to the framing.

b) Framing

The calculation principle is based on reinforced concrete. However, according to the sequence of fabrication and prestressing, the framing mechanism after completion may become quite different from the initial stages of construction. Therefore, design should be carried out considering this fact.

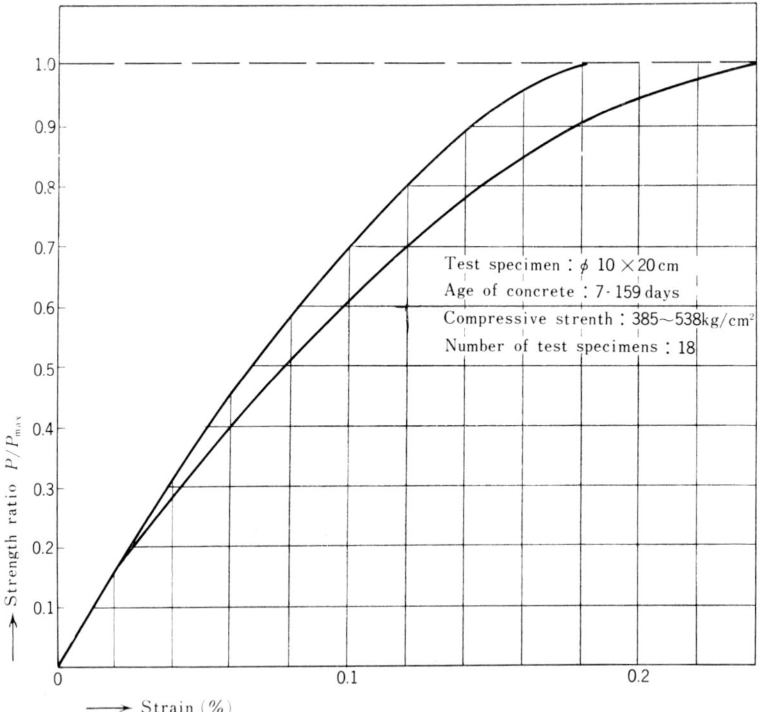

Fig. 9.7 Example of Stress Strain Curve for the Dense Concrete Using Sand and Gravel from Rivers

c) Slabs

Though it is generally assumed rigid, careful consideration is necessary in the design and test results should be referred to.

d) Stress Calculations

Besides the items observed for ordinary reinforced concrete, investigations should be performed on the following items : the stresses in the panel zone of the joint, shrinkage due to seasoning of the member, the stresses due to creep and the effect of stress redistribution. When the approximate analysis which is normally used for reinforced concrete structure is to be applied for prestressed concrete, it is necessary to make sure whether its validity is practically tolerable or not, considering the structural system, its scope and rigidity distribution.

9.4 Remarks on Construction

9.4.1 Concrete

In statically indeterminate structures generally, the required amount of steel tends to be large and consequently it results in heavy reinforcement in the junction between column and beam, making pouring of concrete difficult. Therefore, concreting with good workability and reliable pouring should be adapted.

9.4.2 Tendon and Anchorage

Regarding the prestressing tendon, the official notice No. 223 dated 23 Feb. 1560 by the Ministry of Construction, the AIJ Standards and Code of Practice of Prestressed Concrete and JIS G 3536-1960, Prestressing Wires and Strands, should be referred to and also the following items should be observed:

i) The tendon used in any part which might have a large elongation when the large deformation occurs in the structure, as in the neighborhood of the column base, should have its breaking elongation at least equal to or more than in other parts.

ii) The tendon and anchorage whose safety have been confirmed by testing should be used. Besides, it is necessary to determine the location of the anchorage, to prevent the tendon from getting tangled around the anchorage.

9.4.3 Grouting

Bonding between the prestressing tendon and concrete is achieved by grouting in the post-tensioning system. Grouting plays an important role for protection from rust of the tendon and it is said that well executed bonding performs anothersignificant function of preventing collapse of the whole structure due to propagation of local damage.

It is also said that the breaking strength of the member section is affected about 10% with or without grouting[7]. Regarding grouting, besides referring to the code of practice of prestressed concrete, the following should be noted.

i) If bond is particularly significant, the tendon of the smallest diameter may be used.

ii) The effectiveness of grouting depends on the degree of filling. Therefore, it is dangerous to have uncertain filling using a grout mixture of high strength but of poor workability.

9.4.4 Columns

When prestressing is introduced into columns, the items described in 9.4.2 for the tendon anchorage in the joint should be referred to.

Regarding the shear investigation, the principal tension stresses at the effective prestressed condition are to be calculated and even when the axial force from the summation of external force and prestressing force becomes zero or less, it should be designed so that the shear failure may not take place before bending failure (refer to the Code of Practice of Prestressed Concrete).

The columns are often designed as the ones of ordinary reinforced concrete in accordance with the calculation standards of reinforced concrete structure. This results in different safety against fracture between columns and beams. Therefore, investigation on the safety of columns referring to the Code of Practice of Prestressed Concrete should be performed. As a principle, it is desirable that more safety should be provided in columns than in beams.

9.4.5 Beams

There may be cases where the section subjected to both positive and negative bending moments has ordinary reinforcement for a part of the moments besides the prestressing tendon (Fig. 9.8). In these cases, investigation against failure described in 9.3.2 and also checking the stress in the steel bar should be carried out. It is necessary to reduce the working stresses to an appropriate level considering the residual deformation after earthquakes.

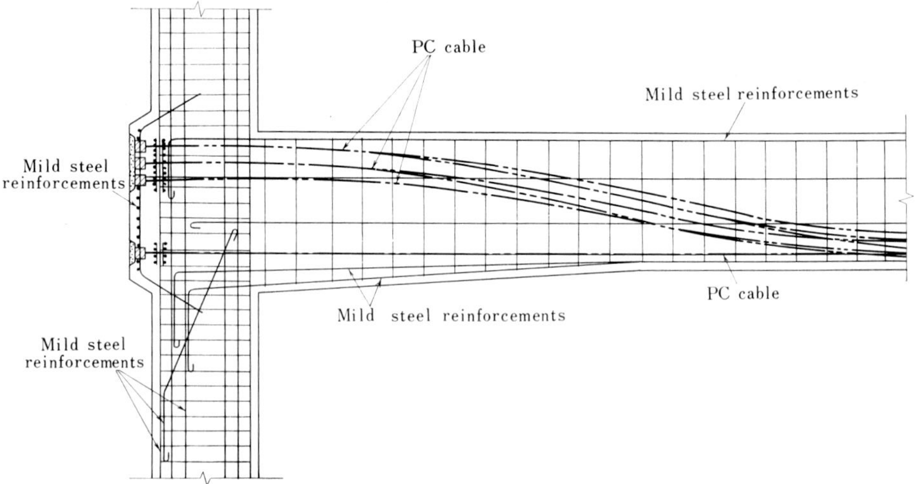

Fig. 9.8 Example Using Ordinary Steel Bars for the Positive Moment at the Beam End

9.4.6 Joints

Since prestressed concrete is often used for large span structures, large stresses occur at the beam ends. Therefore, the joints should be designed to be able to transmit these large stresses reliably. If the strength of the panel zone of the joint is inadequate, cracking may be produced as shown in Fig. 9.9 and the type indicated in Fig. 9.9 (b) is especially dangerous.

When ordinary bars are used for the positive moment (Fig. 9.8), the anchorage of the bar should be carefully checked. In Fig. 9.10 (a) the bond length is sufficient but the stress concentration may occur in the panel zone of the joint. This is undesirable because, once cracking is produced in the panel, an abrupt reduction of strength may be caused. Therefore, the design indicated in Fig. 9.10 (b) is preferable. As for the joint between the girder and beam and the joint between the beam and slab, besides the usual strength calculation, consi-

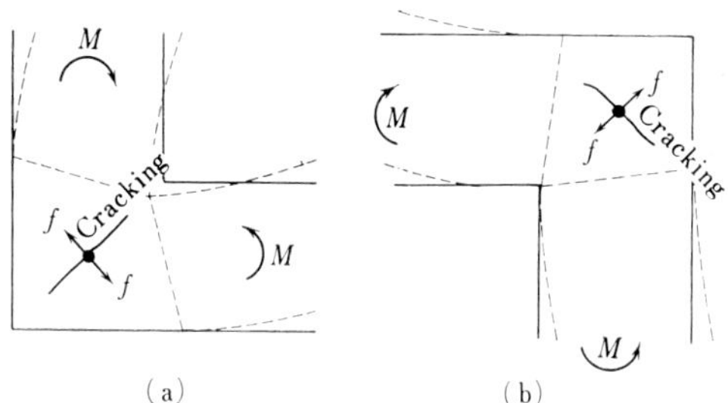

Fig. 9.9 Cracking in the Joint Panel

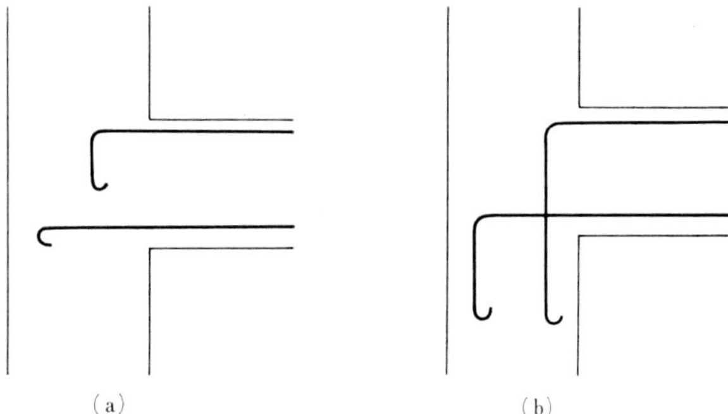

Fig. 9.10 Anchorage of Ordinary Bars in Beam

deration for out-of-plane deformation and investigation on ductility should be carried out.

9.4.7 Miscellaneous

Though walls are often constructed using ordinary reinforced concrete, sometimes seismic resistive core structure may have prestressing introduced. In either case, it is necessary to have a well balanced design considering the difference of deformation character from the framing.

In case of precast slab construction instead of cast-in-place construction, the items described in 9.2.4 are to be referred to.

Foundations are usually cast at the site using ordinary reinforced concrete. The influence of ground conditions on the seismic force given to the superstructure is significant[8]. In case of the building with large spans prudent investigation is very essential.

References

1) W.E. Kunze and others : The March 27 Alaskan Earthquake Effects on Structures in Anchorage. Journal of ACI. June 1965.

2) K. Nakano : Study on Load-deflexion Characteristics of Prestressed Concrete Structures, July 1965, BRI Occasional Report No. 24.

3) Totsuka, Motooka and Nakano : Research on the Design of Juncture of Prefabricated Framing of Prestressed Concrete. Proceedings of Architectural Institute of Japan No. 103 Oct. 1964 (in Japanese).

4) K. Nakano : A Treatise on Design Strength of Prestressed Concrete Structures. Proceedings of Architectural Institute of Japan. Special issue, Sept. 1965 (in Japanese).

5) Totsuka, Umigami and Kawamura : Experimental Study on a Closed Framing with Prestressed Concrete Beam. Journal of Prestressed Concrete Engineering. Vol. 3, No. 3. June 1961 (in Japanese).

6) Japan Prestressed Concrete Engineering Association : Tentative Draft, General Principles for Earthquake Resistant Design of Prestressed Concrete Structures. Dec. 1964.
7) K. Nakano : Experimental Study on Strength Character of Prestressed Concrete Members Jointed by Grouting. Journal of Prestressed Concrete Engineering. Vol. 3, No. 4. Aug. 1961 (in Japanese).
8) T. Hisada, K. Nakagawa and M. Izumi : Normalized Acceleration Spectra for Earthquakes Recorded by Strong Motion Accelerographs and Their Characteristics Related with Sub-soil Conditions. June 1965. BRI Occasional Report No. 23.

10. REINFORCED CONCRETE WALL CONSTRUCTION

10.1 General

10.1.1 Structural System Characteristics

Reinforced concrete construction (hereinafter referred to as wall construction) is a structural system consisting of reinforced concrete bearing walls (walls that can resist both horizontal and vertical loads), roof slabs and floor slabs. This type of construction is widely applied to

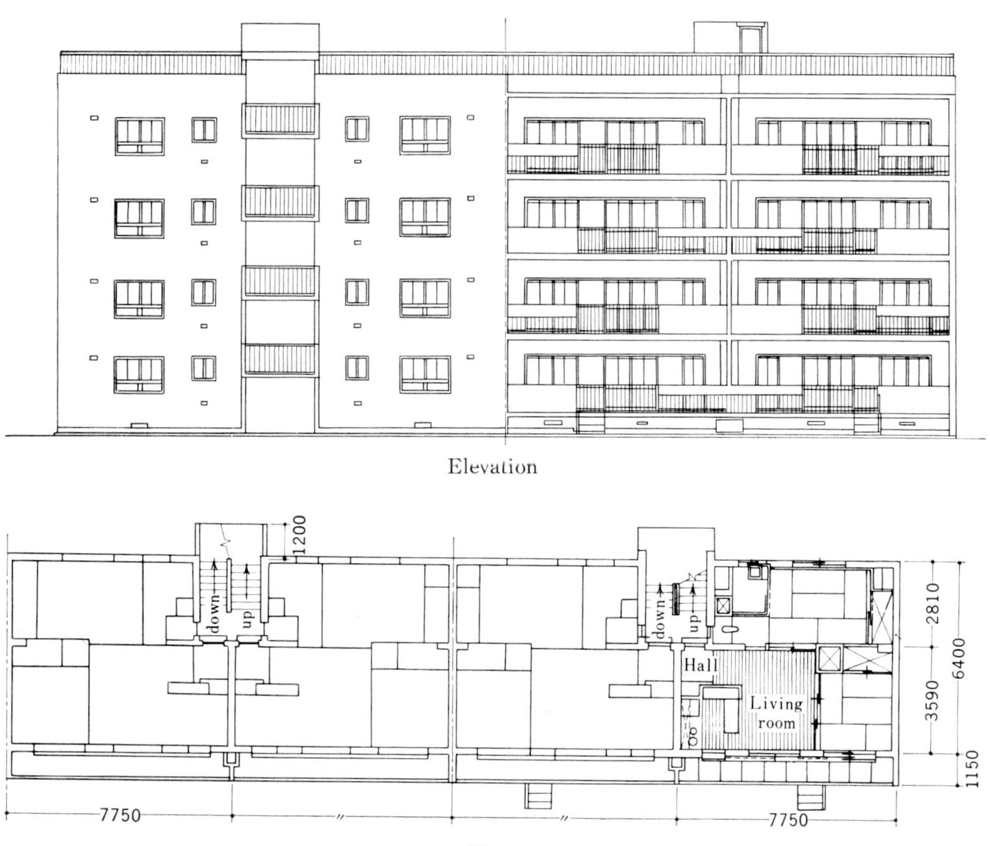

Fig. 10.1 An Example of Apartment House of Wall Construction

apartment houses as shown in Fig. 10.1 for the reason that it is convenient for space utilization and is economical as well in comparison with conventional reinforced concrete frame construction.

The wall construction, theoretically speaking, requires solution of framing taking into consideration the deflections due to both bending and shear and the rigid zone at joints of frames as well, or other complicated analysis because wall beams, wall columns, etc. have large depth compared with member length ; besides, in some cases, their strength become less than those of conventional reinforced concrete construction because their member thickness is small. Consequently, its design is rather difficult. However, in consideration of the fact that its application is mainly for apartment houses, it is a general rule that special analysis is not required provided the design is done in compliance with the "Standard for Structural Design of Reinfoced Concrete Wall Construction" of AIJ. In this chapter, it is intended to give precautions from the aseismic viewpoint for those cases where no special analysis is made.

Photo 10.1 An Aerograph of Apartment Houses of Wall Construction under Management of Niigata Prefecture in the Niigata Earthquake (Offered by Prof. Yoshikatsu Tsuboi, Institute of Industrial Science, University of Tokyo)

Mentioning one other thing, the wall construction treated in this chapter is limited to those using concrete having a 4-week compressive strength of 135 kg/cm^2 or more in case of normal weight concrete and 120 kg/cm^2 or more in case of lightweight concrete, with a value of more than 75 of the strength (kg/cm^2) divided by the specific gravity in dry air. For those cases where other kinds of lightweight concrete are used such as foam concrete, etc., there are standards that apply to them such as "Standard for Structural Design of Reinforced Lightweight Concrete Wall Construction" by AIJ, "Thermo-con Wall Construction Design Manual", by the Fireproof Building Association of Japan, etc.

10.1.2 Earthquake Damage

Buildings of this type of construction suffer less from earthquakes. A few examples of damage due to the Niigata Earthquake of June 1964 are exhibited and every one of the cases may be interpreted as indicating precautions to be taken in design not only for this type of construction but also all other types of costruction.

a) Apartment Houses of Wall Construction under Management of Niigata Prefecture

Photo 10.1 shows an aerograph of the apartment houses and Fig. 10.2 shows the degree of inclination and its direction. These were caused by the liquefaction phenomenon of sandy soil layers and indicates the necessity of precautions in foundation design and construction when the supporting soil layers are soft and weak. Notwithstanding such large inclinations, the superstructure did not indicate such damage as cracks in the building proper.

b) Apartment Houses of Wall Construction

As shown in Fig. 10.3, these are two 4-story apartment houses without basement ; 7.4 m wide and 58 m long, supported by wood piles of 20 cm dia. and 7 m length. Photo. 10.2 shows cracks in the vertical direction in the central part of side walls of the building, and the cause is considered to be attributable to double effects of shrinkage, temperature stresses, etc. and differential settlement of the ground on account of the length of the building. The consideration of the above points are required the design.

c) Bachelor's Dormitory of Wall Construction

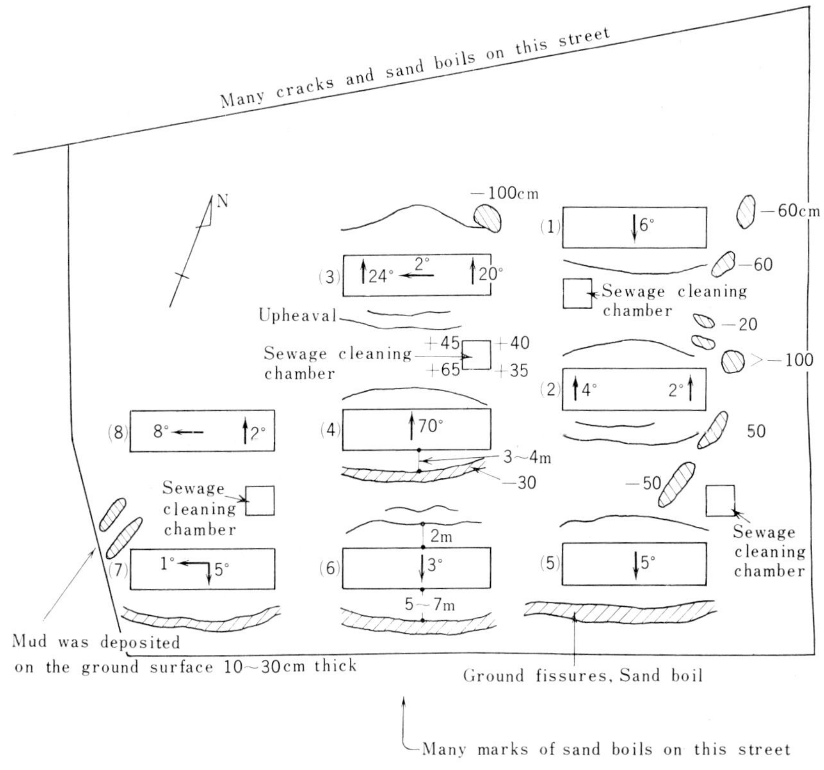

Fig. 10.2 Site Layout, Inclination of Buildings and Condition of Fissures under the Apartment Houses in Kawagishi-Cho, under Management of Niigata Prefecture

Photo 10.2 Cracks in an Apartment House of Wall Construction

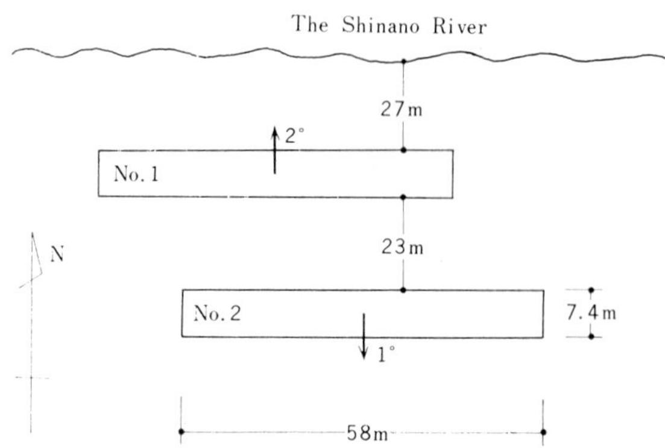

Fig. 10.3 Site Layout and Inclination of Apartment Houses of Wall Construction

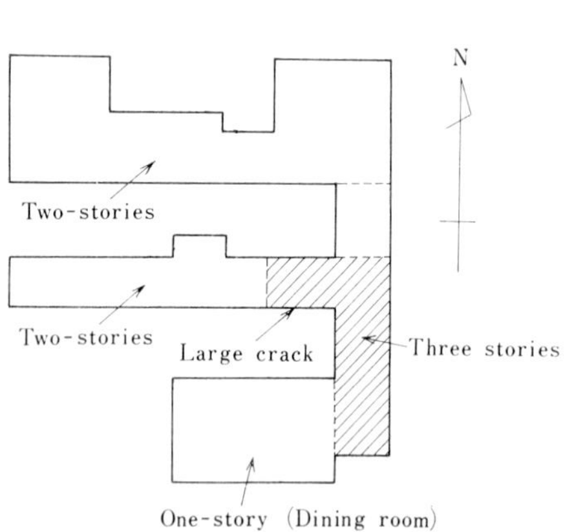

Fig. 10.4 Block Plan of Bachelor's Dormitory of Wall Construction in Akita Prefecture

Photo 10.3 Cracks in Bachelor's Dormitory of Wall Construction in Akita Prefecture (Offered by Prof. Toshio Shiga, University of Tohoku)

As shown in Fig. 10.4, the bachelor's dormitory consists of two units having continuous wall footing but no piles. In one unit which is 3-storied, as seen from Photo. 10.3, such a large crack developed that concrete was stripped off and reinforcing bars exposed at the part of conjunction of the 3-storied and 2-storied portions. This indicates that the part of conjunction between building portions having different number of stories becomes a weak point and requires proper design considerations.

10.2 Structural Planning

10.2.1 Policy

In the aseismic design of buildings of wall construction, it is necessary to pay attention to the following points :

i) The building should be 4 stories or less above the ground and 13 m or less in height at eaves.

ii) Plan and elevation of the building should be so arranged that they do not become too complicated.

iii) Bearing walls should be arranged to be well balanced in the building as a whole so that stresses are distributed as evenly as possible.

iv) It is recommended to use reinforced concrete construction for roof and floor slabs in order to make the building resistant to earthquake forces.

10.2.2 Arrangement of Bearing Walls

In arranging bearing walls, it is necessary to pay attention to the following points :

i) Bearing walls should be placed near the periphery of the building as much as possible, as well as the center of gravity of the building and that of rigidity of the bearing walls should be made to coincide as much as possible so that torsion may not occur in the building.

ii) For effective utilization of the bearing walls, it is safe to place them one over another in the same location in plan throughout all the stories. In case they cannot be so arranged, efforts should be made to provide some structural means that can safely support the reactions of the bearing wall at the lower edge of that wall.

iii) Uniformity in the location, length and thickness of bearing walls throughout the entire building makes flow of forces smooth and from the viewpoint of construction, results in economy as well, because it affords conveniences in the use of metal forms, placement of reinforcing bars, etc.

10.3 Remarks on Stress Calculations

In calculating stresses in buildings of wall construction for earthquakes, the following recommendations may be followed provided the disposition of bearing walls is appropriate :

i) Calculations are to be conducted separately in two principal directions in plan.

ii) It is recommended that the average shearing stress (value of shearing force at the story level divided by respective horizontal sectional areas of the bearing walls in lengthwise and widthwise directions) of each bearing wall due to earthquake should be made 4 kg/cm^2 or less in consideration of stress concentration and other factors.

iii) It is permitted by the standard to calculate the moments in wall column due to horizontal forces on the assumption that the inflection point is at the mid-height of the wall column, and those of wall beam in such a way as to keep balance with those of the wall columns. However, care should be taken where there is considerable difference in the depth of wall column and wall beam.

10.4 Structural Notes

10.4.1 Bearing Wall

a) Length of Wall

In order to make bearing walls as shear resisting members, the wall length l on the horizontal section should be 30% or more of the height of the portion of wall and, at the same time 45 cm or more (refer to Fig. 10.5).

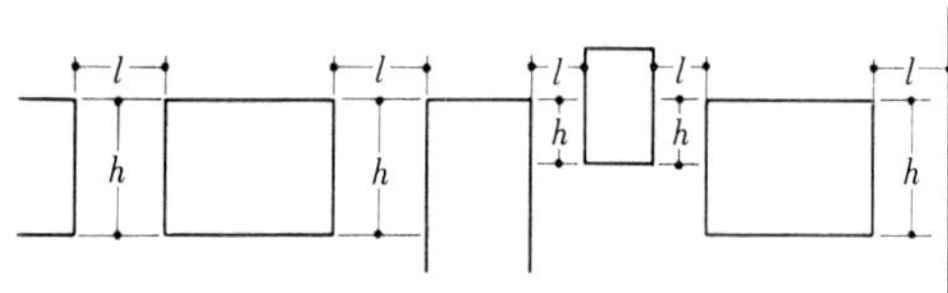

Fig. 10.5 Length of Bearing Wall

b) Thickness of Wall

Thickness of bearing wall t should be made equal to the value t_o shown in Table 10.1 or more in consideration of construction, strength, lateral buckling, etc.

Vertical distance between supports that is critical to the structural strength for the case where the wall is supported by a rigid roof slab or floor slab, is the distance between their respective centers and in the case where the floor of the lowest story is of timber construction, the distance from the bottom of the foundation to the center of the floor of the story above.

Table 10.1 Wall Thickness

Story		Wall thickness t_0 (cm)	Remarks
Stories above ground	One-story building	12 and $h/25$	h : Vertical distance between supports critical to the structural strength
	Each story of a building having number of stories of 2, or top story of a building having number of stories of 3 or more	15 and $h/22$	
	Other stories	18 and $h/22$	
Stories below ground		21 and $h/18$	

c) Amount of Wall

In order to increase the earthquake-resistance of buildings of wall construction, it is most effective to increase the amount of bearing walls. For this reason, as a necessary amount of bearing wall for earthquake-resistance, the wall amount is defined as the value of the sum of the length of bearing wall in lengthwise and widthwise directions respectively divided by the floor area, that is, the length of bearing wall in one direction per unit floor area ; and in case where the thickness of wall agrees with the minimum thickness given in Table 10.1, a wall amount not less than the values shown in Table 10.2 must be provided in the lengthwise and widthwise directions in each story.

Table 10.2 Wall Amount

Story		Wall amount (cm/m²)
Stories above ground	Third story and above counting from the top	12
	First story of 4-story building	15
Stories below ground		20

d) Shear Reinforcing Bar Ratio

Although bearing walls are so proportioned that no shear crack be permitted, shear reinforcing bar ratio of bearing wall should have the value shown in Table 10.3 or more in order to provide for uncounted conditions such as shrinkage, cracking, etc. of the concrete.

Table 10.3 Shear Reinforcing Bar Ratio

Story		Shear reinforcing bar ratio (%)
Stories above ground	One-story building or top story	0.15
	Second story counting from the top	0.20
	Other stories	0.25
Stories below ground		0.25

e) Diameter and Arrangement of Shear Reinforcing Bars

Diameter of shear reinforcing bars should be 9 mm or more, and the spacing should be shown in Fig. 10.6, 30 cm or less (except 45 cm or less for one-story building) in both vertical and horizontal directions parallel to the face of the bearing walls ; and even in the case of staggered

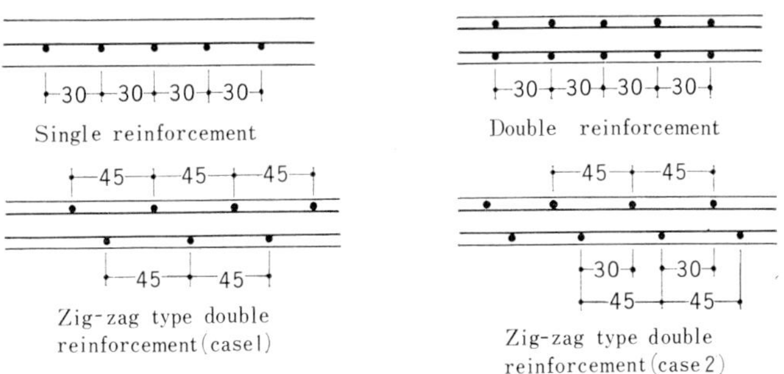

Fig. 10.6 Limitation on Spacing (cm) of Shear Reinforcing Bars (applicable to both vertical and horizontal bars)

arrangement in two layers, it should be 45 cm or less with regard to respective layers. Also, bearing walls more than 18 cm thick should have reinforcing bars placed in two layers.

f) Bending Reinforcing Bars

At structurally important portions such as edges, periphery of openings, etc. of bearing walls, bending reinforcing bars at least equal to those shown in Table 10.4 should be provided.

However, for bearing walls less than 18 cm thick used in the top story and having intersecting bearing wall at right angles, 1–13 φ or 1–D 13 may be used disregarding Table 10.4 in consideration of the resistance of the interesecting wall.

Table 10.4 Bending Reinforcing Bars at Edges, etc. in Bearing Walls

Story	Bending reinforcing bars at edges, etc. of bearing wall	
	$h > 1$ m	$h \leq 1$ m
One-story building	1–13 φ or 1–D 13	1–13 φ or 1–D 13
Top story and second story from the top	2–13 φ or 2–D 13	1–13 φ or 1–D 13
Other stories	2–16 φ or 2–D 16	2–13 φ or 2–D 13
Remarks	h : Height of the edge of opening along which bending reinforcing bar is placed (in case upper or lower part of the opening is of structurally inferior construction to the bearing wall, height of that part of wall should be included)	

g) Diagonal Reinforcing Bars at Corner of Openings

As diagonal cracks occur at corners of openings due to shearing forces, shrinkage, thermal change, etc., diagonal bars having a sectional area more than 1/2 of the reinforcing bars shown in Table 10.4 and 9 mm or larger in diameter should be used in order to prevent expansion of these cracks. However, only in the case where placement of diagonal bars is extremely difficult in execution of concrete placing and other works, diagonal bars may be replaced with additional vertical and horizontal bars.

h) Reinforcing Bars at End of Bearing Wall

At ends of wall or at crossing of walls, and at other similar joints such as that between wall and floor, it is especially necessary, to provide sufficient length for anchorage.

i) Minor Opening near Edge of Wall

In case ventilating opening, other minor openings or ditch are provided adjoining the edge of major opening in wall, wall beam, footing beam, etc., it is necessary to take thorough care with regard to its location and size because the concrete does not spread well and there is the possibility of bad influence on the rigidity, strength, etc. of the member.

10.4.2 Wall Beam

(1) At the top of bearing wall of each story, a continuous wall beam should be provided and reinforced sufficiently against bending.

(2) As for shear reinforcing bars in the wall beam, both vertical and horizontal bars should be equal to or more than those for the bearing wall.

(3) With regard to wall beam that is not made integral with a rigid roof slab or floor slab,

care should be taken so that stiffness of the beam is ensured against lateral bending.

10.4.3 Roof and Floor

Roof and floor should be of reinforced concrete slab and of such construction that it is integrated with the bearing wall or wall beam.

10.4.4 Footing and Footing Beam

(1) For the purpose of preventing uplift, rotation and movement of bearing walls and to develop the full shear resistance, continuous footing should be provided under bearing walls of the story at the base.

(2) Thickness of web of the continuous footing should be more than that for the bearing wall above.

(3) Principal reinforcement of the continuous footing should be determined based on stresses due to contact soil pressure, etc., but other reinforcement should be not less than that of the bearing wall.

(4) For soft and weak soil layers, it is necessary to guard thoroughly against differential settlement by means of piling, soil improvement or other methods.

11. PRECAST REINFORCED CONCRETE WALL CONSTRUCTION

11.1 General

Precast reinforced concrete wall construction (hereinafter referred to as precast wall construction) is such a structural system as shown in Fig. 11.1, in which structural members such as bearing walls, floors, roofs, stairs, etc. of the reinforced concrete wall construction (hereinafter referred to as wall construction) mentioned in Chapter 10, are made into precast reinforced concrete products of large dimensions (room-size) and connected effectively at the construction

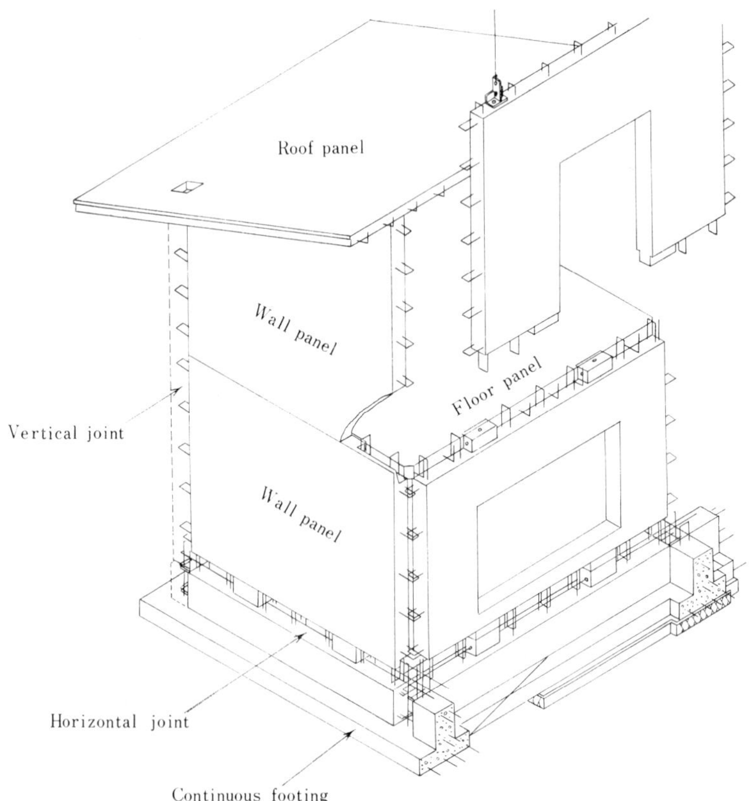

Fig. 11.1 Outline of Precast Wall Construction

site.

As for the advantages of precast wall construction, members of better quality are easily obtainable compared with that of field-cast concrete, weight of members, that is, the dead weight of buildings can be decreased in some degree, shortening of construction period becomes feasible by increasing an early-strength of member giving due considerations to curing method, and so on. On the other hand, the disadvantages are that the stresses of precast members at the time of transportation and assembling must be considered because they are transported to and assembled at the construction site, execution of works becomes difficult if dimensional accuracy of members is not high, it must be designed and executed so that joints of precast members do not become structural weak points.

Especially, earthquake-resistance of the joints of members mentioned above becomes structurally the most important problem for precast wall construction.

The standards for precast wall construction, i.e. "Standard for Structural Design of Precast Reinforced Concrete Wall Construction" and "Japanese Architectural Standard Specification 10" are available, but since there is little difference in principle with the wall construction, description will be made only about those points that differ from those of wall construction in reference to matters important from the aseismic viewpoint, and for the remainder, reference should be made to Chapter 10.

11.2 Structural Design

11.2.1 Policy

(1) As the joint of members is liable to become a weak point, number of joints should be made as few as possible. Therefore, it is desirable to make dimensions of precast members as large as possible.

(2) The method of jointing should be such to produce a joint with strength, initial rigidity, and ductility, and easy construction is desirable, and when adopting a new method other than those already in use, it is necessary to ascertain its characteristics by experiments or by some other means.

11.2.2 Disposition of Bearing Wall

(1) In plan of a building, an area of the portion enclosed by center lines of bearing walls should be made equal to or less than 60 m^2.

(2) In case where basement walls are constructed by jointing precast members, it is necessary to pay attention to the durability of the joints.

11.3 Remarks on Stress Calculations

11.3.1 General

Since precast wall construction is completed by lifting, transporting and assembling the precast members and completely connecting each member, conditions of stresses and deformations at each stage should be calculated in advance under respective conditions and measures should be taken to provide ample safety.

11.3.2 Bearing Wall

As to the supporting conditions and stress conditions of the wall panel, there are differences between those before completion of jointing members and those after, and stress conditions of bearing wall at the time of lateral loading due to earthquakes are quite different depending on the shape of wall panel, conditions of joints etc., and it is necessary to make calculations based on assumptions appropriate to the respective conditions.

11.3.3 Roof, Floor and Stairway

It is necessary to note that in ordinary construction, floor panel, roof panel and stairway are placed on footing, wall panel, etc., and joints between them are made afterwards. Therefore, they are in a condition of simple support at the points where they come in contact with other members when they have been arranged in place, and they will be brought to a supporting condition close to that of fixity when the joints are completed.

11.3.4 Footings and Footing Beams

Footings and footing beams must be given attention because their stress condition changes depending on the method of jointing with the bearing walls.

11.4 Structural Notes

11.4.1 Precast Structural Members

(1) Quality and dimensional accuracy of the precast members should be carefully controlled so that no trouble will be caused with regard to strength and construction.

(2) Quality of concrete is determined in connection with the curing method in such a manner that an early-strength of concrete can be assured so that no trouble will be caused at the time of removal of member from the form and in transportation but compressive strength at 4 weeks should be above 180 kg/cm^2 even in a case where there is no particular designation.

(3) At such times of work as lifting, transporting, assembling, etc. of precast members, exceptional stress condition will be caused in some cases, it is necessary either to reinforce them beforehand or provide anchor lifting hooks in fixed places in order not to induce excessive stresses; or take such appropriate measures as to specify locations of supports when they are laid, and so on.

11.4.2 Bearing Wall

a) Length of Wall

In case where 2 or more bearing walls are effectively connected, it is permissible to consider them as an integral unit in calculating the length and height of the bearing wall.

b) Thickness of Wall

Considering reduction of building weight or reduction of earthquake force attributable to good quality and dimensional accuracy of precast panels, thickness of the wall should be made equal to or more than the values shown in Table 11.1.

Table 11.1 Thickness of Wall

Story	Thickness of wall (cm)
Each story of a building of 2 stories or less, or top story of a building 3 stories or more	12 and $H/25$
Other stories	15 and $H/22$

Note :
H denotes vertical distance between supports critical to the structural strength.

c) Wall Rate

Wall rate should be equal to or more than the values shown in Table 11.2.

Table 11.2 Wall Rate

Story	Wall rate (cm/m^2)
First story of 4-story building	15
Other stories	12

d) Shear Reinforcing Bar Ratio

Shear reinforcing bar ratio of bearing walls should be equal to or more than the values shown in Table 11.3.

Table 11.3 Shear Reinforcing Bar Ratio

Story	Shear reinforcing bar ratio (%)
Each story of a building of 2 stories or less, or top story of a building 3 stories or more	0.2
Other stories	0.3

e) Diameter and Arrangement of Shear Reinforcing Bar

Diameter of shear reinforcing bars of the bearing wall should be 6 mm or larger and the bars may be placed in a single layer.

f) Bending Reinforcing Bars

Table 11.4 Bending Reinforcing Bars at Edges, etc. of Bearing Wall

Story	Bending reinforcing bars at edges, etc. of bearing wall	
	$h > 1$ m	$h \leq 1$ m
Each story of a building of 2 stories or less, or top story of a building 3 stories or more	2-13 ϕ or 2-D 13	1-13 ϕ or 1-D 13
Other stories	2-16 ϕ or 2-D 14	2-13 ϕ or 2-D 13

Note :
h denotes height of opening edge.

The bending reinforcing bars for structurally important portions such as edges, periphery of openings, etc. in the bearing wall, an amount of reinforcement should be provided equal to or more than those shown in Table 11.4.

g) Diagonal Reinforcing Bars at Corners of Openings

At the corners of opening, diagonal reinforcing bars (diameter 9 mm or larger) having more than 1/2 of the amount of reinforcement shown in Table 11.4 should be provided.

h) Wall Beam

Depth of wall beam should be equal to or more than 45 cm.

11.4.3 Roof and Floor

a) Rigidity

For this construction where bearing walls are made by connecting precast panels, it becomes especially important to ensure rigidity of roof slabs and floor slabs. Therefore, precautions are required in the details of the joints of roof panels and floor panels and those of the joints between roof or floor panels and bearing wall panels.

b) Peripheral Reinforcement

At the periphery of roof panels and floor panels, placement of bars for stress transmission is necessary for the purpose of reinforcing against both the bending in the plane of panel caused at the time of horizontal load transmission and the shearing of the panel itself due to the effect of peripheral restraint.

c) End Anchorage of Slab Reinforcement

When the periphery of a floor or roof panel is placed on footing or bearing wall panel, such precautions are necessary as extending the end of reinforcement out of precast panel and either anchor them in the supporting or adjoining members or weld them to the reinforcement of those members.

11.4.4 Joint of Members

a) Rigidity and Strength of Joints

In precast wall construction, it is no exaggeration to say that the rigidity and strength of joints control its earthquake resistance. Consequently, it is necessary to design joints so that they have ample safety against anticipated stresses and in general, to ascertain them by experiments.

It is extremely difficult to decide whether a joint is effective or not based on concrete figures, several examples of connection methods that are presently recognized as being effective by past experience, experiments and theories, and being actually used, are presented in c) for reference.

b) Kinds of Joints

There are numerous connection methods but they are broadly classified into the following categories :

(1) Connection Method Using Cast-in-place Concrete or Mortar

This is a method in which joint is made by filling concrete or mortar in the space between members and it is popularly called a wet-joint.

In this method, quality of filler concrete or mortar should be 180 kg/cm^2 or more in terms of compressive strength at 4 weeks. Besides, reinforcing bars projecting from the adjoining members for joint should be either welded to each other or sufficiently anchored in the filler concrete or mortar. Advantage of this method is that even if there should be errors in member dimensions, it will not matter because the concrete for the joint is cast in place. On the other hand, disadvantage is that it takes time before the concrete for the joint hardens and the jointing surfaces between precast members and filler concrete or mortar are weak against tensile and shear forces and, especially, at the horizontal joint between bearing wall above a floor and that of the floor below, a gap develops along the lower edge of the bearing wall above the floor owing to the settlement of the filler material. With regard to these points, it is desirable to reinforce them paying thorough attention to the work execution, either by placing reinforcing bars against tensile forces or by providing shear connectors in the face of joint against shear forces.

(2) Connection Method Using Welding of Steel Parts

This is a method in which steel plates, or the like that are well anchored to the members to be connected are provided at locations, facing one another, in surfaces of joint of the members and they are connected to one another by welding and it is popularly called a dry-joint.

Advantages of this method are that the joints exhibit satisfactory strength immediately after completion of welding which shortens the construction period and the like. On the other hand, disadvantages are that if the dimensional accuracy of the members is not good, welding will become difficult and the strength also will be affected. Owing to the heat of welding, steel plates, etc. are strained, or strength of the concrete near the welding is reduced being affected by heat and the like. In consideration of these points, it is necessary to pay careful attention to the dimensional accuracy of members, welding method and its sequence, etc.

(3) Mechanical Connection Method

This is a connection method by means of bolts and inserts, and is a type of dry-joint. It is expected that it will be greatly developed and utilized in the future together with the friction type connection by high tension bolts, introduction of prestressing, etc.

c) Examples of Connection Methods

(1) Vertical Joints in Bearing Walls (see Fig. 11.2)

(2) Horizontal Joints in Bearing Walls (see Fig. 11.3 (a), (b))

(3) Joint between Bearing Wall and Floor Panel (Roof panel), and Floor Panel (Roof Panel) and Floor Panel (Roof Panel) (see Fig. 11.4)

11.4.5 Footing and Footing Beam

a) Accuracy

Continuous wall footing or footing beam of cast-in-place concrete is used for the footing and footing beam in this construction in consideration of uniformity in contact with the supporting ground, earthquake resistance of the entire building, durability of joint, etc. However, precautions should be taken especially for maintaining accuracy throughout the building for location of the top surface where contact with the precast panel is to be made and to keeping the top surface level.

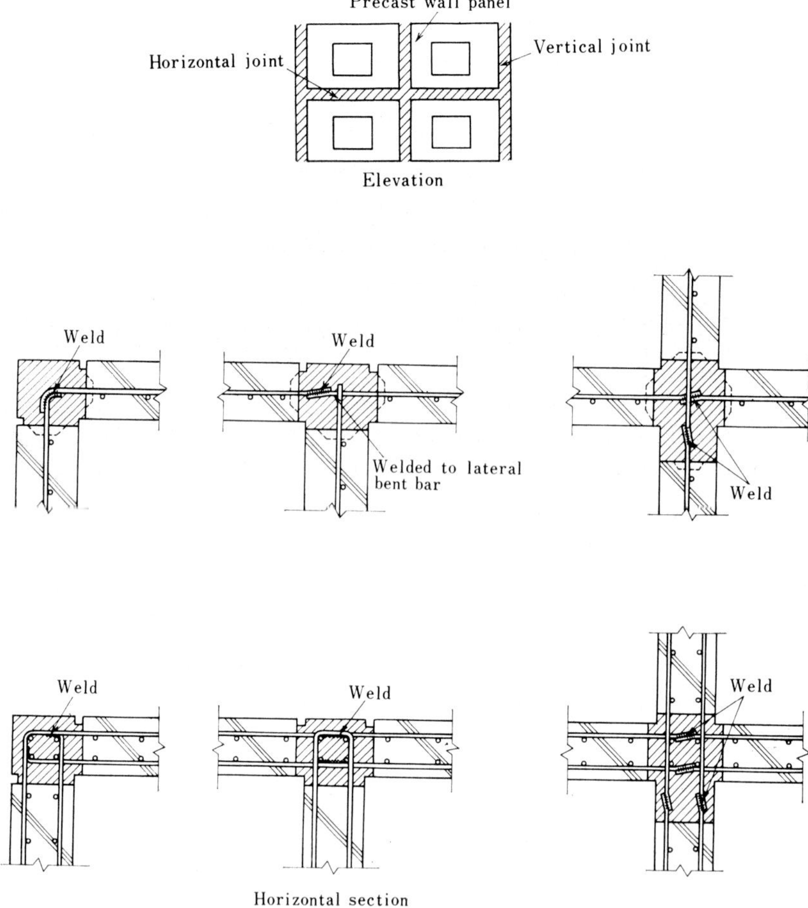

Fig. 11.2 Vertical Joint of Bearing Wall

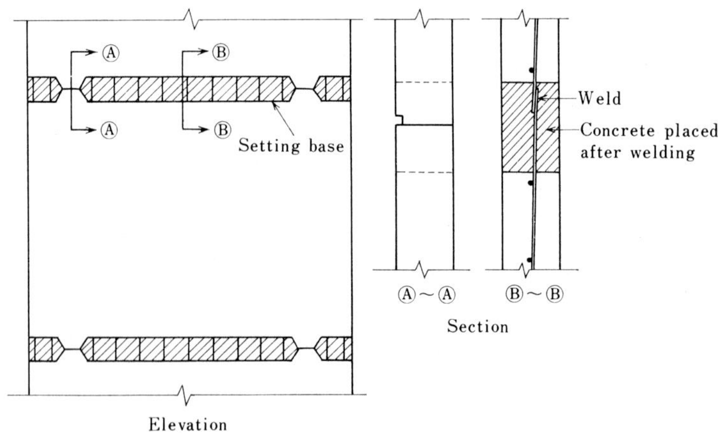

Fig. 11.3 (a) Horizontal Joint of Bearing Wall

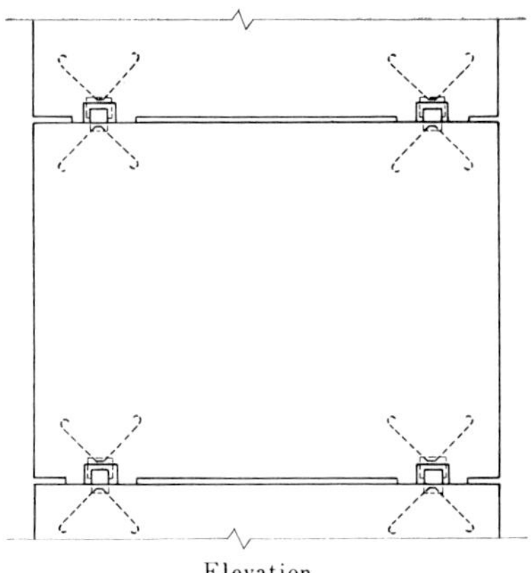

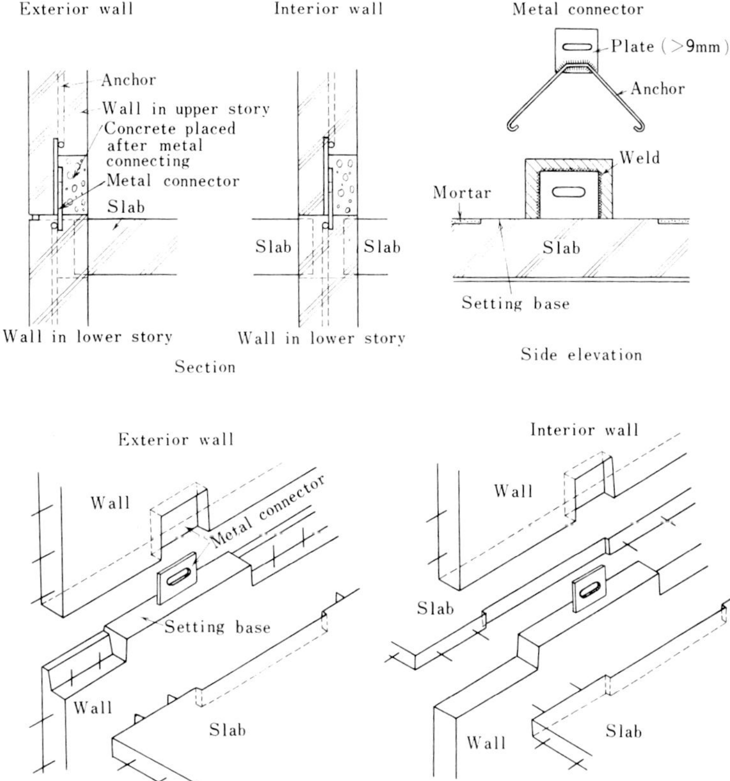

Fig. 11.3 (b) Horizontal Joint of Bearing Wall

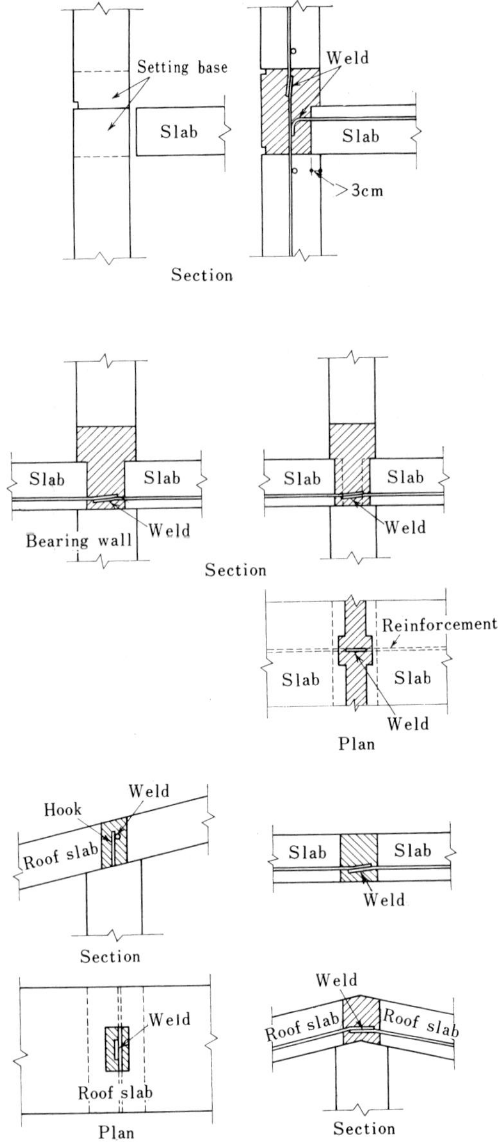

Fig. 11.4 Connection between Floor Slab (Roof Slab) and Other Member

b) Reinforcement

In case the width of the footing is 18 cm or more, reinforcement should be placed in two layers parallel to the surfaces, and reinforcement other than principal bars should not be less than that of the shear reinforcing bars for the bearing wall.

12. PRECAST REINFORCED CONCRETE FRAMEWORK AND PANEL CONSTRUCTION

12.1 General

12.1.1 Structural System and Characteristics

Precast reinforced concrete construction is a structural system in which precast members consisting of rolled steel and concrete such as columns, beams, wall panels, etc. are assembled at the construction site.

In the precast reinforced construction, there are the following types:

i) Pin-joint framework construction
ii) Rigid-joint framework construction
iii) Combination construction of panel-type bearing wall and pin-joint framework
iv) Combination construction of panel-type bearing wall and rigid-joint framework
v) Panel-type bearing wall construction

Among these construction types, popular ones in this country are i), iii) and v), and the standards concerning i) and iii) are the "Standard for Structural Design of Precast Reinforced Concrete Construction" and "Standard for Structural Design of Precast Concrete Floor Construction" by AIJ ; and concerning v), there is "Mass Production Public Management Housing System Precast Reinforced Concrete Construction Design Manual", prepared by the Fireproof Building Association of Japan. In this chapter, matters to be attended to from the aseismic viewpoint with regard to i) and iii) will be described in paragraphs 12.1 through 12.4 ; and as for v), brief description will be given in paragraphs 12.1 and 12.5.

Fig. 12.1 shows an example of construction related to i) and iii), and Fig. 12.2 that related to v). These constructions have as advantages that their stress conditions are comparatively clear and they are suitable for mass production ; but have as disadvantages that in case dimensional accuracy of precast members is below par, rigidity of building or part thereof is reduced and there is structurally speaking, a posibility of stress concentration in particular member or framework, of developing cracks in building finishing work, or of damaging fittings.

12.1.2 Earthquake Damage

It is rare to find buildings of this construction that have suffered damage in earthquakes because only a short time has passed since its use, and the only ones found are those in the

Niigata Earthquake of June, 1964. Even these damages, as described below were considered to be structurally not important.

As an example of the framework construction, a case of railway station (one-story building having a floor area of about 161 m² with continuous wall footing completed in 1950) in Niigata city is as follows : At joint of framework and exterior walls, there were found considerable cracks but in such a degree that no effect on structural strength of the framework was observed. Attention should be paid to the method of installating finishing materials when rigidity of the framework is not sufficient. Besides, there was another 3-story building of the same kind of construction in Niigata city but little damage was observed.

As an example of precast panel construction, a case of buildings used as public servants' quarters (93 m² floor area per building, one-story building for two households, wooden and Japanese type roof framing with continuous wall footing, completed in 1964) is given below. The damage suffered was a slight inclination due to the liquefaction of sand but damage to structural members was not observed.

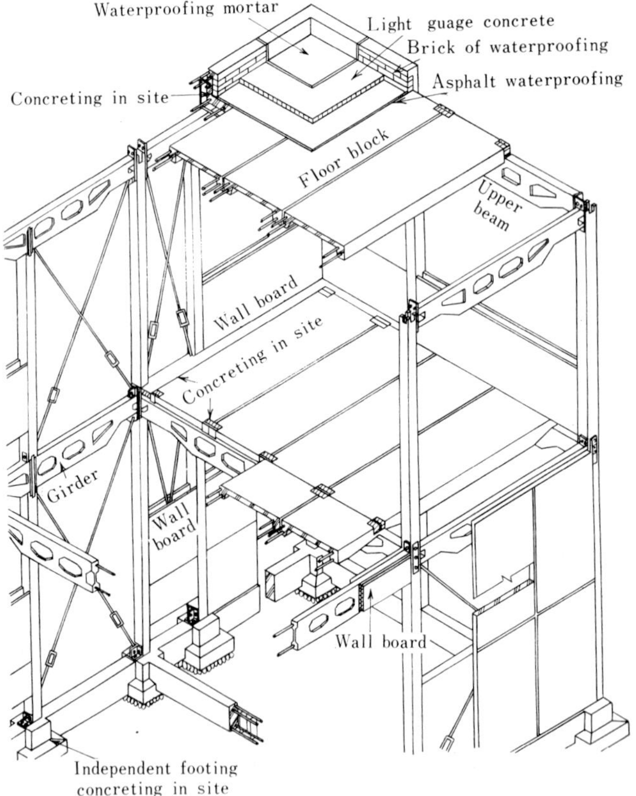

Fig. 12.1 An Example of Precast Reinforced Concrete Framework Construction : i) and iii)

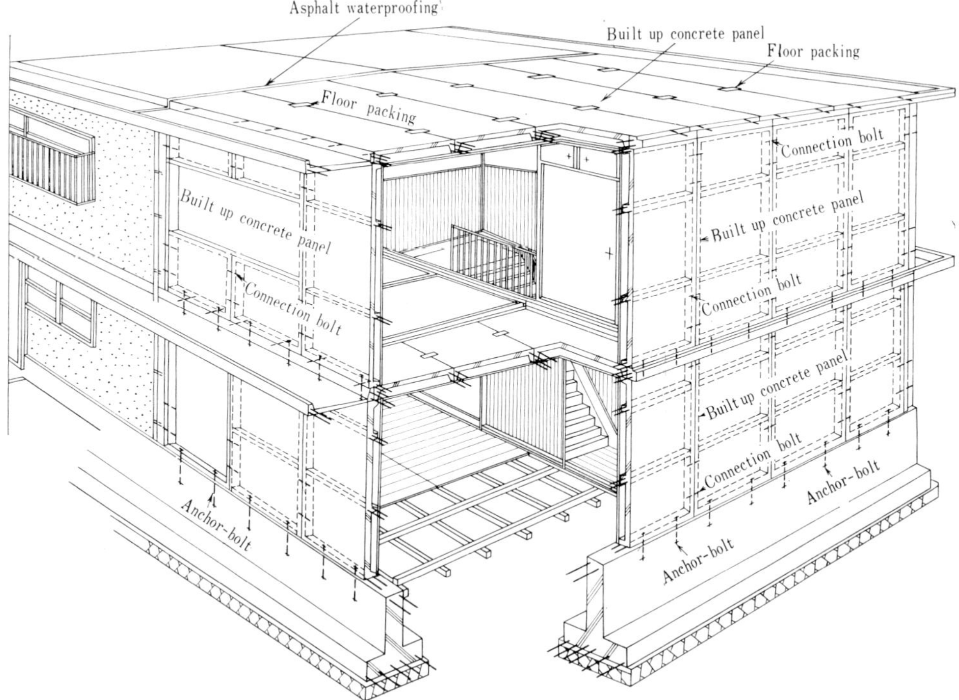

Fig. 12.2 An Example of Precast Reinforced Concrete Panel Construction : v)

12.2 Structural Planning of Framework Construction

(1) Those which have actually been built in this country are 3 or less stories and 14 m or less in eaves height.

(2) As for plans and elevations of the building, those which are as simple as and as well proportioned as possible are recommended and those having orderly disposition of columns, beams, etc. are desirable.

(3) Bearing walls (framings with bracings and panels that can resist horizontal and vertical loads) should be arranged well balanced in the plan of the building. Also, the area of the portion enclosed by center lines of bearing walls should be made equal to or less than 60 m^2.

(4) For the purpose of transmitting horizontal loads to the bearing walls due to earthquakes, roofs and floors should be made to have enough rigidity and strength.

(5) For the bearing walls, the following precaution are required :

As shown in Fig. 12.3, horizontal bracings are to be provided at the top of buttress walls so that lateral buckling of top members are prevented. Bracings as shown in Fig. 12.4 are disadvantageous because they apply a concentrated load to the beam so that a post continuous down to the footing should be provided in such a case.

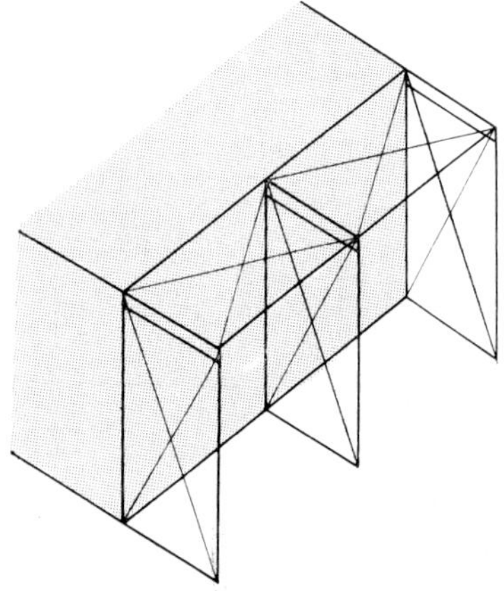

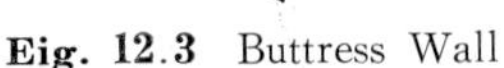

Eig. 12.3 Buttress Wall

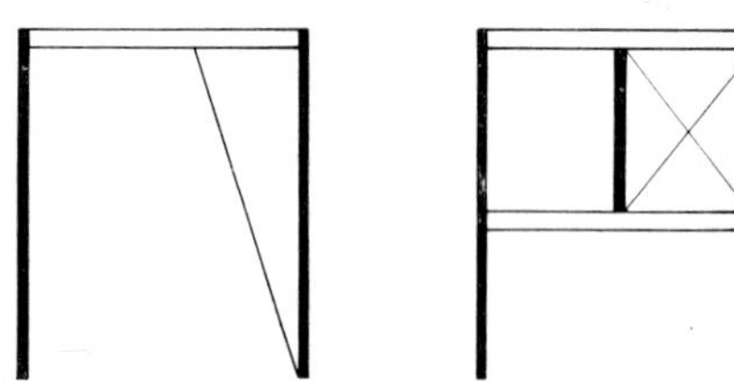

Fig. 12.4 Examples of Bracing placed in Wrong Locations

12.3 Remarks on Stress Calculation of Framework Construction

In calculating the framework stresses, the following assumptions may be used but it is necessary that due considerations be given to the actual conditions.

i) Joints of framework may be considered as pin joints in ordinary cases.

ii) As for bearing walls with bracings, the sum of the strength of respective bearing walls can be regarded as that of the building in anticipation of the yielding of some bracings.

iii) Also the panel-type bearing walls in which column, beam and wall panel are made into one body, can be treated in the same manner as the bearing walls with bracing provided their strength and rigidity have been ascertained. It is desirable, however, to avoid a mixed use of the framework with bracings and the panel-type bearing walls because the latter, generally speaking, compares unfavorably with the former type in many cases with regard to their ductility.

12.4 Structural Notes for Framework Construction

12.4.1 Precast Members

a) Quality

Such principal precast members as columns, beams, floor-blocks, etc. must be of either class 1 or class 2 products as explained below :

i) Class 1 product is defined as the product that is manufactured at a plant or at the construction site and the concrete is made dense by vibrators, etc., using accurate, solid and well-equipped forms made of steel, etc., and cured sufficiently and the concrete has a compressive strength of 200 kg/cm^2 or more at 4 weeks.

ii) Class 2 product is defined as the product using forms and manufacturing method similar to Class 1 product with uniformity in curing and strength is well-maintained. The concrete should have a compressive strength of 300 kg/cm² or more at 4 weeks.

b) Allowable Unit Stresses

In case where forming equipment, curing equipment and other plant equipment are fully equipped, inspections of materials and products are positively conducted and the quality of the product has uniformity ; it can be assumed, for the products of that particular plant only, and based on the results of experiments, that the allowable unit stresses of concrete for the permanent loadings are up to 100 kg/cm² for compression and 10 kg/cm² for shear and tension, and as for those of combination with temporary loadings, twice the above-mentioned values respectively.

The allowable unit stresses of mortar in joints and parts where members make contact otherwise can be assumed for permanent loadings to be 40 kg/cm² for compression, 4 kg/cm² for shearing, and for combined loadings with temporary loading twice the values of permanent loading respectively. Especially, with regards to mortar that is well mixed, the allowable unit stresses may be increased up to 1.5 times the values mentioned above.

12.4.2 Bearing Walls

With regard to widthwise and lengthwise directions in each story of the building, values (wall rate) of the sum of bearing wall length (distance between the center lines of columns at the both ends of the wall framing) divided by a floor area of that story should be equal to or more than the values shown in Table 12.1.

Table 12.1 Wall Rate

Story	Wall rate (cm/m²)
One-story building or top story	12
Second story counting from the top	12
Third story counting from the top	18*

Note :
* In case where the sectional area of bracing is sufficiently provided and the rigidity at connecting point of each member is large enough, this can be reduced to 16.

Lengths of bearing walls to be used in calculating the wall rates are determined as follows:

(1) In case one bracing can take tension or compression only :

i) As for the bearing wall having bracings crosswise within the same wall framing, the full length is taken.

ii) In case where there, in the same plane of wall framing, is a pair of bearing walls that have one-way bracing to resist the horizontal forces, the length of the shorter bearing wall may be taken.

(2) In case one bracing can take both tension and compression :

i) As for the bearing wall having bracings crosswise within the same wall framing, 1.5 times the length is taken.

ii) As for bearing wall having one-way bracing, its full length is taken.

However, length of the bearing wall should be 90 cm or more because rigidity is reduced if an inclination of bracing in the bearing wall is too large.

As for bracing to be used in the bearing wall, its size should be such that it has a sectional area equal to or more than those shown in Table 12.2.

Table 12.2 Sectional Area of Bracing

Story	Diameter of steel bar bracing (mm)	Size of steel angle bracing (mm)
One-story building or top story	13	40×40×3
Second story counting from the top	16	50×50×6
Third story counting from the top	16	50×50×6

Steel bar bracing must be kept under tension by turning the turnbuckle several times without fail. Besides, it is necessary to take precautions as it is apt to be deformed at the connection. One shown in Fig. 12.5 (a) is not satisfactory as the bend is stretched at the part marked by X. Another one shown in Fig. 12.5 (b) is unsatisfactory because the bracing gets loose if the idle space marked by Δ is not eliminated. The arrangement shown in Fig. 12.5 (c), is recommended.

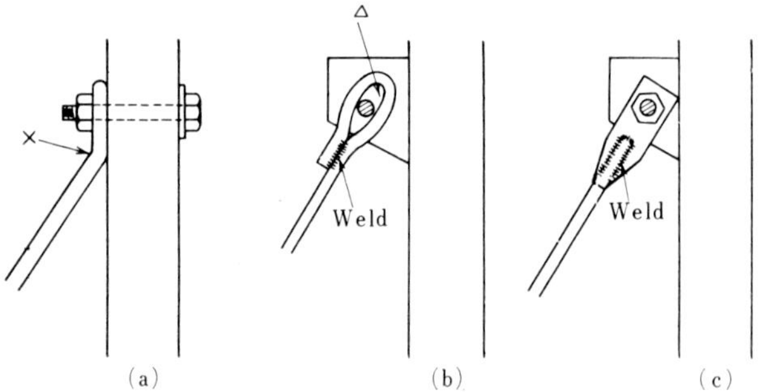

Fig. 12.5 Connection of Steel Bar Bracing

12.4.3 Columns

The least thickness of column should be equal to or more than the values shown in Table 12.3. For principal bars, a minimum of four bars having minimum diameter of 9 mm should be used.

Table 12.3 Least Thickness of Column

Story	Least thickness (mm)	Remarks
Third story counting from the top	18* and $h/20$	h : Vertical distance between supports critical to the structural strength
Other stories	15 and $h/24$	

Note :

* In case where columns of class 2 products are used, two or more columns having the minimum thickness of 15 cm and more than $h/24$ cm may be used by being joined together.

12.4.4 Beams

According to the standard, the width of the beam must be 12 cm or more in for class 1 products, and 9 cm or more in case of class 2 products, but it is desirable to apply an effective finish for fire-resistance on the sides of beam, as the 9 cm thickness is not considered to be satisfactory for fire-resistance.

12.4.5 Connection of Wall Framing Members

Connection methods of wall framing members are classified roughly into two kinds ; one using poured-in-place concrete and the other using bolts. In either case, it is desirable, however, to have such a connection that can positively transmit stresses, is not subject to looseness and deformation, and can properly restrain development of secondary local-stresses.

In case connection is made by bolts, deformation due to lose-nut, clearance between bolts and bolt-hole, etc. is expected and it is not considered to be completely rigid. Besides, diameter of bolt to be used for principal connections should be 16 mm or larger in case one bolt is used for the connection and 13 mm or larger in case two bolts are used.

12.4.6 Assembled Floor

a) Rigidity

In order to transmit horizontal loads to bearing walls, floors that have adequate rigidity and

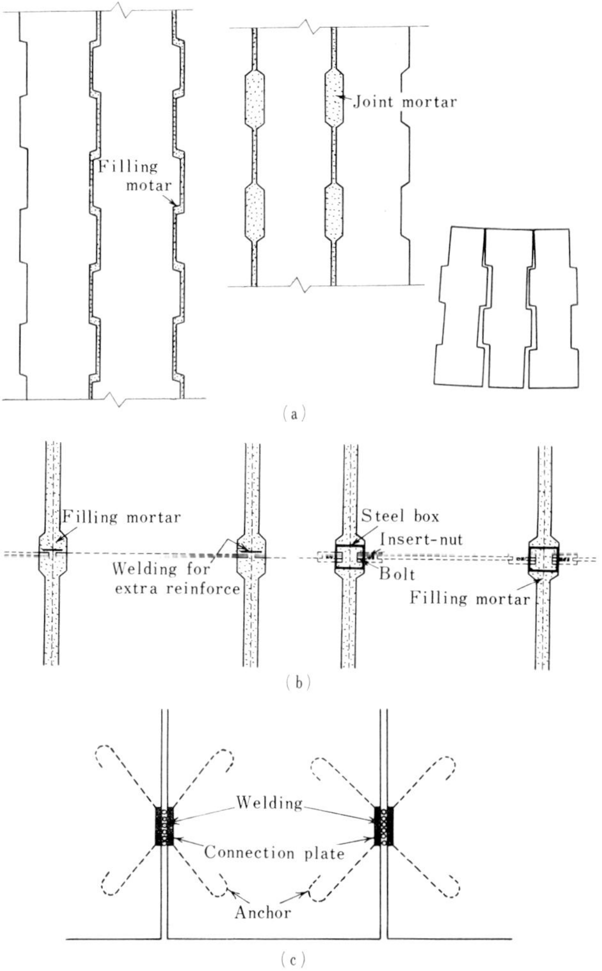

Fig. 12.6 Connecting Method of Assembled Floor

strength are necessary. In case where the floor is made up of concrete blocks or any other precast members (hereinunder referred to as assembled floor), it must be so constructed that the floor blocks and wall framing are made into one body by making connections of the floor blocks to the adjoining members by means of mortar, concrete or bolts. Intention of one shown in Fig. 12. 6 (a) is in insuring rigidity and strength by means of toothed joint and filler mortar; another one shown in Fig. 12. 6 (b), by connecting distribution bars; and the other one shown in Fig. 12. 6 (c), by welding embedded pieces of rolled-steel.

b) Connection of Floor Edge to Beam

As for the connection of end of floor block in assembled floor, principal bars of the floor should be tightly bound by such proper means as anchorage of bars into the beam, or bolt connection, etc., taking precautions so that they cannot be separated.

c) Thickness

Thickness of floor block of assembled floor should be 4 cm or more for class 1 products, and 3 cm or more for class 2 products.

12.4.7 Footings

It is preferable to construct continuous footings or footing beams of castin-place reinforced concrete, in consideration of uniformity in contact with the soil, earthquake resistance, durability of the connections, etc.

Also, the width of the continuous footing or footing beam should be equal or more than the thickness of column; the depth should be 1/12 of the height of the building or more and 60 cm (45 cm for one-story buildining) or more and they should be provided with double reinforcement.

12.5 Precast Panel Construction

The main points of difference from that of framework construction, are the following:

12.5.1 Structural Planning

(1) Those which have actually been constructed in our country are 2 or less stories and 7 m or lower at the eaves height.

(2) It is desirable to arrange bearing walls in L-shape or in T-shape at corners of the building.

(3) It is desirable to place bearing walls of the upper story right above those of the story immediately below.

12.5.2 Structural Notes

a) Precast Members

(1) The member must be a product manufactured at a manufacturing plant by a vibration equipment or some other equipment by which concrete is densely cast, using accurate, solid and well-equipped forms made of steel or some other material; cured in steam or in water; and of which the concrete has a compressive strength of 300 kg/cm^2 or more at 4 weeks.

(2) As for the dimensional accuracy of members, thorough cares should be taken so that there is no trouble in assembling, in executing work and in structural strength.

b) Bearing Walls

(1) An example for shape and dimensions of bearing wall panel (precast panel) that is being used in this country is shown in Fig. 12.7.

(2) Bearing wall rate should be 12 cm/m² or more for each story of 2-story building and 10 cm/m² or more for one-story building, as to the width-wise and the length-wise directions of the building.

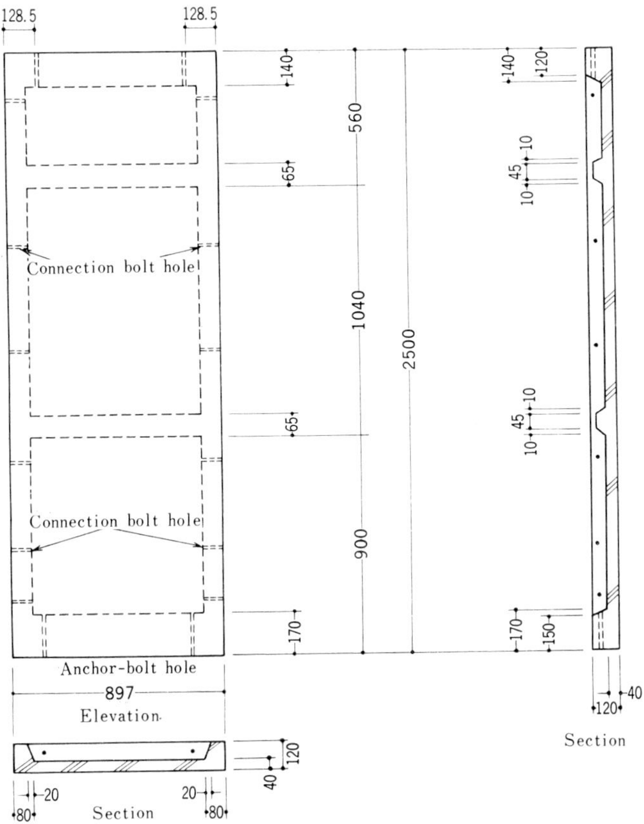

Fig. 12.7 An Example for Shape and Dimensions of Precast Panel (unit : mm)

c) Roof and Floors

(1) Roof structure may be of precast concrete construction, steel construction or timber construction.

(2) Floor construction of the second story should be of precast concrete construction, as a general rules.

d) Collar Beams and Hanging Walls

(1) In case where a roof or a floor of the second story is of a construction other than that of precast concrete and having no rigidity, an effective collar beam must be provided at the top of the bearing wall in order to obtain stiffness of the bearing wall in the direction normal to

the plane of the wall.

(2) In case where a hanging wall receives a large load from the roof framing, floor or beam, attention must be paid to the way in which stresses are transmitted.

e) Footings

(1) The width of the continuous footing should be equal to or more than the thickness of the bearing wall and its depth should be 60 cm or more for two-story buildings and 45 cm or more for one-story buildings.

(2) The continuous footing should have double reinforcement. For the principal reinforcement, at least 2-16 ϕ should be used both in top and bottom for two-story buildings, and at least 1-16 ϕ both in top and bottom for one-story building. For the web reinforcement, bars of 6 mm or more in diameter should be placed at 30 cm or less.

(3) Size of anchor bolts to be used for connecting the lower edge of the bearing wall should be as shown in Table 12.4.

Table 12.4 Size of Anchor Bolt

Story	Design wall rate	Anchor bolt
2-story building	1.5 times the specified wall rate or less	19 ϕ
	More than 1.5 times the specified wall rate	16 ϕ
1-story building	1.5 times the specified wall rate or less	16 ϕ
	More than 1.5 times the specified wall rate	13 ϕ

(4) As for the leveling mortar for the top of footing, it is necessary to execute the work with particular care in order to obtain accurate horizontal level so that there will be no trouble in assembling the wall panels.

f) Connection of Members

Examples of member connections are shown in Fig. 12.8.

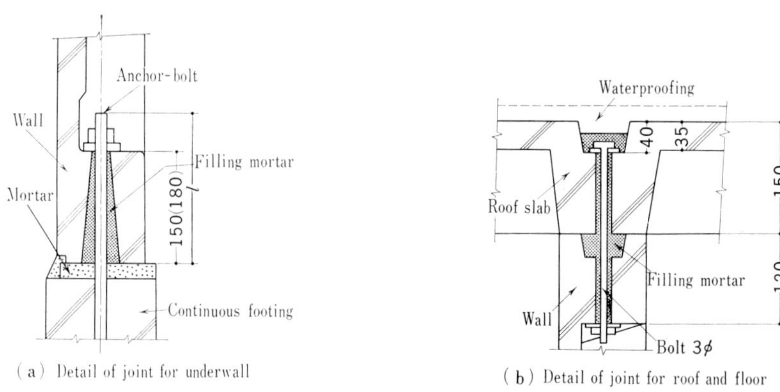

Fig. 12.8 Examples of Member Connection (unit : mm)

(1) Connection bolts to be used for assembling structural members should be fastened after the members have been aligned and adjusted, and care should be taken so that fastening strength

of each becomes uniform.

(2) It is necessary to fill mortar thoroughly in places which are important both structurally and otherwise.

i) Around anchor bolts of footings.

ii) Between wall panel and roof panel, and around the connection bolts for wall panel in the story below and floor panel and wall panel in the story above.

iii) Joints between floor panels and between wall panels.

iv) Around connection bolts of collar beam and wall panel.

13. MASONRY CONSTRUCTION

13.1 General

13.1.1 Structural Characteristics

Masonry construction refers to structures that have walls that are constructed by laying brick, stone, concrete blocks, etc. with joint mortar and it excludes those in which other members support principal loads as in the case of steel framed brick construction. For construction which is reinforced by inserting steel bars in the wall in order to supplement the strength of wall, requirements will be explained in Chapter 14.

Masonry construction resists earthquake forces by shear resistance of the walls in the direction of the plane of wall in the same way as the reinforced concrete wall construction, but while it has great weight and large compressive strength, its strength for tension, bending, shear, etc. is less for its rigidity. Besides, if the work is poorly executed, joints connecting each unit becomes structurally weak points. In the light of these facts, masonry construction is disadvantageous against earthquakes. Since this construction had been developed originally in countries where there is no concern about earthquakes, it has suffered considerable damages in past earthquakes. Observing from the standpoint of earthquake damage, the following facts have been observed :

Fig. 13.1

(1) Parts projecting from the body of wall are apt to be destroyed. Chimnies, towers, gable walls, parapets, etc. come under this heading and destruction of them causes cracks in the wall (Fig. 13.1).

(2) When the roof and floors are of timber construction, wall is destroyed as a bending member loaded in horizontal direction and tension cracks occure vertically at the center, ends or corners of the wall. Longer the wall and larger the openings, and in the upper stories, more prominent is the damage (Fig. 13.2).

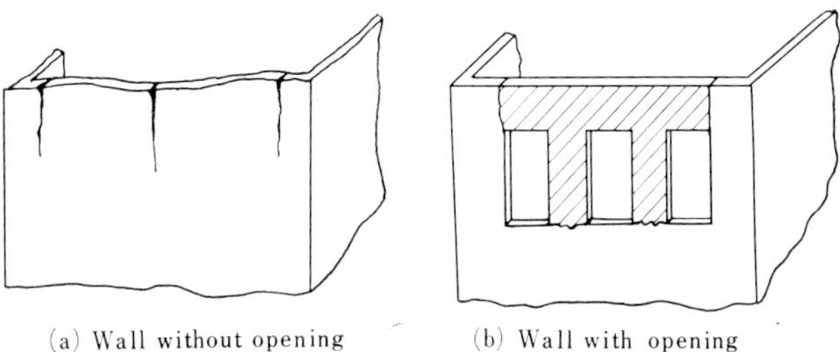

Fig. 13.2

(3) Shear cracks occur diagonally in walls. Subjected to vibrations parallel to the plane of the wall, cracks occur in the lower story in staggered steps diagonally along the joints. Although these occur even in a rigid floor, they are especially noticeable when the work is poorly executed (Fig. 13.3).

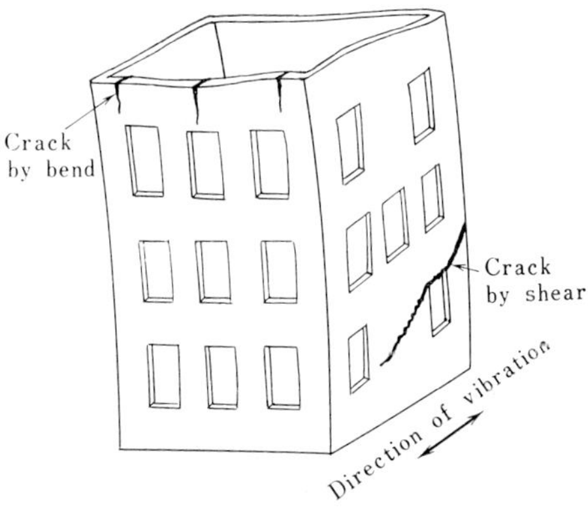

Fig. 13.3

(4) In some cases, the entire wall falls down as a fence wall does, producing horizontal cracks near the base as in the case of a cantilever (Fig. 13.4).

To avoid these damages, there are such means as an increase of wall thickness, alteration of plan, etc., but these have limitations within themselves and it is difficult to raise the degree of earthquake resistance of masonry construction as one wishes. Consequently, it is not feasible with masonry construction to design an aseismic building that above a certain level. In such a case, it is recommended that reinforced concrete block construction, reinforced concrete

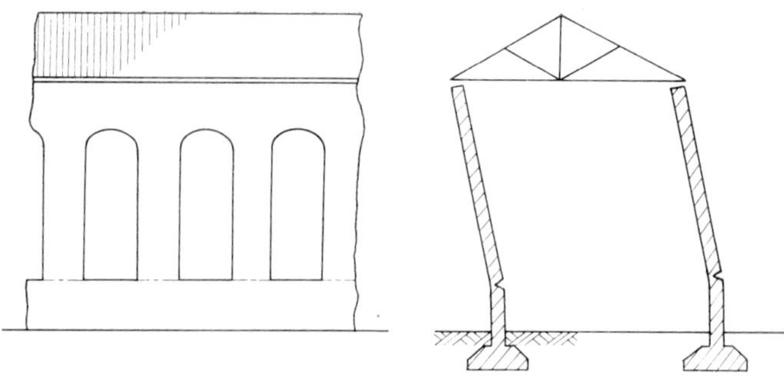

Fig. 13.4

construction or steel construction be used instead.

13.1.2 Earthquake Damage

The use of masonry construction in this country extends about one hundred years since the beginning of the Meiji era, and it suffered considerable damage at the time of the Kanto Earthquake of September 1, 1923. Examples are shown in Photo. 13.1, 13.2, and 13.3. Patterns of damage indicate such characteristics as has been explained in the preceding paragraph.

In structural planning of masonry construction at present considerations based principally on the experiences of past earthquake damages, are given so that it may attain earthquake resistance.

Photo 13.1 Damage Example of a Brick Warehouse at the Time of the Kanto Earthquake in 1923

Photo 13.2 Damage Example of a Brick Building at the Time of the Kanto Earthquake in 1923

Photo 13.3 Damage Example of a Brick School Building at the Time of the Kanto Earthquake in 1923

13.2 Structural Planning

13.2.1 Principles of Structural Planning

To make masonry construction secure against earthquakes, the following principles are indispensable :

i) Masonry construction be used only for small scale buildings as far as possible;

ii) The shape of the building, disposition of walls, etc., should be made in a balanced manner, and the stresses are distributed as uniformly as possible to the whole structure;

iii) That tension, bending and shear developed in walls be as small as possible (for example, such as by proper limitation of the wall height, length, opening, etc.);

iv) That the work should be properly done so that the strength of joints may well be secured.

There are the "Standard for Structural Design of Masonry Construction" and "Japan Architectural Standard Specifications (JASS) 7 and 9" as the standards for design and execution of the masonry construction and Articles 51 through 61, Chapter 3 of the Building Standard Law Enforcement Order also apply. The following construction is regarded in this country as being earthquake resistant masonry construction.

Roof and floors may be of timber construction or of reinforced concrete construction. However, timber construction does not possess high earthquake resistance so that all floors and roof of buildings of two stories or more should be made of rigid reinforced concrete.

At the top of walls of each story, cast-in-place reinforced concrete collar beams (refer to par. 13.3.2) should be provided. Its purpose is to reinforce the upper part of the wall and make it resist earthquake force acting perpendicular to the wall. Consequently, it may possibly be omitted in case there is a cast-in-place reinforced concrete floor. As for footings, continuous footing of cast-in-place reinforced concrete should be provided under the wall. In this manner, the walls of masonry construction are reinforced from both top and bottom, by the floor slab or collar beams at the top and continuous footing at the bottom. It is so constructed that the building become one solid body. As for the walls, it is necessary to give consideration so

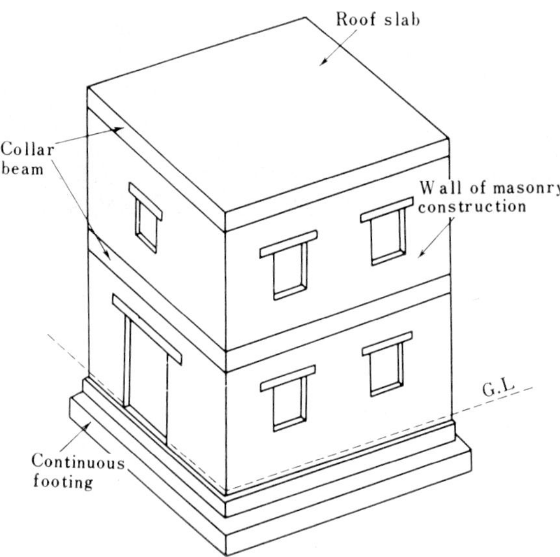

Fig. 13.5

that the building become an earthquake resistant structure, by limiting the size of openings and the thickness of the wall (Fig. 13.5).

Also a structure in which small reinforced concrete columns are built at each corner of the wall of a building is conceivable and masonry construction after completion of wall, thus enclosing entire building by reinforced concrete frame and restraining walls*. This is a kind of mixed use of reinforced concrete frame construction and masonry construction wherein most of the forces are resisted by the masonry wall while a slender framing increases the effectiveness of the walls. This is an example where the weak point of masonry construction is effectively reinforced.

13.2.2 Kinds of Masonry Construction

Masonry construction is classified by the kind of material that constitute the wall, and called stone construction, brick construction, concrete block construction, etc. The concrete block construction is classified into two kinds, A and B, by the kind of concrete blocks used. Their classification based on the strength is shown in Table 13.1. A stone is strong as a unit but bonded joints become the weak point, therefore, it is treated as being equal to class A concrete block. Brick shows, a strength in general, of 100 kg/cm^2 or more by the standard test and is considered to be equal to class B concrete block. Therefore, the masonry construction is roughly classified into two kinds for their strength as shown in the table, and the former is considered as being somewhat inferior.

Table 13.1

	Kinds	Material	Remark
I	Stone construction	Stone	
	Class A concrete block construction	Block having a compressive strength of 60 kg/cm^2 or more for gross sectional area at 4 weeks	
II	Class B concrete block construction	Block having a compressive strength of 100 kg/cm^2 or more for gross sectional area at 4 weeks	Hollow block is not included
	Brick construction	Brick	

Also, in case where different kinds of material are used in combination in a structure, it is interpreted as being that of the weaker material because the strength of the building is dependent on the weaker material. For example, a structure of brick construction in which stones are used in combination is treated as being of stone construction.

13.2.3 Size

Although limitations on the height of buildings established by the Building Standard Law and by the AIJ Structural Design Standard are as shown in Table 13.2, it is recommended that the number of stories be made two at the most, because masonry construction is not earthquake

* Refer to Par 6.2.3 a) of Chapter 6 Reinforced Concrete Construction.

resistant.

Table 13.2

	Height limitation	Height limitation in case roof truss is used
Stone construction Class A concrete block construction	6 m (9 m in case the wall is especially thick*)	9 m, eaves height 6 m
Class B concrete block construction Brick construction	9 m	13 m, eaves height 9 m only for the case where roof truss is on the reinforced concrete roof slab

Note :
 * Refer to par. 13.3.1.

13.2.4 Disposition and Length of Wall, Enclosed Area, etc.

a) Disposition

It is a rule to make the building resist earthquakes as an unit by connecting walls of masonry construction at their top by collar beam and floor slabs. Therefore, it is necessary that the walls be arranged uniformly and appropriately on the building plan. If the distribution of walls is one-sided, a distance between center of mass and center of rigidity of the building become very large ; consequently, large stresses will be caused in the walls on the other sides, or the building will be subjected to torsion and become dangerous. These are the common phenomena in all of the wall construction and an appropriate judgement of the design engineer is necessary.

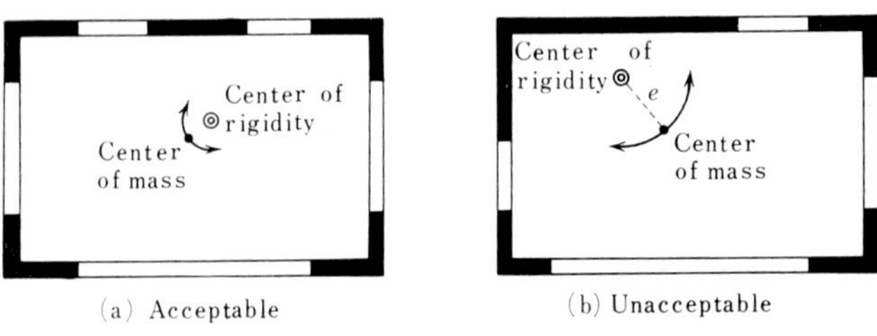

Fig. 13.6

b) Enclosed Area

As masonry construction is a structure like a box, the smaller the box, the stronger it is ; and also in case it is large, the more partitions are provided, the stronger it is. For this reason and for the purpose of preventing unbalanced distribution of walls, the maximum limits of room areas enclosed by walls, for the building of two stories or above, are established as shown in Table 13.3. For buildings of large floor area, they must be partitioned by walls and each area of partition must be so proportioned that they come within these limits (Fig. 13.7).

Table 13.3

	Maximum partitioned area (m^2)
Stone construction Class A concrete block construction	40 (but, 60 in case the wall is especially thick*)
Class B concrete block construction Brick construction	60

Note :
 * Refer to par. 13.3.1.

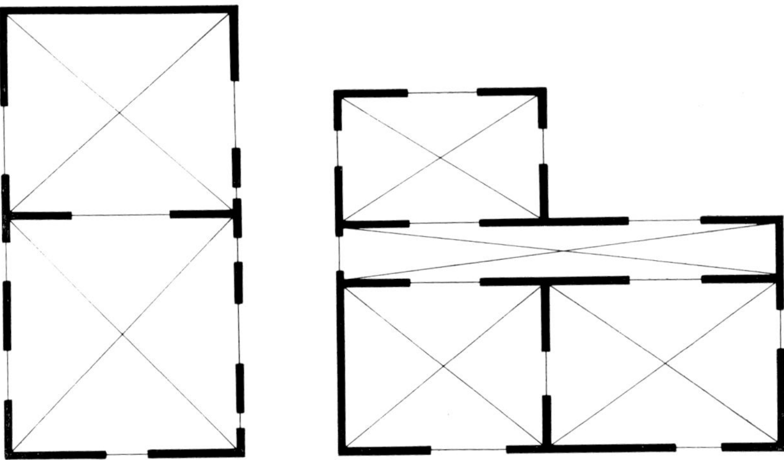

Fig. 13.7 Method for Dividing into Sections

c) Maximum Length of Wall

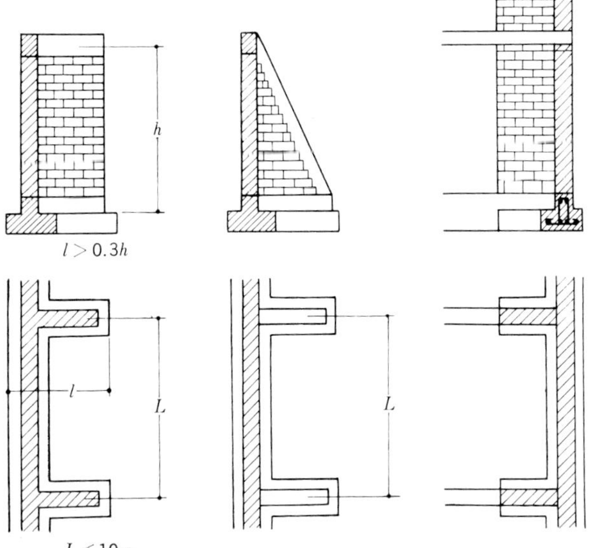

Fig. 13.8 Buttress and Adjoining Wall Details

It is not desirable to have long walls because they are weak against bending and twisting not in the plane of the wall if length of each wall is very long. In this county, the maximum length is limited to 10 m. The length of a wall is defined as the distance between two other walls ("adjoining walls") that are joined to the wall at right angle and facing each other. In case there is need to make a wall long, it must be divided into sections of 10 m or less by providing buttresses in between as shown in Fig. 13.8.

13.3 Structural Notes

13.3.1 Walls

a) Thickness

As for the wall thickness, it is necessary to have a thickness equal to or more than the value shown in Table 13.4 correspondingly to the wall lengths and number of stories. These thicknesses do not include the finish.

The wall thickness should also be 1/15 or more of the height of the wall in that story. As the partition walls in general have less openings, a wall thickness 10 cm less than that of ordinary walls is permissible for them. However, it should not be under 20 cm. For the boundary wall of a multi-unit building, it is desirable to use the wall of the same thickness as that of the exterior walls.

Table 13.4 Thickness of Wall

Length of wall / Number of stories	5 m or under	above 5 m
Building of 2-stories or more	30 cm	40 cm
One-story building	20 cm	30 cm

The definition of "in case the wall is especially thick" mentioned in Table 13.2 or 13.3 refers to the case where the thickness of the wall is made equal to 1.5 times the values shown in Table 13.4 or more. This is a special bonus given because of the increased thickness of the wall. In case double walls are used, either one of the walls must meet the above requirement. The wall, in general, should not be made thinner than that of the story above.

b) Openings

Openings should be kept as small as possible for keeping earthquake resistance of a building unimpaired, because shear resistance of a wall is decreased if it has openings. For each one of the divided partition of a building as shown in Table 13.3, sum of the width of opening should be made less than or equal to 1/2 of the sum of the wall length in respective directions of the length and the width of the building. Also, for the whole building, total sum of width of opening of each story should be made less than or equal to 1/3 of the total sum of the length of walls. The length of the wall between adjoining openings, and that between the edge of opening and center of the "adjoining wall" should be made more than or equal to twice the wall thickness (Fig. 13.9).

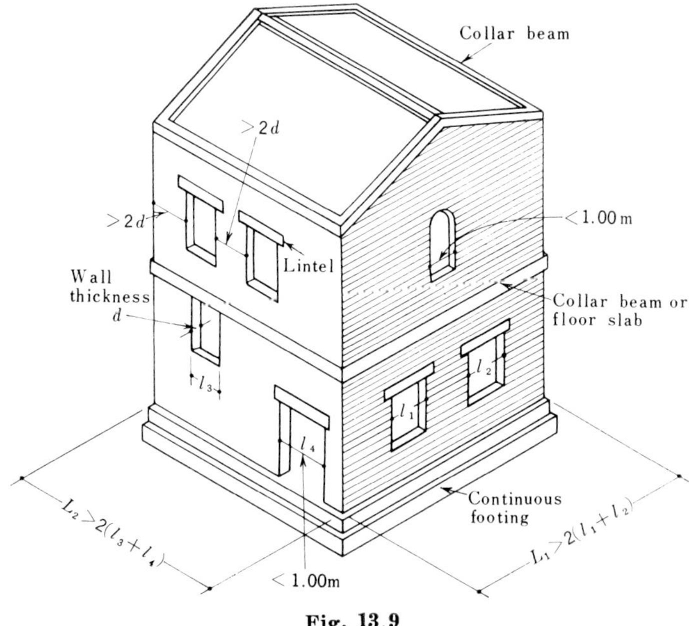

Fig. 13.9

Extruding window or balcony of masonry construction is dangerous, and such must be either made of reinforced concrete or reinforced by steel framing.

c) Reinforcement of the Upper Part of Openings

In the upper part of opening, brick arch or stone lintel has been used in the traditional construction ; but as these are apt to get cracks and even fall down in case of earthquakes, their use should be limited only for small windows having opening width of 1 m or less and they should be sufficiently reinforced with metal pieces. For the upper part of opening having a width more than 1 m, lintel of precast or cast-in-place reinforced concrete should be used and it is necessary to provide sufficient bearing area in the wall of both sides of opening (Fig. 13.10).

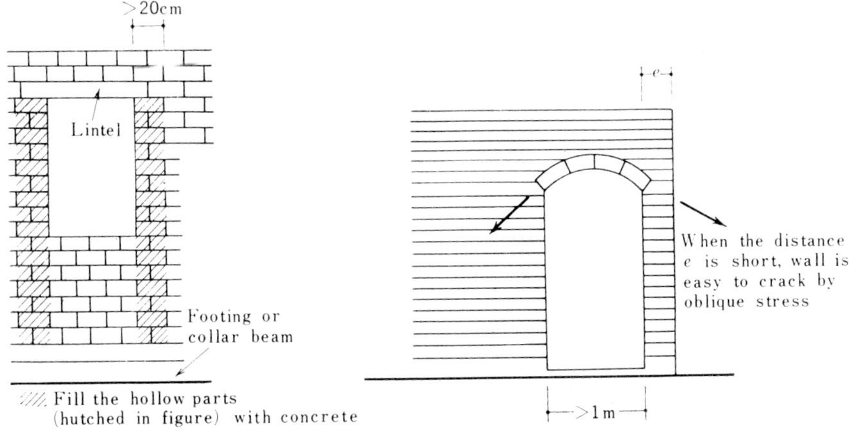

Fig. 13.10

13.3.2 Collar Beams

At the top of wall in each story, continuous beams of cast-in-place reinforced concrete are provided. These are called collar beams. As the collar beam serves to reinforce the wall top, transmitting positively earthquake forces normal to the plane of wall to the "adjoining walls", and making the building resist as one body, it should be provided without fail for masonry construction which is weak in bonding strength. However, in case where cast-in-place floor slabs are provided adjoining the top of walls, collar beams can be omitted as the slabs serve to maintain rigidity of the top of wall and take over the transmission of horizontal forces. Although the law and regulations permit the omission of collar beams for one-story building when the thickness of the wall is equal to or more than 1/10 of the height of wall, and even when the length of the wall is 5 m or less, it is desirable to have them when horizontal force due to earthquake is large. Collar beams should be of reinforced concrete construction having reinforcements in both top and bottom faces, and also on both sides (Fig. 13.11).

Section of collar beam is determined by the distance between "adjoining walls" and also controlled by the vertical load in case there is a large opening immediately below. Even when no collar beam is required, steel or reinforced concrete bearing piece must be provided at the top of the wall for the concentrated load of roof truss, etc. in order to distribute that load.

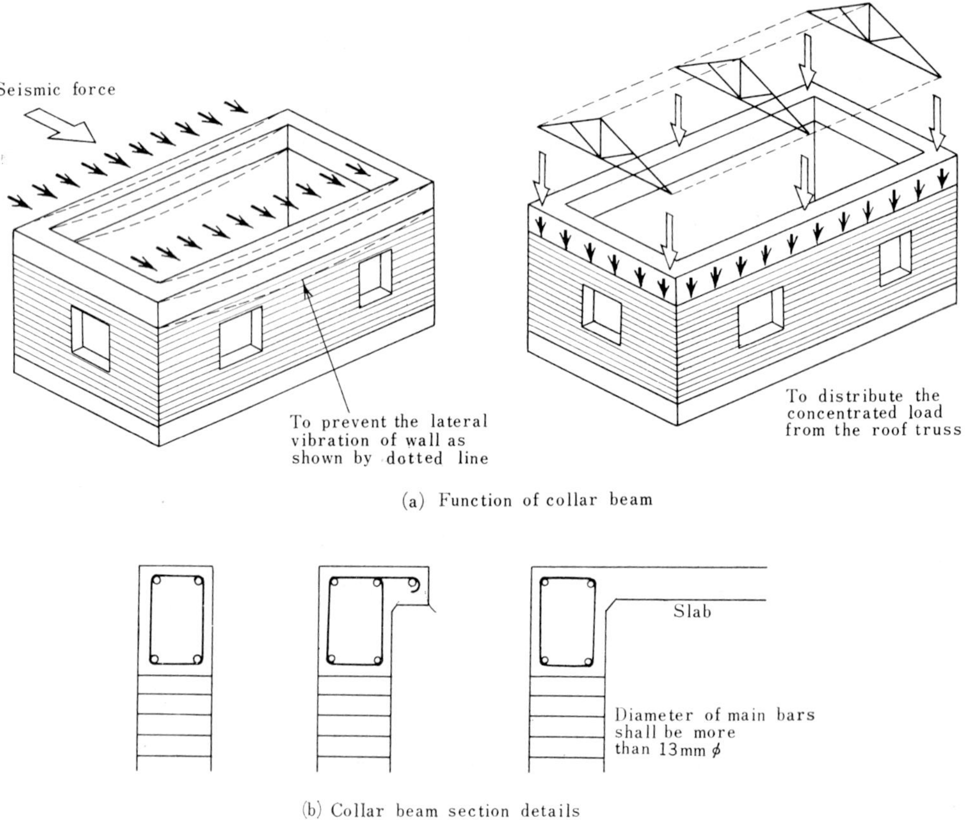

(a) Function of collar beam

(b) Collar beam section details

Fig. 13.11

Projecting walls above the uppermost collar beam or floor slab such as gable wall, parapet, etc. receive large earthquake forces, and it is prohibited under the law and regulations to make them of masonry construction. In this case, however, nothing interferes provided the top of these wall is further reinforced by reinforced concrete collar beams.

13.3.3 Floor and Roof

Floor slabs having high rigidity increase earthquake resistance of the building by distributing horizontal forces due to earthquake uniformly to each wall and bringing the shear resistance of each wall together. For this purpose, for the roofs and floors of a building of two stories or more which are made of masonry construction, it is recommended that they have slabs of cast-in-place reinforced concrete construction or of rigid precast reinforced concrete construction except for the base story floor. It is necessary for these slabs that they be designed and constructed so as to be well bonded with the walls (Fig. 13.12).

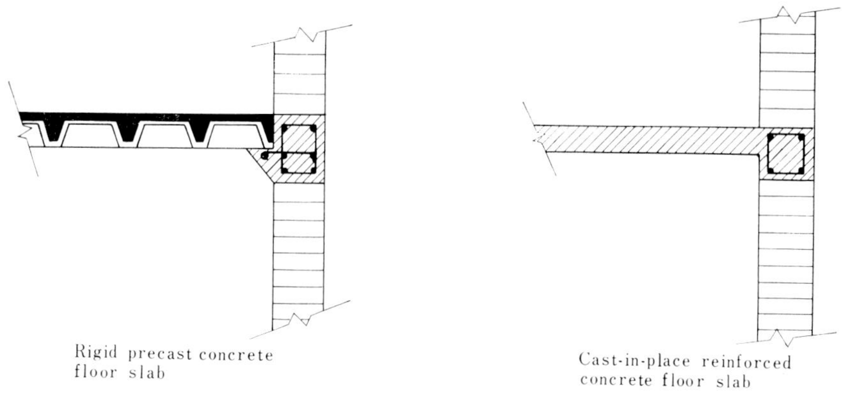

Fig. 13.12

13.3.4 Footings

For the purpose of making the masonry construction earthquake resistant, it is most appropriate to use rigid and continuous footings or footing beams and to connect the bottom of each wall into one body. This has been proven by great earthquakes in the past and by full scale model experiments[1)2)]. Continuous footings are effective not only in case of earthquakes but also against the vertical loads under ordinary conditions in preventing development of cracks due to irregular settlement of the building. Continuous footings should ordinarily be made of cast-in-place reinforced concrete, provided continuously under the walls, and connected by footing beams where openings occur in the above. Continuous footing should be reinforced both in the top and the bottom faces ; the width of footing base should be made wide enough so that the contact soil pressures are made as uniform as possible ; and the depth of footing under ground level should be determined in consideration of effects at the time of earthquake, frost heave, and other conditions. In case the building is of one-story and the supporting soil is good, continuous footing of plain concrete is also acceptable. In case where opening or entrance in the wall is large and clear down to the ground level, the section of the footing beam or the continuous footing should be so determined that it is adequate against the stresses due to the contact soil pressure and earthquake forces. The height of continuous footing and footing beam should be equal to or more than 1/18 of the building height but not less than 40 cm, except for one-story buildings in which it should not be less than 30 cm (Fig. 13.13).

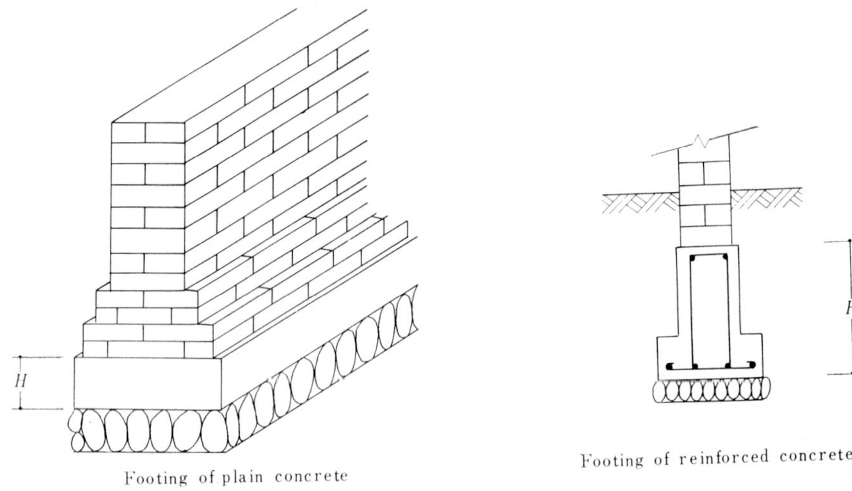

Fig. 13.13 Footing Details

13.4 Remarks on Structural Calculations

The masonry construction is a type of wall construction and its stress calculation is not simple. Also as the body of the wall which is made by bonding is not homogeneous, it is not clear if the stresses are transmitted as calculated. Consequently, a design standard is established based ordinarily on examples of the past or rough calculations on proper assumptions ; and if the articles and numerical values in tables of the standard are applied, a generally reasonable design is obtained. An outline of these requirements have been described, but one should have knowledge of structural calculation to some degree in order to be able to cope with particular cases which may possibly develop. Especially, as the design standard is made for buildings whose scale is not too large, some checking by calculation is necessary in case of designing a large scale building (such as those having large building area or large story height).

13.4.1 Method of Calculations

a) Rough Calculation by Wall Rates

In general, the calculation method for wall construction, calculations are made for the permanent stress due to vertical loads and for the temporary stress, which is the sum of the permanent stress and that of the earthquake force, for every wall, exclusive of non-bearing walls, divided by every one of openings in it is defined as the bearing wall* and if these are

* Refer to Par. 10.1.1, Chapter 10 Reinforced Concrete Wall Construction.

found to be within the allowable unit stresses mentioned in the following paragraph, it is considered to be all right. In this case, with regard to the temporary stress, only an average shear unit stress in the wall is checked for.

For increasing the earthquake resistance of a building, the longer the total length of the bearing wall of the building is, the more advantageous it is. Although another way is to increase the thickness of wall, as the wall in its principal nature resists horizontal forces in the direction of the length of wall, for those walls having the same horizontal section area, the longer the wall is, the greater becomes the earthquake resistance.

For that reason, in wall construction, the values of the respective sums of bearing wall length of each story in widthwise and lengthwise direction divided by the floor area of the story is called the wall rates, and its numerical values are made the bases for judging earthquake resistance of the building. That is to say, the wall rate is the length of wall in a certain direction per unit floor area. If an investigation of the above-mentioned shear unit stress in a wall is to be shown by use of the wall rate, the required wall rate l_0 is expressed by the following formula :

$$l_0 \geqq \frac{Q}{f_s \cdot t} \cdot \alpha$$

where

Q : story shearing force per unit floor area of the story concerned

f_s : allowable shearing unit stress of wall, explained in the next paragraph

t : thickness of the wall

α : concentration coefficient of shearing stress

Concentration coefficient of shearing stress increases the story shearing stress in expectation of possible stress concentration in particular walls as the result of unbalanced wall arrangement. If there is no unbalance in the arrangement of walls, $\alpha=1.0$, but ordinarily it takes values of 1.5~2.0. Also, in case where thickness of each wall is not the same, the method of application of the above formula changes slightly but the formula can be used generally by taking the thinner wall thickness.

b) Alternate Method

In this case, the frame calculation is carried out treating the above-mentioned bearing wall as a kind of wall-column. Based on the shearing force distribution coefficients to be calculated from stiffness ratio of each wall-column, shearing stress, bending moment and axial stress are calculated. Method of stress calculation is similar to that of the reinforced concrete wall construction of chapter 10 (refer to paragraph 10.3). As for these stresses, it is ascertained that they are respectively within the allowable unit stresses for shearing, compression and tension. However, as the tensile resistance of the wall of masonry construction is especially small, it is recommended that it be designed so as to cause no tensile stress in the section if feasible.

For such portions of reinforced concrete construction as collar beams or continuous wall footings excepting the wall itself, sections are determined based on the calculated stresses mentioned above and in accordance with the design of reinforced concrete members.

13.4.2 Allowable Unit Stresses

As to the strength of masonry construction, there are differences in each one of masonry units, and it is usual that there be considerable difference depending on quality of jointing work and type of bonding. Besides, in masonry construction, it is characteristic that the strength of bonded body as a whole is remarkably low compared with that of the masonry unit. The ratio of decrease in strength because of bonding is called bonding factor and regarding this, various experimental studies[4)5)] have been made, and the strength is generally reduced to the neighborhood of one half or thereabouts. Taking these facts into consideration, a larger safety factor for allowable unit stresses must be taken compared with those of the other types of construction. The allowable unit stresses of the wall fixed by the standard of this country are shown in Table 13.5 in which, for example, such figures as 1/8 or 1/12 of the strength of the masonry unit for the allowable compressive unit stress for the permanent loading may seem to be too small, but it cannot be avoided considering the character and dependability of execution in masonry construction[3)].

Table 13.5 Allowable Unit Stresses of Masonry Construction Wall

Kinds \ Allowable unit stresses	Allowable unit stresses for the permanent stress		Allowable unit stresses for the temporary stress
	Compression	Tension & Shear	Compression, Tension & Shear
Brick, Solid concrete block	1/8 of the compressive strength of the unit, but not more than 15 kg/cm^2	1/80 of the compressive strength of the unit, but not more than 1.5 kg/cm^2	1.5 times, the values for the permanent stresses
Hollow brick, Hollow concrete block	Using net sections, 1/12 of the compressive strength of the unit, but not more than 10 kg/cm^2	Using net sections, 1/80 of the compressive strength of the unit, but not more than 1.5 kg/cm^2	

13.5 Precautions for Workmanship

In masonry construction, workmanship is especially important. Precautions are required as there are not a few examples in which great damage was suffered at a time of disaster in spite of good design practice, because the expected strength of the building was not attained on account of poor workmanship.

13.5.1 Materials

a) Bricks

As to the common bricks used in masonry construction, a standard of quality and size is given by JIS R 1250. The strength of a given unit is 100 kg/cm^2 or more. There are two kinds, ordinary burnt and high grade burnt, the latter having a better quality. The size of $210 \times 100 \times 60$ mm^3 is made the standard. For ones to be used for arches and cornices, special shape bricks matching the design are available.

b) Concrete Blocks

As for hollow concrete blocks, there is a standard given by JIS A 5406* ; in which 3

* Refer to Par. 14.2.2, Chapter 14 Reinforced Concrete Block Construction.

kinds of concrete blocks—A, B and C in order of increasing strength beginning from the weakest—are available, but it is desirable in masonry construction to use (Class C) ones having a compressive strength of 100 kg/cm² or more as in the case of ordinary bricks.

c) Joint Mortar

For the mortar for bonding, a mix of 2.5~3 sand to 1 cement (volume ratio) is used. For finishing joints, a special mix of 1 : 1 is used. As the strength of mortar has great influence on the strength of bonded member, mortar of leaner mix than this should not be used. As the mortar mixed with lime reduces the strength, precautions are required when it is used for structural purposes.

13.5.2 Laying

a) Wetting

As bricks, stones and concrete blocks have water absorbability, their joint surfaces must be thoroughly washed with water prior to bonding in order to give moisture. If it is not wetted sufficiently, moisture in the joint mortar is absorbed and the hardening condition of mortar becomes unfavorable and separation of joint surface may result. Precautions are required also to avoid excessive wetting, because excessive water will under such a condition cause bleeding of joint mortar.

b) Joints

In laying units, bedding, buttering and pouring of mortar should be done with sufficient care and precautions should be taken especially to the vertical joints so that no void is left in the joints. For the case using hollow concrete blocks, refer to the chapter on reinforced concrete block construction. For the thickness of joint, 10 mm is the standard.

c) Laying

Straight joint should be avoided because it reduces considerably the strength of bonded member (Fig. 13.14).

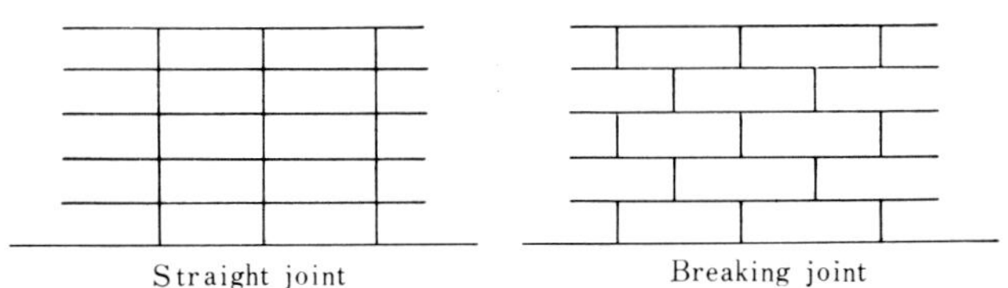

Fig. 13.14

Although there are various kinds of bonds, those in which straight joint is avoided and units may be laid with regularity are recommended. For brick construction, English bond, Flemish bond, etc. are often used (Fig. 13.15). Each part should be laid equally as much as possible. A day's laying height should generally be limited to within 1.2 m and unfinished edge should be raked and toothed edge should be avoided.

13.5.3 Others

Buildings of masonry construction have reinforced concrete collar beams, continuous wall footing, lintels, etc. Also there are cases where steel frame is used in part for reinforcement.

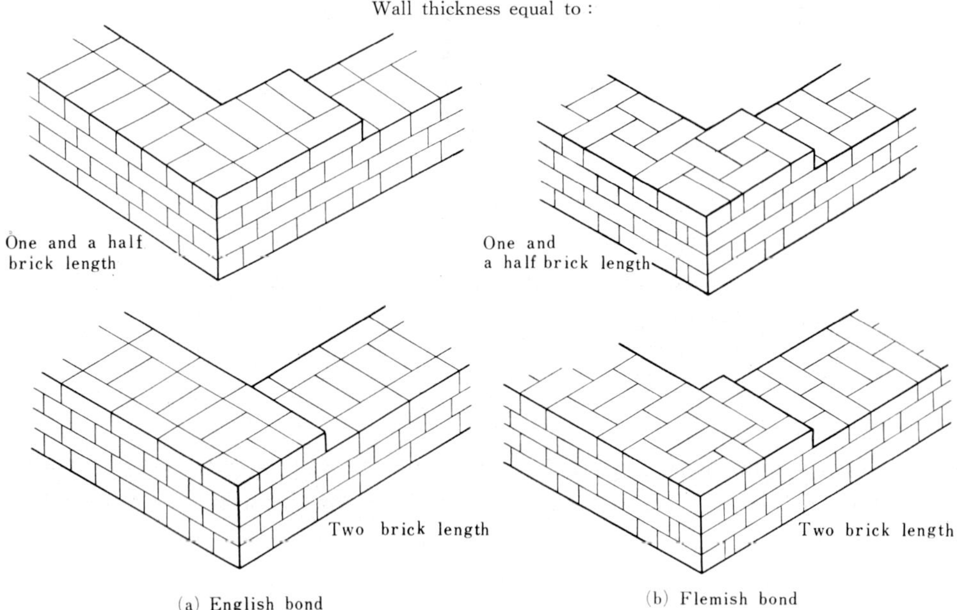

(a) English bond

(b) Flemish bond

Fig. 13.15

In each one of the construction methods, those precautions required for the execution of respective construction should be followed. In regard to masonry construction, however, precautions should be taken so the masonry construction are well connected because these connections made with other types of construction are apt to become weak points. In some cases, metal reinforcements are sometimes used.

14. REINFORCED CONCRETE BLOCK CONSTRUCTION

14.1 General

14.1.1. Structural Characteristics

Reinforced concrete block construction is similar in appearance to masonry construction but it differs in that it is a construction which is reinforced by inserting steel bars in the body of the wall. This construction method has been especially developed in this country and it is an improvement over ordinary masonry construction so that it is able to resist the external forces due to earthquakes. In this construction, wall made by bonding hollow concrete block units is a principal element of structure ; and concrete is filled in long and narrow hollows in the body of wall running both vertically and horizontally, after steel bars have been inserted in them so that the body of the wall is made into one unit. Its reinforcing method is based on the theory of reinforced concrete. Consequently, it is a kind of reinforced concrete construction. However, the body of wall, unlike that of ordinary concrete, is an assemblage of individual small block units jointed by mortar ; therefore it has the same weak points as that of the masonry construction. That is, a bonded body of concrete block unit has less resistance to compression, tension, shear, etc. in comparison with that of monolithically cast concrete ; and if jointing work is poorly executed or concrete filling in the cavity is inadequate, not only the resistance is further reduced, but also the reinforcing effect of steel bars becomes less and the advantages of this construction are lost.

Although this type of construction has, the characteristics of a masonry construction that has been made earthquake-resistant, cases occur where its earthquake-resistance is not reliable at all, depending on the design and work execution ; and it is necessary to give thorough considerations to those precautions to be taken of which mention will be made in this chapter.

14.1.2 Earthquake Damage

Generally speaking, the reinforced concrete block construction has been popularized since the end of World War II, and most of the buildings of this construction are less than ten years since their completion. Consequently, there is little earthquakes damage experience. At the time of the Niigata Earthquake in 1964, there were found a few examples of damage. The distinctive feature of damage due to this earthquake was caused by liquefaction phenomenon of sandy layers resulting in considerable inclination of buildings as a whole and cracked wall caused by differential settlement. The damage to most of the superstructure of reinforced

Photo 14.1 Damage Example of an Apartment House of Reinforced Concrete Block at the Time of the Niigata Earthquake in 1964

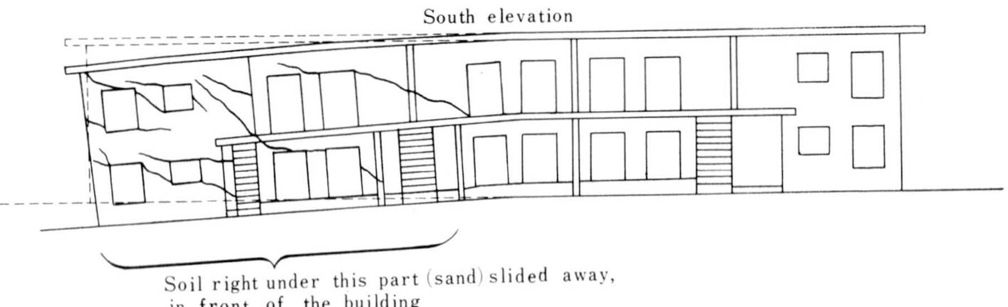

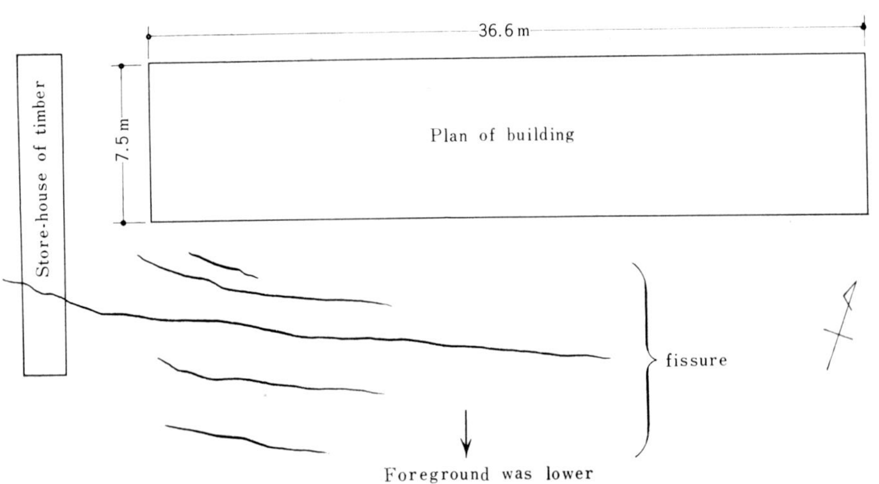

Fig. 14.1

Table 14.8 Length of Splice and Anchorage for Reinforcing Bars

Structural part / Kinds of steel bar		Round bar	Deformed bar	
Splice	Anchorage		Without hook	With hook
In principal places such as parts of joints and corners of bearing wall, periphery of opening, etc.	In cases where vertical bars at the principal place are embedded in collar beam, floor slab or footing beam, and where horizontal bars at the top and bottom of an opening are anchored in the wall on both sides of the opening	40 *d*	40 *d*	30 *d*
In other locations	In the rest of cases*	25 *d*	25 *d*	20 *d*

Note :

d : diameter of steel bar ; in case of splicing, diameter of the smaller bar.

* In case anchor length of horizontal bars cannot be taken sufficiently being close to the end of wall, the end of the horizontal bars may be hooked on the vertical bars in the part where they are to be anchored.

b) Concrete Protection

Durability of reinforced block construction is dependent upon the thickness of concrete protective covering for the reinforcing bars. Required thickness of the concrete protective covering is shown in Table 14.9. As the concrete blocks are generally porous and rain water easily infiltrates into them, thickness of concrete block itself is not included in the concrete protection. Accordingly, steel bars placed in the hollow part of bearing wall must be given required thickness of concrete protection only by the concrete filled in the hollow part.

Table 14.9 Thickness of Concrete Protective Covering

Parts of structure	Thickness of concrete protective covering (cm)
Bearing wall	2 (Thickness of concrete block should not be included.)
Collar beam	3*
Footing	4
Bottom surface of footing	6 (Thickness of leveling concrete should not be included.)

Note :

* For those parts facing inside and having an effective finish for the durability of reinforcing bars, 2 cm may be acceptable.

14.4 Notes on Structural Calculation

As the reinforced block construction is a wall construction, its structural calculations are made in the same manner as those for reinforced concrete wall construction. However, as the body of wall is not a homogeneous one as in the case of masonry construction, safety of the material itself must be well considered as to the allowable unit stresses, etc.

Stress calculations for the wall construction are not simple. However, for the design of small scale concrete block buildings such as dwelling houses or ordinary office buildings, design can generally be made without calculation, based on provisions and values in the tables of the "Standard for Structural Design of Reinforced Concrete Block Construction". But structural calculations are necessary for those portions for example, reinforcements for collar beams and continuous footings, or in cases of especially large scale buildings when special design is required.

14.4.1 Principles of Calculation

As reinforced block construction has no such limitation as to the width of openings as is the case for masonry construction, an elevation with a large opening is generally made possible. In case where an opening width is small (for example, not more than 2 m), nothing more is required for the bearing walls if the dimensions, wall rates, reinforcements, etc. have been confirmed to meet the standard ; and as for reinforced concrete collar beams, footing beams, etc., which are located at the top and bottom of the opening, it may be enough to calculate them as a beam fixed at both ends and subjected to vertical loads and horizontal forces in direction normal to the plane of the wall, or as that subjected to vertical loads only.

If opening width is made large, wall elevation comes to have a resemblance to that of a frame structure, and the bearing walls, collar beams, etc. will be subjected not only to the vertical loads but also stresses of rectangular frame in the plane of wall due to horizontal loads ; consequently suitable calculation for sections will become necessary (Fig. 14.18). These stress calculations are complicated, but ordinarily simplified methods of frame calculation are used. The methods are similar to those for reinforced concrete wall construction. For details. special reference works should be consulted.

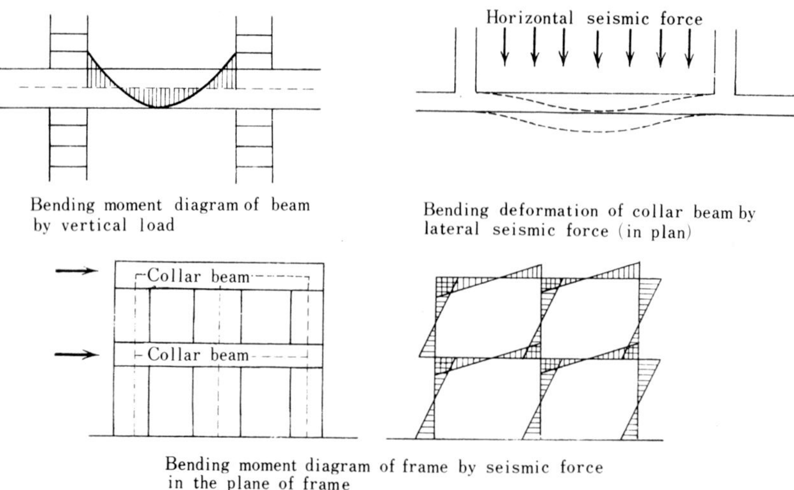

Fig. 14.18

As for various kinds of roof trusses and floor slabs of reinforced concrete construction which are parts of reinforced block construction, calculations are made in accordance with the calculation methods for the respective constructions.

14.4.2 Allowable Unit Stresses

Allowable unit stresses of reinforced concrete block wall are shown in Table 14.10[3].

As for concrete block wall, even if the concrete block has considerable strength as a single

Table 14.10 Allowable Unit Stresses of Reinforced Concrete Block Wall (kg/cm²)

Kinds of construction \ Allowable unit stresses	For permanent stress		For temporary stress	
	Compression	Tension & shear	Compression	Tension & shear
Class A concrete block construction	6	0.6	Twice the value for the permanent stresses	
Class B concrete block construction	8	0.8		
Class C concrete block construction	11	1.1		

unit, strength of the wall drops when bonded. Although this cannot be determined definitely as it depends on joint materials, bonding methods, quality of work, etc., it is concluded based on the available past experimental data[4),5)] that it will become about one half of the strength of a single unit block.

Reduction ratio of the strength of wall to that of a single unit is called bonding factor, and the allowable unit stress of concrete block wall is shown by the following formula :

(Allowable unit stress of wall) = (Strength of concrete block) × (Bonding factor)
÷ (Safety factor of material)

Where

Strength of concrete block : Average strength of dry concrete block based on gross sectional area

Bonding factor : 0.6 for class A concrete block construction, and 0.5 for class B and C constructions

Safety factor of material : 1.5 for permanent loads and 3.0 for temporary loads

14.5 Precautions for Workmanship

Execution of work for reinforced block construction is especially important as also in the case of masonry construction as it has significant influence on the strength. Since almost all of the reinforced block construction that has suffered damages from the earthquakes in the past had been poorly executed, both designers and builders should take heed in this regard. As for timber construction work and reinforced concrete construction work which may be combined with this construction, precautions in the related chapters should be referred to.

14.5.1 Materials

a) Concrete Blocks

Concrete blocks to be used are the hollow concrete blocks prescribed in JIS A 5406. As for ceramic blocks, those prescribed in JIS A 5210 should be used.

b) Mortar and Concrete

Mortar and concrete mixes to be used in reinforced block construction are as shown in Table 14.11.

As hydrated lime mixed in joint mortar serves to increase the stickiness but decreases the structural strength, it should not be used excessively. Size of gravel to be used for filling concrete should not be too large. As the hollow parts of block are small and almost all of

Table 14.11 Mortar and Concrete Mixture (Volume Ratio)

Use \ Construction		One-story or two-story building	3-story building	Remarks
Mortar	Jointing	1 : 0.2 : 2.8 (or same as below) (cement) : (lime) : (sand)	1 : 0.2 : 2.5 (or same as below)	Thickness of joint : 10 mm
	Filling	1 : 3 (cement) : (sand)	1 : 2.5	Consistency shall be determined in consideration of water absorption of concrete block
	Joint finishing	1 : 1	1 : 1	Proper amount of waterproof agent shall be mixed
Concrete	Filling	1 : 2.5 : 3.5 (cement) : (sand) : (gravel)	1 : 2 : 3	Slump : 20～23 cm
	Structural	1 : 2.5 : 3.5	same as left	Slump : 15～20 cm
	Leveling for the footing bed	1 : 3 : 6	same as left	minimum amount of water in order to obtain workability should be used

them are to have reinforcing bars in them it, ordinarily the size should be 1/5 of the diameter of the hollow part or less.

14.5.2 Arrangement of Reinforcement

As for the arrangement of reinforcement for continuous footings and collar beams, they are omitted here as they are similar to that for reinforced concrete construction ; only those points requiring precaution as to the arrangement of reinforcement in the wall itself will be given here.

Vertical bars should be raised from the continuous footing, extended througth the one story height without being spliced and anchored in the collar beam. This means that the reinforce-

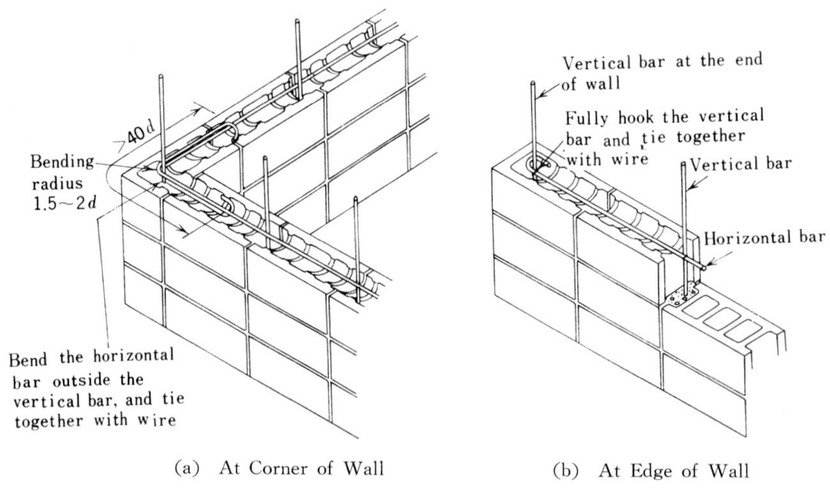

(a) At Corner of Wall (b) At Edge of Wall

Fig. 14.19 Arrangement of Reinforcement for Wall

ment must be arranged and imbedded at the same time when the concrete of the footing is poured. Therefore, it must be definitely ascertained that the bar spacing is sufficiently accurate ; should it be inaccurate, no correction will be possible after the concrete hardens. They should be anchored with sufficient anchorage length.

For horizontal bars, it is necessary to hold them exactly at the center of the hollow part. It is not good to pour the concrete while the bars are lying in contact with the bottom of the hollow part. For the anchorage of horizontal bars at a corner, a sufficient length of anchorage should be provided in the joining wall. Where a wall is not joined at the end with other wall, it should be hooked on to the vertical bar at the end and bound with binding wire (Fig. 14.19).

14.5.3 Laying

a) Wetting

As the concrete blocks have water absorbability, it is recommended to give proper moisture to their joining surfaces. If there should be no moisture, water in the joint mortar will be absorbed resulting in poor binding quality.

b) Joints

Joint mortar should be applied to the entire surface to be bound. Often application is made to the portion of both sides (portion of face shell) only, leaving the portion of web at the center undone but this is not good practice (Fig. 14.20).

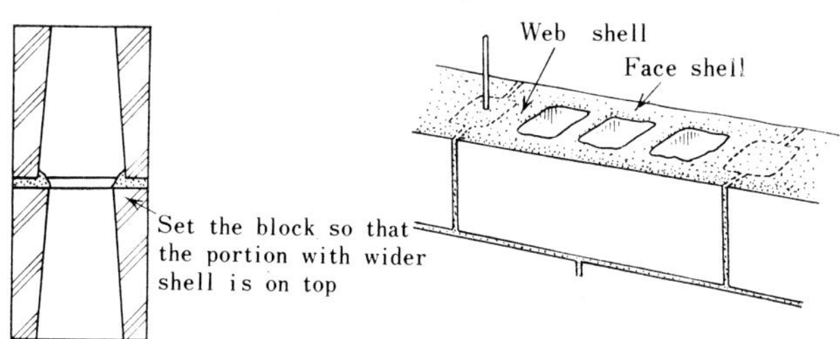

Fig. 14.20 Application of Joint Mortar

In case of laying the first block course on the top surface of reinforced concrete collar beam or continuous footing, mortar of rich mix should be used for the joint in order to obtain good bonding.

c) Laying

For reinforced block construction, straight joint is often used so that vertical reinforcing bars can be put through straight at the location of the joint. Staggered joint makes the work execution a little harder. Height of laying for a day should be made about 1.2 m (6 courses) or less.

d) Mortar or Concrete Filling

Hollow parts where vertical bars are inserted or hollow parts adjoining to vertical joints should always be filled with concrete or mortar. Hollow parts are not wide and when steel bars are inserted in them, filling become especially difficult. Therefore, tamping rod or the like must be used so that concrete may be well packed. If this is not well done, steel bars will have exposed portions and be corroded, causing complete failure of the reinforced block construction. Pouring should be made at approximately every two courses. Jointing of successive pours should be made about 5 cm below the concrete block surface. This is done so that the weak point of the jointing of successive pours would not coincide with the horizontal joint (Fig. 14.21).

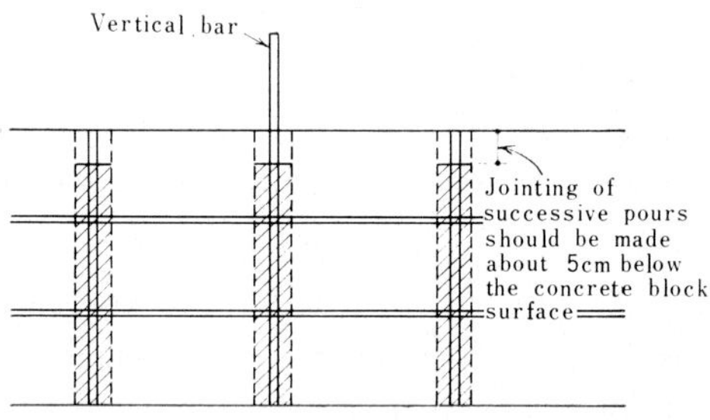

Fig. 14.21

14.5.4 Waterproofing

Concrete block construction is often criticized beceause rain water infiltrates througth the exterior wall. This may be the case of exterior walls of exposed concrete blocks. Nevertheless, there is no doubt that it would cause trouble if rain should infiltrate and the infiltrated water corrodes the steel bars. Therefore, the matter of waterproofing of the wall surface is an important problem in reinforced block construction. Rain water is considered to infiltrate through either the exterior surface of blocks or poorly worked joint. Joints should be worked carefully and either the concrete block surface should be applied with waterproof agent or waterproof concrete blocks made by mixing waterproofing admixture should be used. If exterior surface is finished with waterproof mortar instead, there will be no fear of such troubles.

References

1) T. Kawano, K. Matsushita and T. Hisada; Vibration Test on Full Size Building Sample of Brick Construction, Architectural Institute of Japan, Research Report No. 14, October 1951.

2) S. Takenouchi and Y. Nakagawa; Strength Test on Building of Reinforced Block Construction, Architectural Institute of Japan, Research Report No. 27, May 1954.

3) Architectural Institute of Japan; Commentary on Standard for Structural Design of Reinforced Concrete Block Construction (1964).

4) S. Asano, Y. Nakagawa and S. Takenouchi; Experiments on Strength and Rigidity of Wall of American Standard Type Concrete Block, Architectural Institute of Japan, Research Report No. 22, May 1953.

5) K. Matsushita; Strength of Brick Masonry Construction, Architectural Institute of Japan, Research Report No. 7 and 9, August and November 1950.

15. CONSTRUCTION OF PARTS OF BUILDINGS

15.1 Aseismic Design of Minor Part Construction

In contrast with the principal members that are designed as the resisting members of the structure, those not regarded as the principal resisting members are referred to in general as minor part construction. For example, fitting fixtures, exterior and interior finish, etc. are so called. Needless to say, some of these minor part construction are checked as to their strength by experiments and calculations but, in many cases, the dimensions have been determined based on experience of the past and it has come be accepted to have the necessary strength for ordinary use.

In regard to the aseismic problem, the minor part construction differs from the principal construction in the following points :

i) There are many cases where more importance is attached to the function rather than to the strength.

ii) In many cases, earthquake damage occurs to minor parts due to damage to the principal members to which they are attached.

iii) Generally, it has smaller mass compared with that of the principal part, therefore in the upper stories, it suffers earthquake damage because the earthquake motion already amplified through the building is amplified again in the upper part.

iv) There are cases where unfavorable conditions for vibration occur even for the same construction, depending on the method of attachment.

v) Finishing materials, etc. have high rigidity compared to their strength and at the time of earthquakes, many of them are destroyed before the forces have been transmitted to the main members. Consequently, even in the case of medium or minor earthquakes where no damage is expected in the main construction, it is possible that the minor constructions are immediately destroyed causing death and injury.

Consequently, the design of minor parts is not confined to the mere investigation of strength of those parts but considered to be a problem to be treated carefully at the time of design of the whole building. If possible, construction method of minor parts should be determined in relation with the planning of the period of vibration and interstory deflection of main structure ; and in cases where different construction methods must be used in combination despite their same purpose, it can be stated that precautions are required especially with regard to the design and work execution.

However, for instance, as in the case of fireproof finish, the function of which is the protection of the main structure at the time of disaster, it is important that they be left unbroken at the time of the earthquake in order to prevent fire which possibly may break out immediatly after it. But, on the other hand, there is the possibility where policy can be established to prevent the main structure suffering from damage, by making those parts other than the main structure suffer more or less damage, and thereby increase the damping of the building, provided it is clear that it does not cause loss of human lives even if they may be ruptured. In short, in order to prevent damage to such minor parts that have the possibility of being connected directly with great loss of human lives and the building proper at the time of earthquakes, it is essential to investigate their earthquake resistance in relation with the main structure.

15.2 Roof and Parts Projecting from Roof

15.2.1 Roof

With regard to roof finishes or roofing, clay tile, slate or sheet metal roofing are popular and straw thatched or shingle roofing are decreasing. From the viewpoint of aseismic requirements, light roofs are desirable, and in view of this, those of sheet metal roofing and asbestos cement slate roofing are advantageous. Except for those cases where the strength of the building itself has been extremely reduced owing to decay or some other reason, sheet metal roofing and asbestos cement roofing can be said, even in the case of wooden buildings, to possess earthquake resistance to some degree by themselves. Besides, even if it should collapse, percentage of survival of the people staying inside would be high because of the lightness of roof and roof framing members. Moreover, there is almost no case of casualties caused by the collapse of the roof alone in past earthquakes. However, as for the stones placed on cedar bark roofing against wind storm or snow accumulating on sheet metal roof or shingle roof, danger of their falling down in earthquakes is always present.

With regard to asbestos cement plate roofing, as they are attached to thin sheathing board, proper precautions to prevent nails from loosening must taken.

In some cases, corrugated asbestos cement boards are used without sheathing for roofs of factories, etc., and when the rigidity of this part of the possibility, it is necessary to guard against the building is insufficient, forces will first act upon this part, and rupture will be caused originating from the holes for fastening metal pieces to purlins and the roof may fall down.

Besides, asphalt roofing and the like are of light weight and advantageous in the light of earthquake resistance as in the case of sheet metals, but its durability is a problem as in the case of straw thatched or shingled roof.

On the other hand, with regard to clay tiles, slates, straw thatched roofing, etc., all are of large weight and are undesirable from the aseismic viewpoint, as they cause the collapse of buildings, loss of human lives due to collapse, casualities due to rupture of roofs only, etc. Of course, when they have been designed taking these points into consideration, they become

Photo 15.1 Fallen Roof Tiles

earthquake resistant.

In the case of clay tile roofs, tiles are apt to slip down at the time of earthquakes. The old fashioned clay tile roofs having clay between the tiles and sheathing are especially dangerous. Clay is not being used at present because of the necessity for reducing the weight, and as to tiles themselves, those of hook-on type are being used for the most part, and moreover they are fastened by nails or wire. Therefore, an improvement has been made in regard to this point but, owing to the moisture absorptive sheathing board and hooking strip or corrosion of fastening metals themselves, these get loose and a lump of roof tiles falls down in many cases in an earthquake.

Although the weight of slate roofing is comparatively light, as the slates are nailed to sheathing board as in the case of asbestos cement slate roofing, they become loose and fall down sometimes in an earthquake.

Cement roofing tiles can generally be said to be one of the most undesirable roofing materials from the aseismic viewpoint because their weight is great and in many cases, they are fastened without hooking strips to thin sheathing boards by one or two nails per unit.

With regard to earthquake proofing of roofing materials, attention must be paid so that no casualty is caused by their falling down, consequently, their fastening method to the sheathing board is a problem. In many cases, they have resisting strength at the time when they have been just constructed or repaired, but they get loose owing to repetition of thermal changes for many years or corrosion, and suffer damage at the time of earthquakes. Therefore, it is recommended that those materials such as copper, stainless steel, etc., for which rusting is not expected to progress, be used for the fastening metals and sufficient margin for strength be provided. Parts of roof suffer great damage also at the time of strong winds, therefore it is never futile to pay attention especially to the fastening method at the tip of eaves and around the ridge. Besides, it goes without saying that the sheathing should have sufficient strength to support fastening metals and it is desirable that attention be paid especially to its rigidity in case where those roofing materials of large size such as corrugated asbestos cement board are used.

15.2.2 Parts Projecting from Roof

Those projecting parts from the roof such as parapets, smoke stacks, penthouses, etc. have small mass as mentioned before compared with that of the building. Consequently, they vibrate at the time of earthquakes in such a manner as if their vibratory base were at the top of the building. For this reason, it is ruled by the Building Standard Law that the projecting parts from the roof of buildings, the height of which is 30 m or less, be designed with a seismic coefficient of 0.3 or more. Of course, considering the safety factor, etc. of the material, it may be expected that those parts can actually resist against an earthquake force stronger than this but precautions are required because should the building be subjected to an earthquake having an acceleration of about 100 gal, it would be 300 gal at the top of the roof and for the parts projecting from the roof, it would be probable that it becomes 2 or 3 times this value in the worst case. In addition, if these projecting parts are located so that their rupture is directly connected to their falling down from a high place, there will be added danger of their inflicting injuries on persons.

In cold districts, there are many cases where large cracks have developed in parapets and the like because of snow and freezing. As to smoke stacks, there is a decrease in strength due to corrosion after many years' use. Therefore it is desirable that these matters be taken together into consideration at the time of design. As for tall and long parapets, it is necessary to reinforce them either by providing lateral supports by means of buttresses or by providing such bond beams as used in masonry construction.

Mentioning others requiring precautions among the parts projecting from the roof, there are water tanks, cooling towers, advertisement towers, etc. On the premise that each one of them will sometimes to be subjected to an acceleration as great as several times that for the main structure, not only its the structural strength but also the influence of their acceleration on the main structure of the building as forces acting upon it, such as torsion, etc. should be carefully investigated and scrupulous design and work execution, especially as to the construction method of those connecting portion, are strongly stressed. Sometimes they are installed on an existing old building and in this case, it is clear that such structural investigations as in the case of extension of building superstructure are required. Besides, as many of them are exposed to the weather, it goes without saying that precautions for maintenance must be taken constantly so as to prevent the decrease in strength.

15.3 Ceilings

With regard to ceilings, those in which the strength of the support is especially weak and the weight large present a problem. It is good from the aseismic viewpoint that those heavy ones such as plaster ceiling are on the decrease lately, being replaced by new materials. In case of plaster ceilings, lap of the plaster on the lath will become short owing to the contraction of the latter due to drying and besides, in plaster itself cracks are caused, or leakage of rain water, etc. will overlap on these to make favorable condition for their falling down. In case of

Photo 15.2 Fallen off Ceiling and Wall Plaster Finish

suspended ceilings, where relatively light materials such as wood board, fiber board, etc. are used, it will not easily fall down so long as there is no defect in the fastening method of suspenders. Furthermore, damage will be minor even if they should fall down.

15.4 Walls

Walls as well as ceilings are vulnerable and many examples of plaster falling off in earthquakes occur. Moreover, stone faced walls are more dangerous. Stones are, in many cases, attached to the unfinished wall surface by means of metal anchors and except for those particular ones such as marble, etc., mortar is poured in the space between the wall for attachment. The poured mortar, if the stone has no moisture absorbability, will have separated from the stone and there are many cases where the fastening is accomplished merely by the metal anchors. Depending on the method of work execution, even the mortar will not sufficiently flow in the back side, and in many cases, the metal anchors come out or are broken causing falling down of the stones at the time of earthquakes. Especially for those marble stones attached leaving hollow space behind, metal anchors and the mortar around them alone will have to support the heavy stones in case they have slipped off the edge of the stone below on account of an earthquake. Where the main structure is of construction having little rigidity, it is desirable to avoid the application of plaster finish or stone wall facing upon it because those parts having high rigidity will first be ruptured. By the same reason, if those having high rigidity but little strength are used for partition walls and exterior finish walls, these parts will also be susceptible to rupture.

Generally speaking, for walls which anticipate an interstory deflection of more than 1/300, it is recommended not to use stone wall facing. Not only facing stones, but also large size tiles and terra-cotta must be sufficiently tied to the wall body by means of metal anchors.

Moreover, by the same reason as in the case of the parts projecting from the roof, the parts projecting from the wall must be properly protected. Especially in the case of an exterior wall, precautions are required because the fall of the projecting parts has the danger of inflicting injuries on persons.

As for the wall and its finishes, selection must be made taking into considerations whether the main structure is flexible or rigid.

For rigid structures of reinforced concrete construction and steel framed reinforced concrete construction having reinforced concrete walls, such finishes as stone facing mentioned above can be utilized if precautions are taken in the method of attachment. In these cases, it is proper to install them, without much regard to interstory deflection as the masonry walls and concrete panels have sufficient reinforcement against forces in and out of the plane of the wall. However, in flexible structures such as ordinary steel framed construction, it is desirable to make it a general rule that those which can adapt themselves to deflections of the main structure are used as walls and their finishes.

In carefully designed curtain walls, those properties such as air-tightness, heat shielding, heat insulation, etc. are not affected by deflection but considerations must been given so that no damage to glass panes occur. For high rise buildings of steel framed construction in this country, motions in interstory deflection amounting to 1/150 are anticipated but techniques to cope with these conditions have already been developed. However, in case where they are attached to the end of cantilevered floors or similar locations, precautions required especially at the parts of attachment.

For walls, especially for exterior walls, falling off of the finishing materials has been treated so far but damage to and falling down of window glass panes also require precautions. There are designs that have canopy, balcony or lean-to roof at the low story in order to provide against this case. Moreover, investigation of the strength of walls against strong winds and

Photo 15.3 Fallen Plaster Finish of Earthen Wall of DOZO-Construction

Photo 15.4 Fallen off Wall Facing Stones

earthquake forces in the direction normal to its surface is required.

On the other hand, there are partitions, walls, etc. of the part for which mere restoration is sufficient even if they are damaged, and there is no need to use flexible construction except in special cases. Needless to say, there are cases where damping effect of the building as a whole may be increased by allowing more or less damage on purpose to the anticipated portions.

As for cantilever beams of reinforced concrete construction, precautions are required because there are many cases where top bars have been lowered owing to poor execution of the work thereby decreasing the resistance against vertical motions.

15.5 Fittings and Fixtures

Among fittings, glazed windows hurt persons when they are broken. Moreover, even in the case where it seems at a glance that there are no damages to buildings by an earthquake of minor or medium intensity, damages to window glass panes are caused. Therefore, it becomes one of the items which requires speciall attention.

As a clearance is provided between glass panes and sash bars, glass panes can ordinarily adapt themselves to deflections of the building. However, in fixed windows of which putty has hardened, allowable amount of deformation is very little, and some of them are considered to be about 1/1000 rad or possibly in the order lower than that. Consequently, in case where the building itself is flexible and only the glass panes are rigidly fixed, for example, those buildings of reinforced concrete construction having few walls and being fully glazed by means of fixed window sash, there should be caused damages to the glass panes so long as no

Photo 15.5 Damages to Fitting and Glass Panes

consideration has been made to provide sufficient clearance for the glass panes.

The problem of clearance for glass panes is solved with comparative ease by inserting as packing, as in the case of curtain wall mentioned above, those elastic materials (for example, thiokol, neoprene, etc.) that change little by aging. At the time of great earthquakes, it is conceivable, that it will be ruptured in the direction normal to the plane, depending on the glass pane size and the condition of attachment.

Precautions are required because among the fully glazed sashes mentioned above, there are such ones for which considerations have not been given sufficiently as to the deformation in the direction normal to the plane.

It cannot be said that there is not at all such construction method in which fittings are made so as to have earthquake resistance and its reinforcing effect to the main framing is expected but it is natural that generally many of fittings are manufactured placing emphasis on operational function in ordinary times and habitability. Also, in general, there are many cases where exterior fittings are designed based mainly on strength for the wind forces and it also covers the strength for the earthquakes. However, as for those which are dangerous when damaged, which have a possibility of overturning, which have especially a large weight, investigation should be made once from the viewpoint of earthquake resistance. With regard to this, it is the same for other fixtures.

15.6 Stairs, Cantilevers and Overbridges

As for stairs, a sweeping statement cannot be made because of the variety in their relationship with the main structure, construction method of stairs themselves. Those stairs constructed with the same material as that of the main structure and made integral with the main structure will in many cases increase the resisting strength of the main structure but this is at the same time an indication of the possibility that the part of stairs will first be destroyed. Or, having vibrations different from those of the main structure, the part of stairs in some cases is ruptured at the portion of their support. In case where floors or beams are cantilevered from the main structure, it is well to anticipate that those parts will vibrate greatly. Moreover, in cases where fittings are attached thereto, damage to the fittings is sometimes caused on account of these vibrations.

In case of stairs, considering that they become routes of egress, it is desirable to use such construction method that is not susceptible to damage. But in many cases they are located at the dividing part of the building, or in such part where the resisting strength of surrounding floors has been reduced on account of many openings for ducts, pits, etc., or in such locations where concentration of stress will occur. Therefore, thorough precautions are required in the design of these parts.

Also the stairway have vibratory properties different from that of the main structure itself as in the case of an overbridge spanning between two buildings, and there are such cases where there is an extreme change in the spans and therefore, there is great danger not only of being damaged but also of moving and falling. In order to cope with these, such considerations are required such as to provide sufficient bearing at the part of support and in addition to make one end movable depending on the case.

15.7 Gates, Fences and Retaining Walls

Gates, fences and retaining walls of small scale are made of concrete blocks, stones and sometimes of brick construction. As for those of a larger scale, they are generally dependable because almost all of them are now made of reinforced concrete construction and majority of them are made into earthquake resistant construction with supporting structural calculations. Those suffering most damages are the ones of small scale about 2 m or under in height which are not subjected to severe surveillances under the laws and regulations. Structures of this kind are used everywhere and injuries inflicted on persons can not be disregarded from their number in case they should have collapsed completely. In general, the footing work is simply done and there are many footings whose width is too narrow and the thickness of structural way and the reinforcement are insufficient. Consequently, that many of them suffer damages due to inclinations and cracks at the time of earthquake even when it is a light one and those of overturning and destruction when it is a severe one. It is desirable that even those of small scale be considered as objects of aseismic design.

Photo 15.6 Damage of Concrete Block Wainscot Wall

16. VESSEL STRUCTURES, STEEL TOWERS AND STACKS

16.1 Vessel Structures

16.1.1 General

Vessel structures to be treated in this chapter are the vessels and their supporting structures of self standing elevated water tanks, oil tanks, gas tanks, silos, etc. Water tanks installed on the roof of buildings are not included.

High pressure vessels and elevated water tanks of special shape, etc. are the ones which have been developed in foreign countries and brought into this country through the introduction of technical know-how and precautions are required because some of them have not been thoroughly investigated as to their vibratory behavior due to earthquakes.

In case where the content in the vessels is liquid, responses of these structures to earthquake waves indicate characteristics that differ from those of other structures. This is the point which requires precautions in the design of vessel structures.

16.1.2 Structural Planning

Vessel construction is divided between the part of vessel that is functionally needed and the supporting structure that supports the above. Also, there are some which are installed directly on the ground without the supporting structure and those installed underground.

a) Vessels

Vessels can be classified depending on their services, into those of liquid, gas, solid, or powder. Further, depending on their uses, they are divided between pressurized ones and

Table 16.1 Vessl Classification

Content	Pressure	Nomenclature	Law, regulation or standard
Liquid	No	Petroleum oil tank	JIS B 8501 Petroleum Oil Tank API[1]
Liquid, Gas	Yes	Pressure vessel	Pressure Vessel Safety Regulation (The Ministry of Labor), Pressure Vessel Structural Standard (The Ministry of Labor, JIS), ASME[2]
Gas	Yes	High pressure gas container	High Pressure Gas Control Law
Liquid, Gas		Vessels related to nuclear power	Atomic Power Generation Standard

non-pressurized ones. A summary of these classifications are shown in Table 16.1; and as the respective vessel part has applicable laws, regulations and standards as well, these requirements must also be satisfied at the time of structural planning. For vessels other than the above, design can be based on the Standard for Structural Design of Vessel Construction by AIJ.

b) Supporting Structure

As for supporting structures for vessel, many of them are directly attached to the vessel and it is difficult to distinguish between the container part and the supporting part. Accordingly, in case of those vessels as shown in Table 16.1 and having supporting structures, it is hard to determine strictly the applicable limit of the related laws, regulations and standards; but those parts directly attached to the vessel shell by welding or the like are regarded as a part of the vessel and applications of these laws, regulations and standards are made. As for supporting structures other than these, the Standard for Structural Design of Vessel Construction may be used.

16.1.3 Structural Notes

a) In plan, it is desirable to make it as symmetrical in shape as far as possible. For vessel constructions, there are many vessels for which stress analyses are difficult. Therefore, it is recommended that the plan be arranged in such a manner that it is made into a symmetrical shape as far as possible in order to simplify the design calculations.

b) As for the details at the part of vessel support, considerations should be given so that the welding work can be done with ease. In addition, in case where a vessel is supported by the side plates or the bottom plates, precautions are required so as not to cause eccentricity to the side plates as far as possible because a large force will act locally upon the comparatively thin shell plate.

c) In case where the connections of main body of vessel are made by riveting or bolting and for which watertightness or airtightness is required for its service, a proper selection as to their diameter, pitch and guage must be made.

d) Welded joint of the body of vessel should be made by butt welding. However, fillet welding may be used in the following cases:

i) in case where the thickness of plate is 4.5 mm or under,

ii) in case of lapped joint to be welded on one side only for the bottom plates that rest directly on the foundation, and

iii) in case of lapped joint to be welded on one side only for the roof plates that do not resist stresses.

e) As for agricultural silos of unreinforced block construction, there are many examples of earthquake damage that they have suffered. In regard to those matters to be noted from the seismic viewpoint, refer to respective chapters of masonry construction and reinforced concrete construction in this book.

16.1.4 Notes on Structural Calculations

a) Some vessels are subjected to leak and pressure tests using gas or liquid depending on the kind of vessel. In case where gases are used for a test, their weight will not be a problem whereas in case where hydraulic test is performed for these vessels to be used for storage of gases, design against the weight of water at the time of test become very important in many cases. Although there is a question whether seismic loads should be concurrently considered at the time of such hydraulic test, ordinarily such condition is not considered in the design.

b) As for the seismic forces to be applied to vessels and their supporting structures, in addition to those cases where they are based on the Building Standard Law Enforcement Order, their determination can be made in consideration of dynamic characteristics of the whole container structure, regarding it as one vibratory system. However, in that case, with regard to such content in the vessel as liquid which will have influence on the vibration, dynamic characteristics of the content must also be taken into consideration.

c) There is an example in the Niigata Earthquake where a petroleum oil tank having no content failed by buckling (Photo. 16.1). An investigation is required also for the case where the vessel is empty.

Photo 16.1

d) In case where vessels are of reinforced concrete construction, their bending stresses reach from the edge deep into the central part. Consequently, for this construction, stresses must be determined based on the bending theories. In addition, design of sections must be made in accordance with the "Standard for Structural Calculation of Reinforced Concrete Structures". And in case where the radius of curvature of vessel shell plates is large, the ordinary design formula for slab (Art. 16) of the "Standard for Structural Calculation of Reinforced Concrete Structures" can be used for the design of sections. Besides, an investigation is required especially for the concrete against tension at the tension part of the vessel. As for parts of a vessel of reinforced concrete construction, where the thickness is thin and the reinforcing bars are especially crowded, use of deformed bars is recommended in order to increase the bond strength.

e) In case where a vessel of thin plate is analyzed by the membrane theory, stress disturbance to be caused at locations where the profile of the vessel changes abruptly or at the edges must be established by experiments or by some other means and sufficient reinforcement should be provided at these portions.

f) Stresses in the vessel at the time of earthquake vary depending on whether the type of content is gas, liquid or solid. Besides, investigations must be made as to thermal stresses of the vessel itself in case where the content generates heat and to settlement or differential settlement of the vessel due to its content in case where specific gravity of the content is large.

g) In case where a vessel is installed underground, as the stresses due to soil pressure or underground water pressure increase when the content is emptied, investigations for these conditions should not be neglected.

h) A center upon which horizontal forces act at the time of earthquake may be considered to be at the center of gravity of the vessel with its content.

16.2 Steel Towers

16.2.1 General

Steel towers, depending on their services are of such kinds as television tower, radio tower, microwave tower, or the like, which are used for transmitting and receiving electric waves; and illumination tower, advertisement tower, etc. ; and all the requirements for structural calculations of these tower-like structures are compiled in the "Standard for Structural Calculation of Steel Tower Structure" by AIJ. As the steel towers and steel masts for electric power transmission are ruled by the "Electrical Facilities Code"[3)] of the Ministry of International Trade and Industry Ordinance, the Building Standard Law is not applicable to them.

As for structural types of steel towers, there are two kinds ; the steel mast and the steel tower.

Steel tower is defined as a comparatively large type having respective footings for every one of the main column members of the tower ; and the steel mast, as a small scale type for which distance between the main column members is short and the whole tower is considered to be one column and the foundation is made in one body.

Generally speaking, steel towers are self standing cantilever type fixed to the supporting soil layer but there are also guyed masts in which tall mast-type tower body is supported by wire ropes.

16.2.2 Structural Planning

a) As for steel towers related to electric waves or the like, there are cases where deflections and deformations are limited because of their services and therefore selection of structural type is important.

b) It is necessary to make structural planning taking into consideration the field erection method based on scale and height of the tower.

16.2.3 Structural Notes

a) For ordinary steel structures, use of bolt is severely limited by Article 67 of the Building Standard Law Enforcement Order. But for steel towers, use of bolts is permitted as a rule for the principal structural members also.

As for nuts, it is necessary to use double nuts or provide locking device in order to prevent their loosening from vibrations.

For steel towers, it is common to apply galvanizing but it is not desirable to subject galvanized bolts to tensile forces. The reason is that more or less looseness is provided between the bolt and nut before coating in anticipation of the thickness of zinc coating and those nuts have a tendency of slipping out more or less easily compared with those of ordinary bolts.

b) As the footings are separated from each of the principal column members of the tower, tie-beams of sufficient section are required.

16.2.4 Notes on Structural Calculations

a) As for stress caluculation of ordinary steel towers, wind force is the predominant load, but as to those towers having heavy object at the top, calculation of stresses due to earthquake is required.

b) Stresses in the body of tower, in principle, should be calculated as those of a space frame but with regard to trussed steel tower of ordinary construction, calculations may be made by substituting it with plane trusses.

In case where the inclination of the plane of truss is small, the body of the tower may be treated by projecting it on the vertical plane. In case where the inclination of the plane changes, influences due to such changes should be taken into consideration.

16.3 Chimneys

16.3.1 General Description

As for types of chimney structure, a majority is those of self standing reinforced concrete construction so far. However, as the problem of public hazard becomes the subject of much discussion in recent years, height of chimney become higher and higher and as a matter of fact those having a height of more than 100 m are not rare. Consequently, new types of structucture such as self standing steel chimney, steel chimney with supporting tower, and combined chimneys in which several chimneys are interconnected and made into one unit, etc. have been developed.

As reinforced concrete chimneys have inevitably heavy weight and deteriorate rapidly causing cracks due to thermal stresses in them, they are greatly influenced by earthquakes. Even in earthquake disasters in the past, there are many examples of collapse of concrete chimneys.

16.3.2 Structural Planning

a) It is necessary to select an economical type of structure depending on the scale of the chimney and conditions of the site.

Designs for the respective types of structure may be made in accordance with the "Standard for Structural Calculation of Reinforced Concrete Structures" and the "Standard for Structural Calculation for Steel Chimney."

b) As large type chimneys are constructed on reclaimed land or in areas having weak supporting soil layers in many cases, thorough precautions are required in planning foundation construction. As for high chimney exceeding 50 m, there is lately a tendency to use steel construction in order to decrease the weight and reduce the force of earthquake.

16.3.3 Structural Notes

a) For the chimney lining, common bricks, gunite, and other capable heat insulating materials are used and it is desirable to attach them firmly to the chimney shell so that they do not fall down due to chimney vibrations.

b) As for combined chimneys and chimney with support, considerations for thermal stress due to difference in temperatures of each part are necessary and appropriate expansion devise must be provided.

c) Connecting parts between flue and chimney and those between base and foundation of steel chimney are apt to become a point of structural discontinuity, it is necessary to design them with ample safety taking repetitive loads to a certain extent into consideration.

16.3.4 Notes on Structural Calculation

a) As the vibratory behavior of chimney is different from that of ordinary building, it is desirable to use a large horizontal seismic coefficient. In the current design standard, it is stipulated that the horizontal seismic coefficient of 0.3 or more should be taken for self standing chimneys. As for foundations, however, it is permitted to use 0.2 (Art. 139, Building Standard Law Enforcement Order).

It is presumed from past earthquake damage to chimneys, that the portion at the height 1/2 to 2/3 from the base is the weak point when designed by the current statical method of

Photo 16.2 (Offered by Prof. H. Kobayashi, Tokyo Institute of Technology)

calculations, precautions in the design are required (refer to Photo. 16.2).

b) In case of a chimney having a supporting tower, structural calculation of the tower can be done in a similar manner as that of the steel tower but as the chimney itself has considerable rigidity, design must be done in such a way that the deformation of the tower is limited within an allowable amount of deformation of the chimney. In addition, it is necessary to reinforce sufficiently parts of support and the neighboring parts and also to consider the movement due to differences in temperature.

References

1) API : Specification for All Welded Oil Storage Tanks.
2) ASME : Unfired Pressure Vessels.
3) JEC-127 : Steel Tower Design Standard for Power Transmission.
JEC-128 : Steel Mast Design Standard for Power Transmission.
JEC-144 : Steel Tower and Steel Mast Design Standards for Radio Communications for Power.

17. TEMPORARY STRUCTURES

17.1 General

As for temporary structures constructed to be used for a short period of time, there are in addition to temporary buildings and temporary structures ruled under Article 85 of the Building Standard Law and Clause 2, Article 147 of the Enforcement Order of the same law, those such as scaffolding for construction work, temporary supports, wale framings, towers, etc. Temporary buildings and structures are of many kinds ranging from those occupied by unspecified number of people and those related to construction operations such as work shops and warehouse used at construction site, and some may be of large span and multistoried to those of small scale, light weight, etc.

Comparing these temporary structures with permanent structures, they have such features as that their period of use is short and many of them are those which can be moved and used repeatedly by assembling and dismantling. Accordingly, for structural members such as columns, beams, roof frames, wall units and their connection methods, assembling systems and connection methods of their own are used for many of them in addition to those used in usual permanent structures. Also, their footings usually are of a simple construction.

There are but few reference materials as to earthquake damage to temporary structures but valuable lessons were learned from the damage examples to temporary structures at the time of the Niigata earthquake.

Based on the above material, points to be considered in the aseismic construction of temporary structures are the following :

i) Although the period of use of temporary structures is short, the same considerations as those for permanent construction should be given to them. The influence of earthquakes is great for temporary structures loaded with heavy objects.

ii) Ordinarily, from economic requirements, the safety factor for the structure is taken low in many cases so that there is little margin in strength left and many structures have a low degree of statical indeterminacy. It is necessary, however, to plan so that they have sufficient ductility at the time of earthquakes.

iii) In addition to welding, riveting and bolting, metal fittings, and specially devised ones are used. However, as many of them allow large deformations, it is necessary on consideration of their characteristics, to plan framing plan and increase the rigidity.

iv) Simple footings are used in many cases and they are apt to be carelessly made. However, they must be planned prudently so as not to cause trouble to the superstructure.

In the Building Standard Law and Enforcement Orders of the same law, for the requirements as to the material and construction temporary buildings and structures are mostly excluded from the application of the provisions relating to permanent structures. In the design of temporary buildings and structures of steel or timber construction, it is desirable to design them refering to the provisions of the "Standard for Design and Construction of Temporary Steel Buildings and Structures" and the "Standard for Structural Design of Timber Structures" by AIJ. In the "Standard for Design and Construction of Temporary Steel Buildings and Structures," structural requirements are differentiated by classifying those for public use and of large scale into class A and the rest into class B.

Essential points in aseismic planning of temporary buildings and structures are described below.

17.2 Notes on Structural Planning and Stress Calculations

a) Structural Planning

Although the basic principles of aseismic planning of temporary structures are not different from those of permanent structures, it is necessary to consider the characteristics of temporary structures in the planning.

i) Type of framing to be used should be as simple as possible and the mode of stress transmission should be clear.

ii) Earthquake forces may be distributed to braces in wall framing, bearing walls, shoring struts, guys, etc., but their arrangement should be in good balance so that torsion will not be caused.

iii) Connections and arrangement of structural members should be planned so that the structure as a whole will have ductility under earthquake conditions.

iv) Footings should be designed as not to cause differential settlement, lateral movement, rotation, etc.

b) Earthquake Forces

When earthquake forces are treated as static horizontal forces based on the seismic coefficient method, seismic coefficients stipulated in the Building Standard Law Enforcement Order should be used.

c) Materials to be Used and Allowable Stresses

Timber should conform to the "Ministry of Agriculture and Forestry Standard for Timber", and steel materials to JIS.

For the allowable stresses of timber, values for permanent loading given by the "Standard for Structural Design of Timber Structures" should be increased depending on the period of use as follows :

One week or less	1.3 times the allowable stresses for permanent loading
One week to one month	1.25 times the allowable stresses for permanent loading

One month to 3 months 1.20 times the allowable stresses for permanent loading

For the allowable stresses for the two classes of steel having tensile strength of 41 kg/mm^2 and that of 50 kg/mm^2 or over, values corresponding to the respective allowable stresses for temporary loading should be used without distinction for permanent and temporary loadings. For the allowable stresses of steels, those of welds, allowable buckling stresses, allowable strengths of wire rope, stranded steel wire, hard steel wire, connection metal fittings, etc., standards by AIJ should be used.

In designing temporary steel structures, the allowable stresses used are so high that they are almost close to the yielding point so the safety factor against loading must be considered depending on use and loading condition of the structures.

17.3 Structural Elements

a) Footings

The footings, in addition to those of concrete, there are simple footings made of natural stones, bricks, concrete blocks, etc., footings in which column bases are directly joined to piles, and so on. Footings that will cause trouble to the superstructure such as through settlement, lateral movement, rotation and uplift are not permissible under earthquake conditions. At the same time, footings should be designed and constructed giving due considerations to selection of safe supporting soil layer, its improvement, determination of the depth of footing in relation to the soil bearing capacity and frost damage, landslide, falling of cliff, ground water, etc., and possible interference with underground installations.

b) Columns and Beams

It is necessary to connect column bases firmly to footings or sills so that uplift and lateral movement are prevented at the time of earthquakes and to interconnect and reinforce them by struts if conditions require. For assembled type columns consisting of elements, they must be either reinforced considering the decrease in rigidity of the connection part and increase in eccentricity or treated taking these effects into consideration. Column head of tower-like structures having guy systems should be uniformily supported by guy wires so as not to cause twisting and in case they are of great height, horizontal displacement should be prevented by providing guy wires or tie-beams at intermediate points appropriately spaced.

For beams, especially those of adjustable span type, attention should be given to the supporting points and the connections at the adjustable part so that dislocation at the supporting point and looseness or slip in the adjustment part will not occur in earthquakes.

c) Braces, Knee Braces, Shoring Struts and Guy Wires

For braces, knee braces and shoring struts of timber construction, refer to the chapter on timber construction. As a general rule, round steel bar braces should be 13 mm ϕ or larger and wire ropes for guys should be 12 mm ϕ or larger. Appropriate initial tension should also be given so that no slackening will occur. But care should be taken as there was a case where a brace was broken due to by a large horizontal force with impact because it had been tightened too much at the time of construction. The part of attachment of guy wire should be

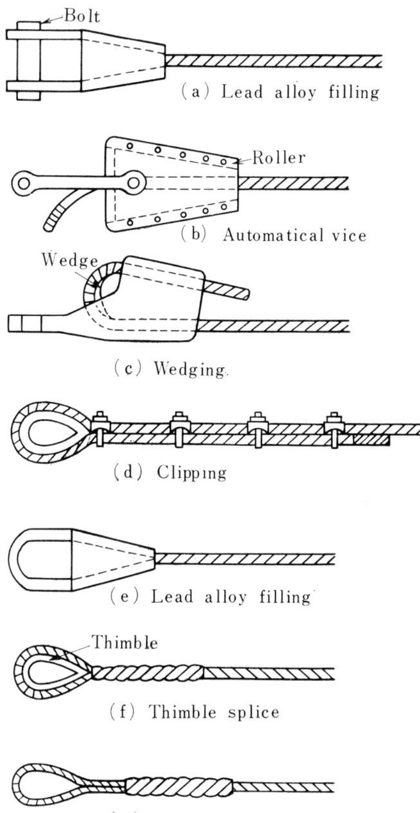

Fig. 17.1 Fastening Method of Wire Rope End

Table 17.1

Rope diameter	9∼16	18∼22	25	26
Number of clips	4	5	5	6
Clip spacing (cm)	7.5	13	15	16

designed with ample safety so that deformation, looseness, slip, etc. at the anchorage are not caused. As for clipping of wire ropes, number of clips to be used should be equal or more than those given in Table 17.1 correspondingly to wire rope diameters and it is necessary to tighten them properly. Their efficiency under properly tightened condition is in the neighborhood of 85% as a maximum and if tightening is insufficient and the number of clips is insufficient, efficiency decreases remarkably (refer to Fig. 17.1).

d) Part of Member Connection

For connection parts of temporary structures of timber construction using nails, connectors, bolts, metal pieces, adhesives, the chapter on timber connection should be referred to and for connections in temporary structures of steel construction using welds, rivets, bolts, friction type high tension bolts, drive-rivets, pins, etc., the chapter on steel construction should be referred to. In addition to the above connection methods, such connections as those using metal fittings, hackbolts, and other various kinds of connection methods are often used. For temporary structures, those methods for which safety with regard to strength and deformation characteristics have been ascertained may be used but compared with connection methods using ordinary bolts or rivets, some of them cause large elongation, shrinkage and rotational deformation in spite of being within the limit of safety with respect to strength. Consequently, it is necessary to plan the framing so that the rigidity or ductility of the structure as a whole is not decreased due to the decrease in the rigidity of connecting parts of columns and beams owing to these deformation characteristics.

(1) Bolt Connections

In bolt connections, it is important that clearance between the diameter of the bolt hole and that of the bolt is small but in temporary structures, hole diameters usually are apt to be made or become large, for easy assembling or because of wear of holes due to repeated use. In case of steel construction, fabrication should be done in such a way that clearance between bolt diameter and the hole be limited to less than 1.0 mm. In addition, bolts are used sometimes as connections which are subjected to vibrations and impacts. In these cases, semi-finished bolts should be used and locking for the nuts should be provided. For cases where shear forces are resisted by the threaded part of bolts, their allowable strength should be determined based on the net area of the section. They must be of such length that the bolts project out from the nuts by two screw threads or more.

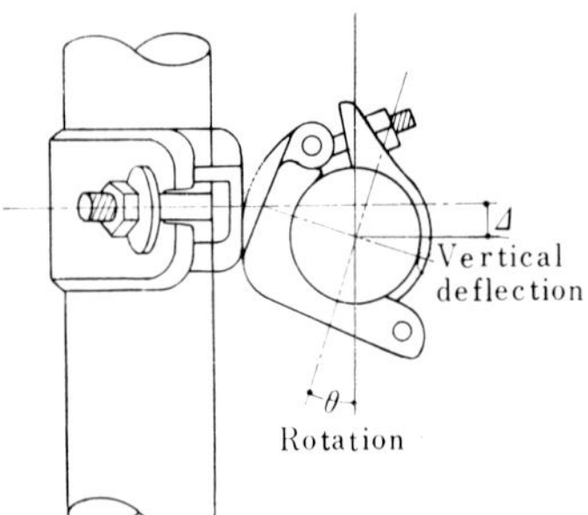

Fig. 17.2 Deformation of Fastner Metal Fitting

(2) Metal Fitting for Connections

As for metal fittings for connection, there are those clamps and joint pins covered by JIS A 8951, and connection metal fittings used for tubular steel structures, etc. clamps eccentrically connect steel pipes accompanying displacements due to deflection and rotation as shown in Fig. 17.2 and 17.3*.

* As for connections using clamps, ordinarily no particular difficulty will be caused by assuming them to be pins, but they should be tightened so that sufficient frictional force is obtained and influences of bending and twisting caused by eccentric connection should be taken into consideration. Framings having clamps for columns, beams, etc. have the tendency when acted upon by horizontal seismic forces, to vibrate in the deformed condition at the initial stage as the origin and to damp out quickly.

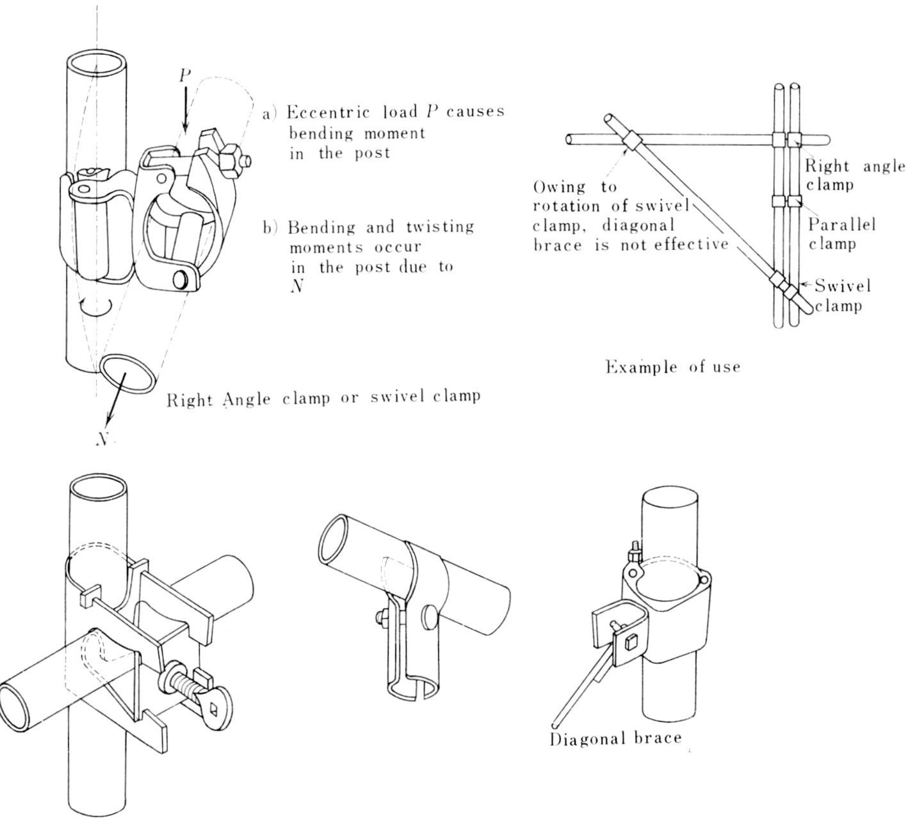

Fig. 17.3 Examples of Fastner Metal Fittings

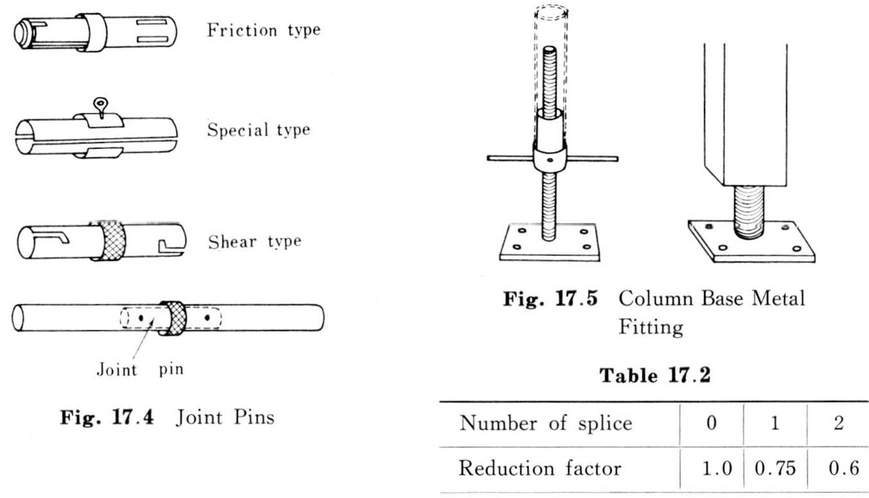

Fig. 17.4 Joint Pins

Fig. 17.5 Column Base Metal Fitting

Table 17.2

Number of splice	0	1	2
Reduction factor	1.0	0.75	0.6

There are friction type, shear type and special type for joint pins. As shown in Fig. 17.4, these take axial forces by means of frictional resistance between the fitting and steel pipe or shear resistance of pins or projections but, as there is a large clearance between the fitting and steel pipe, much elongation or deflection is caused by tensile force or by bending moment.

Splice efficiencies of the connection parts using joint pins are shown in Table 17.2. In compression members using joint pins, bending stiffness of the member is decreased at that part and decrease in buckling strength due to eccentricity is considerable so treatment such as locating the part of connection using joint pins close to the panel point or appropriate reinforcement is necessary.

(3) Column Base Metal Fitting

Column base metal fitting to adjust the column height and pin holes should have strength with ample safety against the applied axial force and in consideration of the reduced stiffness of this part, safe side assumption should be made with regard to the buckling length of the column. Column base is ordinarily assumed to be pin supported (refer to Fig. 17.5).

17.4 Temporary Structures for Construction Work

For temporary structures for construction work, almost no cosideration is given to earthquake forces, because they recieve earthquake forces less frequently during their period of service and many of the framings are of comparatively light weight and receive little influence of earthquake forces. However, the causes of damage to temporary structures in past disaster examples were due to lack of consideration for horizontal forces in almost all cases. In addition, considering the damage examples of scaffolding, temporary supports, wale framings and towers in the Niigata earthquake, earthquake forces can not be ignored. Especially for the temporary supports for heavy objects and wale framings, considerations of earthquake forces are necessary.

a) Steel Pipe Scaffolding

Steel pipe scaffolding is a framing having intermediate supports of ties and the rigidity of wall ties has a great influence on the strength of scaffolding. In scaffolding made by assembling straight pipes, braces are placed in the framing plane but not ordinarily in the plane at right angles to the plane of framing. But, as the deformation of the framing due to earthquake forces in this plane is large, it is necessary to provide diagonal members in the plane of vertical members picked up at appropriate intervals. Besides, the scaffolding using frame units is a framing having comparatively large shear rigidity in both planes in and out of the framing but as each vertical framing element consisting of frame units is connected merely by diagonal braces in the plane of scaffold framing, and the vertical framing elements are independent one from another with regard to the deformation not in the plane of scaffold framing, so it is necessary to make the whole framing into one body by providing horizontal struts with proper vertical

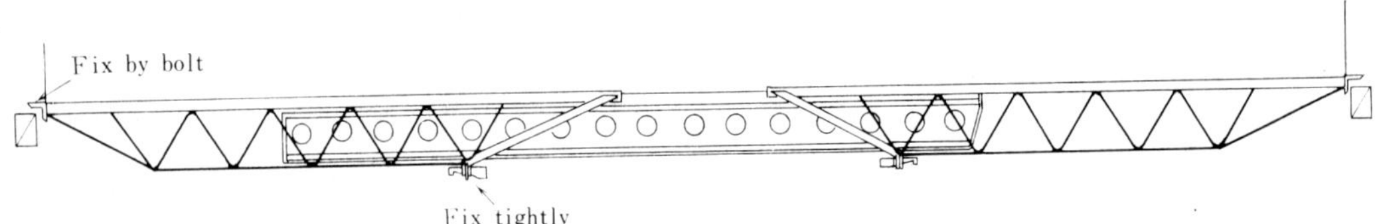

Fig. 17.6 An Example of Adjustable Beam

spacing. As for wall ties, it is necessary to provide them with sufficient rigidity so as not to be broken or bent in earthquakes.

In addition, footings for scaffolding should be designed to be amply safe against axial loads of the verticals (refer to Fig. 17.7).

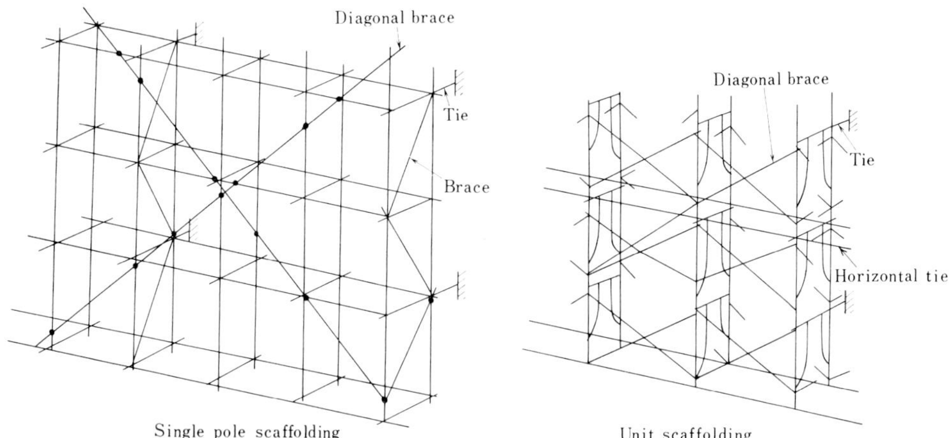

Fig. 17.7 Braces and Horizontal Struts in Scaffolding

b) Temporary Supports

Temporary support is provided by pipe supports, assembled steel columns, framed supports, assembled adjustable span beams, etc. As these supports and beams have been developed independently considerations are often lacking with regard to the methods of interconnection. Therefore, the connections should be provided with ample safety using bolts, metal fittings and the like, and in addition braces, shoring struts, horizontal struts, etc. should be provided against horizontal forces*.

c) Wale Framing

Soil pressures which act against the wale framing in earthquakes have not been made clear but when there is the possibility of increase in soil pressures, it must be taken into consideration. The wales, struts and support piles and their connections should be designed so that they may resist as a whole the soil pressures without being disconnected in earthquakes.

d) Maintenance

Framing should be inspected from time to time when in use and absence of looseness, damage, deformation, etc. of bolts must be ascertained. Also service load must be well controlled to ensure that the design condition is not exceeded.

* It is stipulated in Par. 6, 8 and 9, Art. 107 of the "Labor Safety and Sanitary Regulations" that the displacement should be prevented by providing horizontal struts in five layers or less for steel pide framed supports and at intervals of 4 m or less in case of assembled steel columns.

14.3, Fig. 14.3).

A crack in the ground ran right under the building and the lower part of the building widened and cracks developed in the wall vertically and diagonally from the wall footing to the collar beam. It is thought that the strength of the continuous wall footing and collar beams were insufficient because of poor design.

14.2 Structural Planning

14.2.1 Principle of Structural Planning

In designing buildings of reinforced concrete block construction (hereinafter referred to as reinforced block construction), the following points are indispensable from the point of view of aseismic design:

i) That this type of construction be limited to buildings not requiring too much live load such as dwelling houses, small office buildings, etc.;

ii) That the scale should not be too large and the number of stories should be three or less;

iii) That the shape of building, disposition of walls, etc., arrangement should be made so that they are balanced as a whole and the stresses may be distributed as uniformly as possible to the whole structure;

iv) That unreasonable stresses are not caused in the walls (for example, the height of wall should not be made too great, the length and width of wall should not be made extremely small, etc.); and

v) That the work should be executed with special care so that the strength of joints is ensured and concrete filling of hollow parts, in which steel bars are to be inserted, is made positive.

There are "Standard for Structural Design of Reinforced Concrete Block Construction" and JASS 7 as the standards for design and work execution of the building of reinforced block construction. The provisions in the law are described in Chapter 3 of the Building Standard Law Enforcement Order. A type of reinforced block construction recommended in this country is.

Although it is very desirable to use cast-in-place reinforced concrete slab for the roof and the floors, timber or steel construction is sufficient for buildings of two stories or lower, provided collar beams are used. For buildings of 3 stories, slabs of reinforced concrete construction or of rigid and strong precast reinforced concrete construction are required, except for the floor of the lowest story. At the upper and lower ends of walls, collar beams of reinforced concrete construction should be provided as in the case of masonry construction and continuous wall footing of reinforced concrete construction should be provided at the bottom. Vertical bars inserted in the wall should be completely anchored in to the collar beam and to the continuous wall footing (Fig. 14.4).

Although there are no particular limitations on the size of openings in the wall —excepting those for height, length and thickness of unit wall standing between the openings, for wall rates of the whole building, for maximum enclosed area, etc.— consideration must be given so that

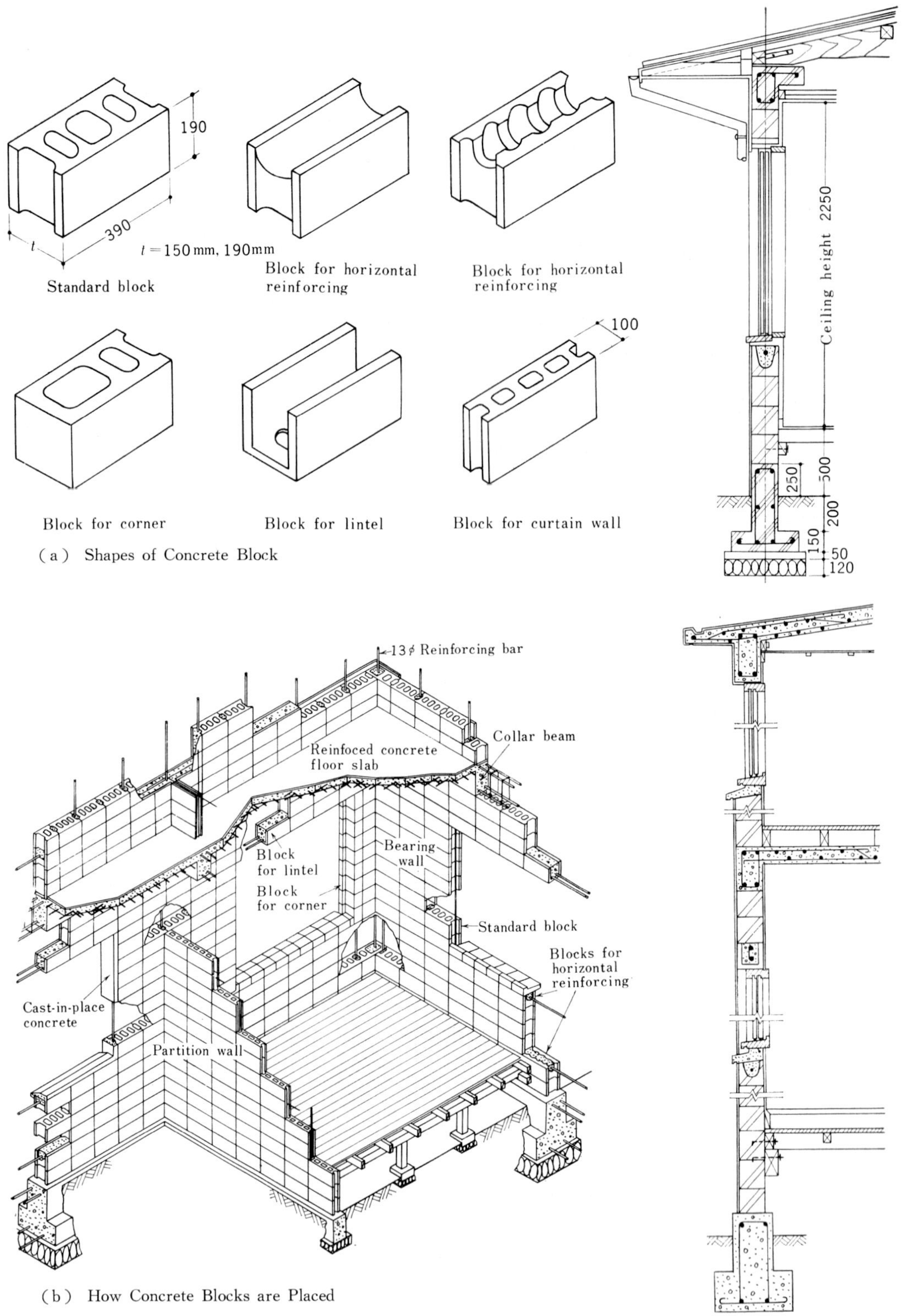

(a) Shapes of Concrete Block

(b) How Concrete Blocks are Placed

(c) Cross sections of Reinforced Concrete Block Structure

Fig. 14.4

the building is made into an earthquake resistant structure as a whole. The standards for workmanship are similar to those for masonry construction but precautions are required in the arrangement of reinforcing bars. Based on these standards, precautions that are required for earthquake resistance will be described below.

14.2.2 Materials

a) Hollow Concrete Blocks

Concrete blocks that are used for reinforced block construction are those hollow concrete blocks standardized by JIS A 5406. Principal features of the standard are the following :

(1) Classification of Aggregates

Concrete blocks differ in property depending on the aggregates used.

i) Those using river sand and river gravel.

ii) Those using river sand and light gravel.

iii) Those using light sand and light gravel.

Light sand and light gravel are those ordinarily called light weight aggregates and there are various kinds such as volcanic gravel, blast-furnace slag, cinders, coal slag, artificial light weight aggregate, etc. Generally speaking, the properties of concrete block product —although it cannot be said absolutely— those using river sand and river gravel are superior to others in strength and water absorption but they are heavier and are more expensive.

(2) Classification by Quality

Concrete blocks are classified into 3 kinds, A, B and C, in accordance with the quality in terms of specific gravity, strength, moisture content, etc. as shown in Table 14.1. Class A and B are generally called light weight concrete blocks and Class C is called heavy weight concrete blocks. Heavier the weight, greater the compressive strength and more suitable for use in buildings having many stories.

Table 14.1 Quality of Hollow Concrete Block

Kinds	Air-dry bulk specific gravity	Compresive strength based on gross sectional area (kg/cm^2)	Reference value
			Ratio of moisture content to the maximum water absorption rate (%)
Class A concrete block	Less than 1.8	25 and over	Less than and equal to 40
Class B concrete block		40 and over	
Class C concrete block	Not less than 1.8	60 and over	

(3) Classification by Shape and Size

Those most often used are the standard concrete blocks ; and special blocks to be used for placing horizontal bar, those having shape suitable for corners, those not of full size, etc. are called special concrete blocks. The shape of hollow part has not been standardized but minimum sizes and the minimum thickness of cell wall are given in Table 14.2 and Fig. 14.4.

Table 14.2 Size of Hollow Concrete Blocks (unit : mm)

Shape	Size			Tolerance	
	Length	Height	Thickness	Length and thickness	Height
Standard concrete block	390	190	190 150 100	±2	±3
Special concrete block	The minimum size for the length, height and thickness shall be 90 mm. The tolerance for those having the same size as that of the standard concrete block such as concrete block for horizontal bar placement and corner concrete block shall be similar to those of the standard concrete				

Note : 180 mm thickness may be substituted for 190 mm in Hokkaido for the time being.

b) Other Kinds of Blocks

Hollow blocks of ceramics called ceramic blocks have lately come to be manufactured so that they can be used generally for the same purpose as ordinary concrete blocks. As to the quality, there is a separate JIS A 5210 standard ; and they can be used without difficulty, depending on the quality, in constructions similar to those of reinforced concrete blocks.

14.2.3 Kinds of Construction and Scale

Although reinforced block construction uses reinforcing steel bars, in the light of the weak point that the concrete assemblage consists of small concrete blocks, it cannot be said that it is absolutely safe for high-storied buildings as in the case of reinforced concrete construction. Especially, as it is only a short time since the popularization of this construction and it has not experienced many great earthquakes, it is stipulated under the present standard that allowable height limit is up to 3 stories.

Reinforced block construction is classified depending on the quality of concrete blocks to be used ; the building for which class A concrete blocks are used is called Class A concrete block construction, and the one for which class B concrete blocks are used is called Class B concrete block construction*. It is recommended to use concrete blocks having a large strength for the buildings having many stories as they are subjected to large stresses due to earthquakes. As for the class of concrete blocks to be used, there are limitations corresponding to the number of stories of construction as shown in Table 14.3.

Table 14.3 Limitation on Kind and Scale of Reinforced Block Construction

Kinds	Limitation on number of stories	Limitation on eaves height
Class A concrete block construction	Up to 1 story	Up to 4 m
Class B concrete block construction	Up to 2 stories	Up to 7 m
Class C concrete block construction	Up to 3 stories	Up to 11 m

14.2.4 Disposition of Walls, Enclosed Area by Walls, etc.

a) Bearing Walls

Not all concrete block walls support the load. Those not having reinforcing bars, those

* This differs from the classification into the class A concrete block construction and the class B concrete block construction of Paragraph 13.2.2, Chapter 13 Masonry Construction

having reinforcement but placed in the wrong location (for example, a wall of the upper story placed above the opening of the story below) or insufficiently reinforced along the periphery (collar beams and continuous footings are not provided), or the like cannot resist not only the horizontal loads at the time of earthquake but also the vertical loads at ordinary times.

Thus only those walls for which every condition for reinforcement is fulfilled and that can support permanent vertical loads and the horizontal loads are called bearing walls ; and in designing the construction, disposition of the bearing wall and its construction are checked. As for the construction of the bearing wall, mention will be made in Section 14.3 ; and the rest of the walls should be considered not to share the loads at the time of earthquake and be regarded as mere partitions.

b) Characteristics of Bearing Walls

Description of the characteristics of bearing walls of reinforced block construction is given below :

(1) Earthquake force becomes the dominant factor for the design of bearing walls.

Although the bearing wall should be designed so as to support safely both the vertical loads and the horizontal loads, it is especially necessary to make them capable of resisting earthquake forces. For this purpose, it is necessary to reinforce the body of the bearing wall by collar beams and continuous footings of reinforced concrete construction.

(2) The stronger the concrete block units, the stronger the structure.

Bearing walls are subjected to both bending and shearing by earthquake forces. Although the steel bars inserted along the edge of the wall resist the bending moment, the concrete block wall itself must resist the shearing force. For this reason, the class B and C concrete blocks having great strength should be used for buildings of 2 or 3 stories.

(3) More the bearing walls, the stronger the structure.

For the purpose of increasing earthquake resistance of the building of reinforced block construction, it is the surest way to increase the amount of bearing walls. For this purpose, either the length of wall or the thickness of wall is increased.

(4) Short walls have little earthquake resistance.

Even in cases of same wall rate*, earthquake resistance is larger in one case where longer walls are placed in key locations than in the other case where there are many shorter walls. In short walls, effect of bending is large; on this account, horizontal cracks easily develop at the upper and lower edges of wall thereby causing deformation of the wall, and the

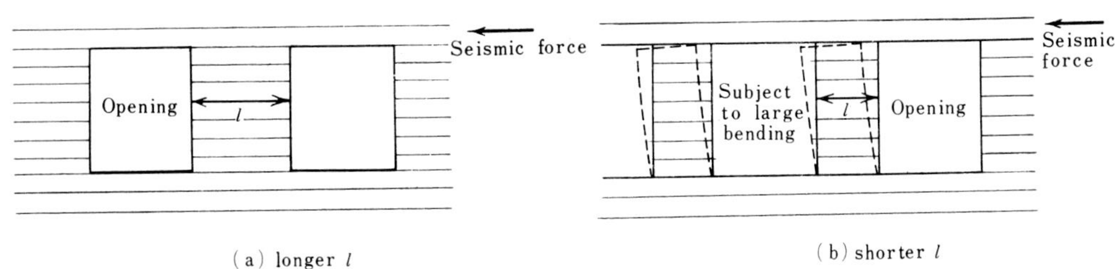

Fig. 14.5

* Refer to Par. 10.4.1, Chapter 10 Reinforced Concrete Wall Construction, and to Par. 13.4.1, Chapter 13 Masonry Construction. Refer also to Par. 14.2.5 of this Chapter.

resistance of the wall against shear also becomes unreliable (Fig. 14.5).

c) Disposition of Bearing Walls

Bearing walls should be proportionately arranged in the plan. This is common to all wall constructions. If the distribution of wall is one-sided, divergence of the location of center of mass of the building from that of rigidity of the walls become large and the building as a whole is twisted at the time of earthquake and dangerous stresses occur. For buildings in this country, there is a general tendency to provide large openings on the south side, therefore the amount of wall on the south side decreases thus causing easily one-sided distribution of the walls and these facts require attention. In case there is an inevitable need of such, it is recommended that all the walls in that direction be made of reinforced concrete frame construction (Fig. 14.6).

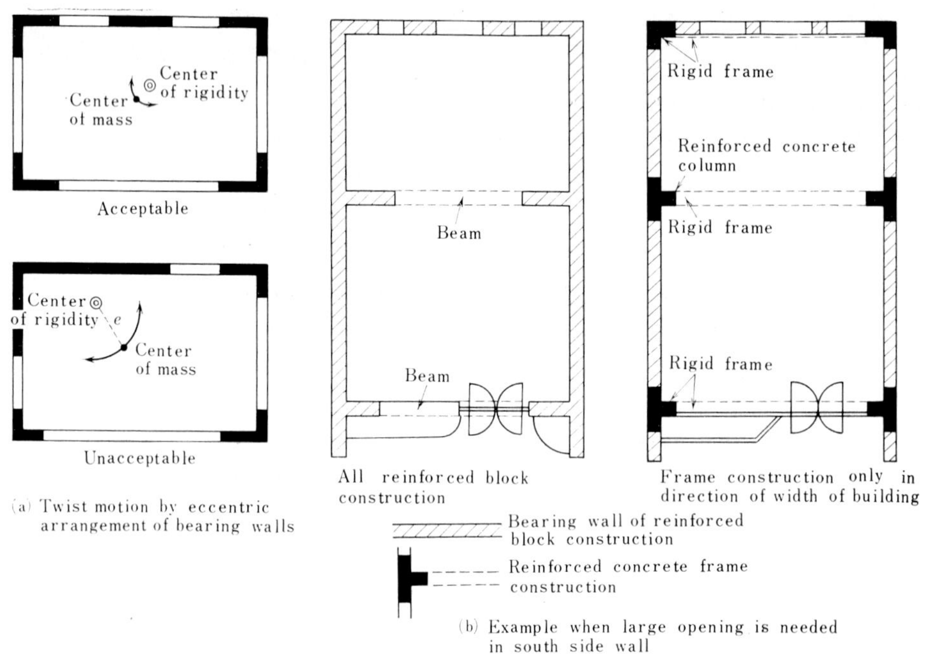

Fig. 14.6

At the corners and intermediate positions of importance, bearing walls should be arranged at right angle so that the plans developed have L, T or cross-shapes. For buildings of 2 or 3 stories, these precautions are especially necessary. As the wall is weak against external forces normal to the plane of the wall, they are combined with each other at right angles and made to reinforce each other. Besides, balance in elevation is also important ; for example, in case of a building of 2 or 3 stories having a large opening in the lower story and a wall right above it, the wall of the upper story will be tilted at the time of earthquakes and cannot maintain sufficient earthquake resistance unless the collar beam between the upper and the lower story is especially strong. Therefore, such an arrangement should be avoided or such wall should not be regarded as a bearing wall (Fig. 14.7).

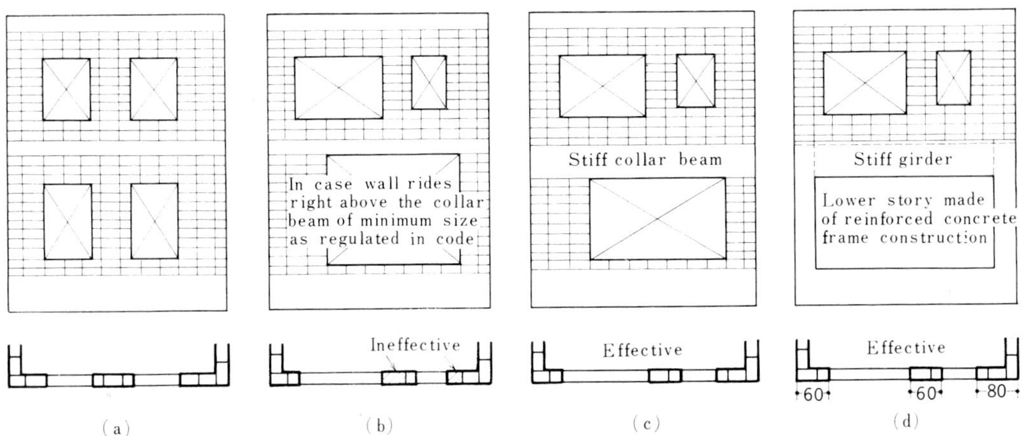

Fig. 14.7 Treatment of Upper Story Bearing Wall

d) Limitation on Divided Areas

In wall construction, smaller the interior divisions are the stronger it becomes. This will prevent, at the same time, unbalanced arrangement of bearing walls. On this account, as in the case of masonry construction, etc., a maximum limit to every one of the divided areas enclosed by bearing walls on the plan is regulated. In an ordinary design, it should be 60 m^2 or less ; and in case roof and floors are of timber construction, it is desirable to make it about 45 m^2 or less. As for each one of the divided areas, it is desirable to make the values as equal as possible (Fig. 14.8).

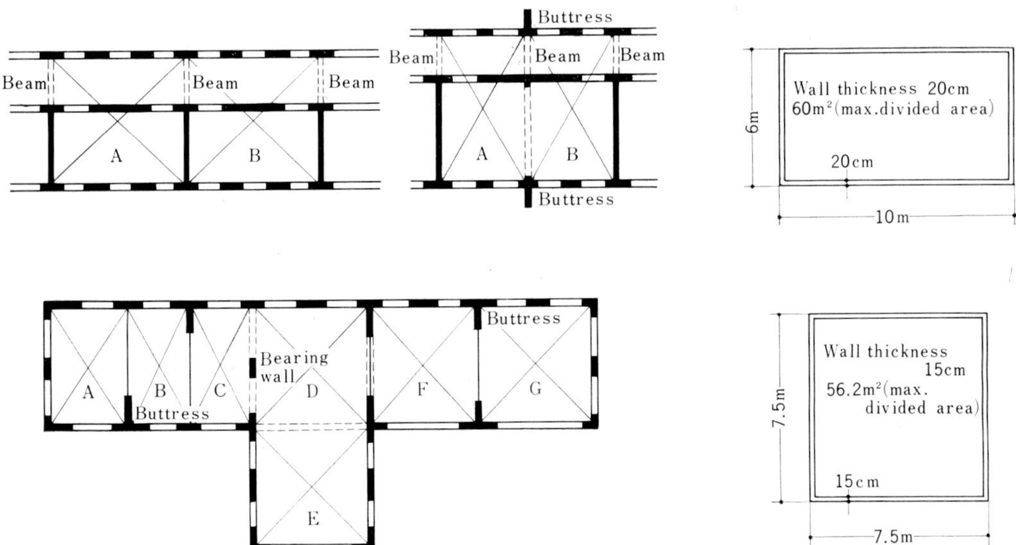

Fig. 14.8 Determination of Divided Area

e) Distance between Adjoining Walls.

If a bearing wall is made extremely long, it become dangerous against bending and twisting

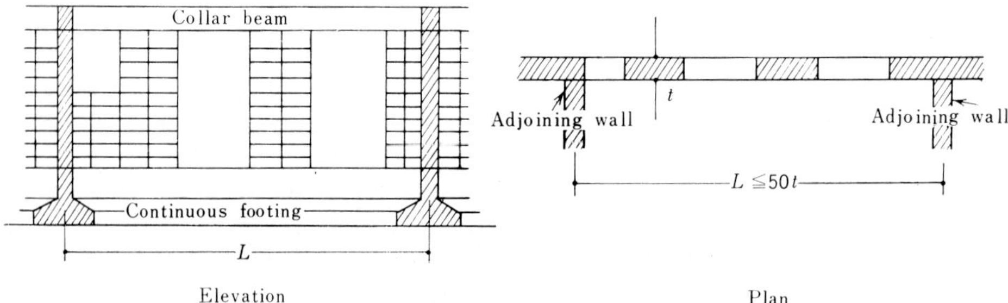

Fig. 14.9 Relation between Wall Thickness and Distance between Center Lines of Adjoining Walls

not in the plane of the wall. In this case, other walls connected to the face of the wall at right angles become a reinforcement against forces normal to the plane of the wall under consideration. Distance L between center lines of the two walls (called "adjoining walls") which are joined to the face of the wall should be made 50 times the thickness of the concrete blocks or less (10 m or less for masonry construction). If 15 cm thick concrete blocks are used, $L \leqq 7.5$ m (Fig. 14.9).

14.2.5 Wall Rate

In order to indicate the proportion of length of the bearing wall that is required for earthquake resistance, "a value of sum of the length of bearing wall divided by the floor area," namely "the length of bearing wall per unit floor area" is especially called the wall rate and used as a numerical value to indicate earthquake resistance of a building (mention has already been made in Chapters 10 and 13). The wall rates are calculated respectively as to the widthwise and lengthwise directions and also to each story. The standard of required wall rate is shown in Table 14.4.

Table 14.4 Standard Value of Wall Rate

Kinds of concrete block construction	Wall rate (cm/m^2)			Remarks
	One-story building or top story	Second story counting from the top	Third story counting from the top	
Class A concrete block construction	15	—	—	Values indicated are for the case of the standard wall thickness, and the values indicated in ( ) are for the case of 15 cm wall thickness
Class B concrete block construction	15	21(25)	—	
Class C concrete block construction	15	15(18)	24	

The above table, however, is based on the standard wall thickness t_0 shown in Table 14.5 and in case the wall thickness t_0 is greater than the standard wall thickness, actual wall rate multiplied by t/t_0 can be made the computed wall rate, provided the actual wall rate is not decreased more than 3 cm/m^2 from the numerical values shown in Table 14.4. Values of wall rate shown in the table are the minimum, therefore, for such special cases as those in which an unbalanced disposition of bearing walls is not avoidable, or the floor load is especially large, etc., the design should be made in such a manner to increase appropriately the wall rates shown in the table.

14.3 Structural Notes

14.3.1 Construction of Bearing Walls

a) Length of Walls

If the length of the bearing wall is short, earthquake resistance is decreased. Therefore the minimum length of the wall that is effective in bearing wall construction should be made 55 cm and not less than 30% of the average height of openings on both sides.

b) Thickness of Walls

The thicker the bearing walls, more earthquake resistance it will have and more stable and the easier the execution of work. Required thickness of bearing walls is given in terms of both the absolute value and the ratio to the height of the part of bonded concrete blocks. Besides, as there is a difference in deformation or in degree of danger at the time of disaster between the wall of the second story from the top of a 3-story building and that of 2-story building, though they are both termed as the second story counting from the top, the standard for the latter is made lenient. These thicknesses do not include finishes because the rule is made for the structural strength.

Table 14.5 Standard Wall Thickness

Story	Wall thickness t_0
One-story building or top story	15 cm minimum but not less than $h/20$
Second story counting from the top	18 cm minimum but not less than $h/16$*
Third story counting from the top	18 cm minimum but not less than $h/16$

Note :

* For the case of 2-story building only, this can be 15 cm minimum but not less than $h/20$. h : height of the part of bonded blocks.

c) Arrangement of Reinforcement

As for reinforcing bars in bearing walls, there are two kinds, shear reinforcing bars and bending reinforcing bars. The former are steel bars of small diameter to be inserted in the body of the wall and distributed both vertically and horizontally ; the latter are steel bars of large diameter to be placed along the edges of the wall and the periphery of openings.

As for concrete block wall, wall rate is determined at least so as not to cause occurance of cracks in the wall against earthquakes of the design magnitude. Since it is a bonded body, however, occurance of cracks due to defects in workmanship or unexpected defects may not be avoided. Shear reinforcing bars are placed so as to take all the shear forces even in such cases, and by this reason, reinforced block construction has tenacity which cannot be expected in masonry construction.

Amount of shear reinforcing bars required in ordinary design is shown in Table 14.6. Moreover, these reinforcing bars, in addition to acting as shear reinforcement, have the other purpose of preventing destruction of the wall due to bending not in the plane of the wall caused by unexpected external forces (earthquake force, wind pressure, shock, etc.) in the direction normal to the wall surface (Fig. 14.10).

Table 14.6 Shear Reinforcing Bars

Kind of bar / Kind of construction	Vertical bars		Hrizontal bars	
	Diameter (mm)	Spacing (cm)	Diameter (mm)	Spacing (cm)
One-story building	9 or above	80 or less	9 or above	80 or less but not more than 3/4 l**
Top story	9 or above	80* or less	9 or above	80 or less but not more than 3/4 l
Second story counting from the top	{ 9 or above 13 or above	{ 50 or less 80* or less	{ 9 or above 13 or above	{ 60 or less but not more than 3/4 l 80 or less but not more than 3/4 l
Third story counting from the top	13 or above	50 or less	13 or above	60 or less but not more than 3/4 l

Note :

* In case of 3-stories, this should be 50 cm or less.

** Disregarding 3/4 l, this can be made 60 cm or above. l is the actual length of bearing wall (cm).

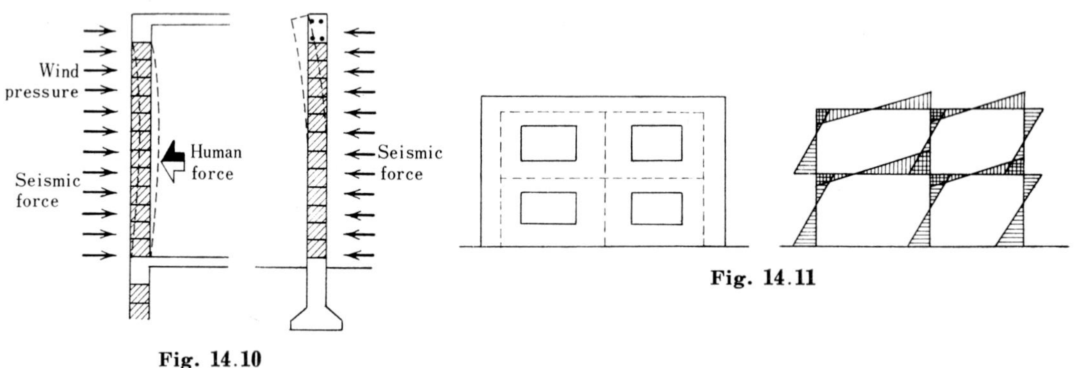

Fig. 14.10

Fig. 14.11

Table 14.7 Bending Reinforcing Bar

Story	Reinforcing bar at the edges of bearing wall		Reinforcing bar at corner jointing part and top and bottom of opening of the bearing wall
	$h > 1.5$ m	$h \leq 1.5$ m	
One-story building and top of 2-story building	1-13 ϕ (1-D13)	1-13 ϕ (1-D13)	1-13 ϕ (1-D13)
Top story of 3-story building, or second story counting from the top	1-13 ϕ (1-D13)	1-13 ϕ (1-D13)	1-13 ϕ (1-D13)
Third story counting from the top	1-16 ϕ (1-D16)	1-13 ϕ (1-D13)	1-13 ϕ (1-D13)

Note :

h : Height of opening on both sides of the wall (in case upper or lower part of the opening is of structurally inferior construction, height of this part is disregarded).

For bending bars, as the reinforced block construction is a type of framing construction if considering the building as a whole, as shown in Fig. 14.11, bending moment is caused in the body of the wall when it is subjected to horizontal forces. Consequently, as in the case of bending members of ordinary reinforced concrete construction, bending reinforcing bars are required at the edges of member and around the opening. The design standard for the bending reinforcing bars which are required in ordinary cases is shown in Table 14.7.

d) Filling of Masonry Joints and Cavities

All of the vertical and horizontal joints, where the individual concrete blocks in the wall touch one another, must be completely bonded by mortar. Besides, the hollow part where the reinforcing bars are inserted must be filled with mortar or concrete so that the reinforcing bars and the wall become one body. As for the hollow parts of the vertical joints where no steel bar is placed, it is necessary to fill, without fail, mortar or concrete. This is done in order to integrate the body of the wall and, at the same time, to make it resist the forces normal to the plane of the wall as well (Fig. 14.12).

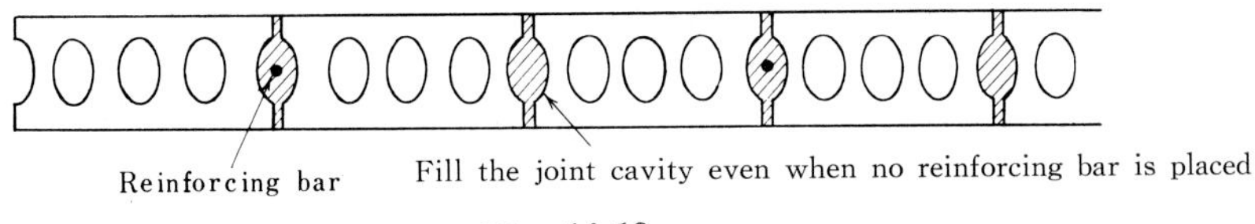

Fig. 14.12

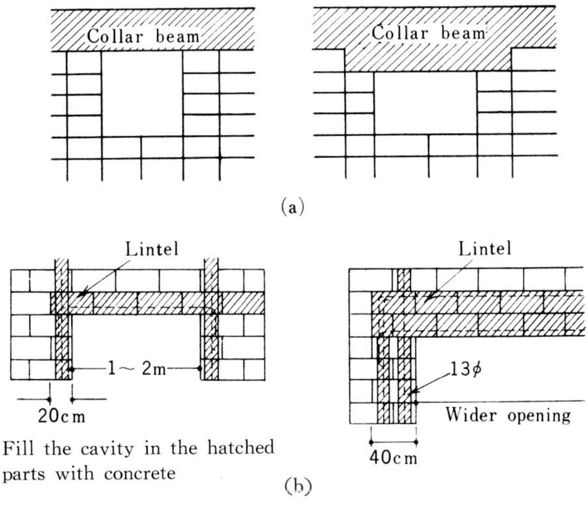

Fig. 14.13

14.3.2 Lintel

Lintel is a member above the opening supporting the portion of the wall above. As construction work of the lintel is troublesome, it is convenient, if practicable, to use construction that has no lintel, in that the upper part of an opening come directly into contact with the collar beam or the collar beam is made hung down at the part of an opening in order that its part of cast-in-place reinforced concrete serves also as the lintel. It goes without saying that the hanging part of the collar beam is provided with steel bar reinforcement (Fig. 14.13 (a)).

If such is not feasible, either a reinforced concrete lintel beam should be provided in order to support the part of the wall above or sufficient bars should be provided in the wall so that the wall itself is safe against its deadweight (Fig. 14.13 (b)).

14.3.3 Collar Beams

In reinforced block construction, as in the case of masonry construction, a continuous beam of reinforced concrete construction is provided at the upper part of walls of each story. This is the collar beam and it serves the purpose, together with the continuous wall footing, of making the wall into a strong one-body structure.

The functions of collar beams are the following :

i) They prevent the wall swaying in the direction normal to the plane of the wall at the time of earthquakes.

ii) Bearing walls are dispersed all over the building and placed in respective locations. It serves for the purpose of a horizontal frame that connects these bearing walls one another, in order to make them all work together and transmit effectively the horizontal forces.

iii) They have the function to anchor vertical reinforcing bars of the wall. Having the anchorage of vertical bars at the top and bottom respectively of the collar beam at the division of each story and the continuous footing at the bottom, these walls become strong, having been provided tightened vertical hoops.

iv) They prevent differential settlement of the building and undesirable effects of local concentration of load.

v) Collar beam above the opening supports such vertical loads as roof loads and floor loads.

However, the collar beam, as its principal role, supports uniformly the horizontal loads that are applied to the whole building. In the case of one-story building having reinforced concrete slab roof, the collar beam can be omitted for this case, because the slab takes over most of the function of the collar beam.

Collar beam must be of cast-in-place reinforced concrete construction having reinforcing bars at top and bottom and both sides, and must have a section sufficiently safe against loads in both vertical and horizontal directions. As the minimum requirement for the section, the depth of

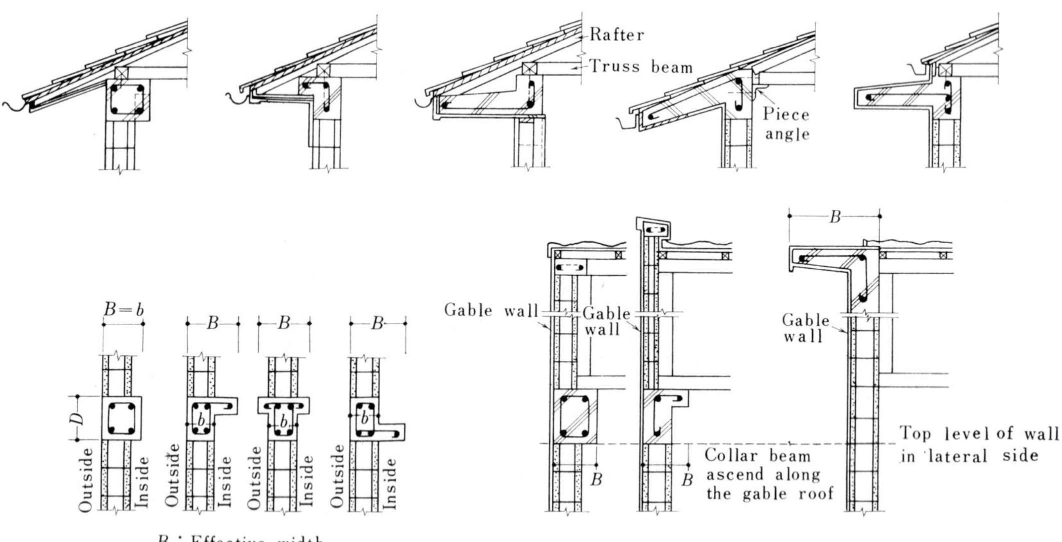

Fig. 14.14 Various Types of Collar Beams

collar beam should be 30 cm or more (25 cm or more in case of one-story building) but not less than 1.5 times the wall thickness and the width should be equal to or more than the width of the bearing wall above and below, and in case the floor is not a reinforced concrete slab construction, it should have an effective width not less than 1/20 of the distance between the center lines of the "adjoining walls" (refer to Fig. 14.14). The necessary amount of reinforcement is calculated by the structural calculation method for reinforced concrete construction.

Also, as the collar beam must have rigidity required for the horizontal frame of the building in case there are no rigid slabs, the corners and juncture on the plan should be so constructed that the joints have sufficient rigidity. The length of anchorage for steel bars should be sufficient and, if required, collar beam is haunched sometimes at the corners (Fig. 14.15).

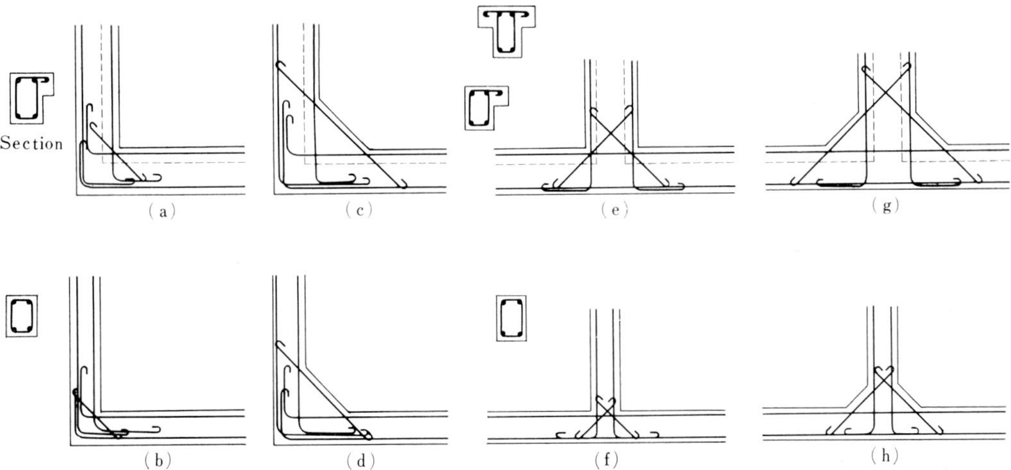

Fig. 14.15 Plan of Collar Beam at Corners, etc.

14.3.4 Floor and Roof

It is best to have rigid and strong floors such as of reinforced concrete slab in order to transmit effectively the horizontal forces to the bearing walls of each story and make every one of the bearing wall work together. Especially, the earthquake force becomes large as the number of stories increases, therefore in 3-story buildings of reinforced block construction, all the roof and floors, excepting the floor of the lowest story, should be made of either reinforced concrete slab or rigid precast reinforced concrete slab.

Also in buildings of 2-stories or less it is desirable from the structural viewpoint to use reinforced concrete slabs if possible.

14.3.5 Footing

At the lowest part of wall, continuous footing or footing beam of reinforced concrete construction should be provided.

The role of continuous footing is to reinforce bearing walls in collaboration with the collar beams and giving fixity to them by connecting each one of them at the bottom of the building so that they are made into one-body. At the same time, it prevents differential settlement and damage to walls due to local reaction, and play the role of transmitting uniformly the distributed load to the supporting soil. Accordingly, it is necessary for the continuous footing to be as deep as possible and strong. It should also be made a part of the frame connecting the bottom of each wall.

The width of continuous footing should be equal to or more than the thickness of the wall. Greater the height, the better ; but it should be not less than 60 cm (not less than 45 cm for one-story building) and at least 1/12 of the eaves height. Once a reinforced concrete construction should undergo a great earthquake, it will be subjected to rocking vibrations and shear but it has been ascertained by experiments[1),2)] that in case the height of the continuous footing is high, the building reacts well as one-body and the damage is minimized (Fig. 14.16).

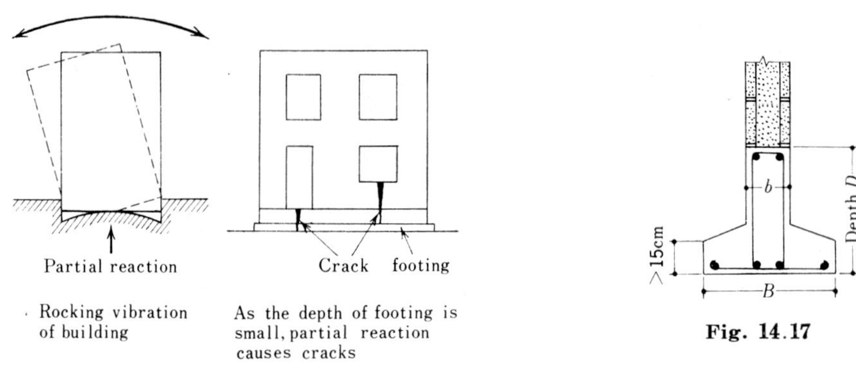

Fig. 14.16

Fig. 14.17

Continuous footing or the footing beam should have reinforcement in both the top and bottom and its section should be so designed that it is safe against loads acting to it. This should be based on the calculation method for reinforced concrete construction. In general, continuous footings are used under bearing walls and small openings while footing beams without spreading are used under large openings. For continuous footings, the footing base should be checked in addition to the beam section (Fig. 14.17).

14.3.6 Splices, Anchorage and Concrete Protection for Reinforcing Bars

a) Splice and Anchorage

As the details of steel bars to be used in reinforced block construction are similar in the main to those in the reinforced concrete construction, only the points that require particular attention are mentioned below.

Vertical bars of the bearing wall should not be spliced in the body of wall except when splice is welded. This rule is made based on the fact that the hollow cell in which reinforcing bar is to be inserted is narrow, therefore it is very disadvantageous to make a splice joint in it ; besides, as there are collar beams edging concrete block walls at every division of stories, anchorage of bars is easily accommodated in them. As for the length of splice and anchorage in other locations, Table 14.8 is to be referred to in ordinary cases.